DURATION AND CHANGE
Fifty Years at Oberwolfach

DURATION
AND
CHANGE

Fifty Years at Oberwolfach

Edited by
Michael Artin Hanspeter Kraft
Reinhold Remmert

Springer-Verlag
Berlin Heidelberg New York
London Paris Tokyo
Hong Kong Barcelona
Budapest

Michael Artin

Department of Mathematics
Massachusetts Institute
of Technology
Cambridge, MA 02139
USA

Reinhold Remmert

Mathematical Institute
University of Münster
Einsteinstraße 62
D-48149 Münster
Germany

Hanspeter Kraft

Mathematical Institute
University of Basel
Rheinsprung 21
CH-4051 Basel
Switzerland

Gedruckt mit Unterstützung des Förderungs- und Beihilfefonds
Wissenschaft der VG WORT

Frontispiece (p. II): *The guest house at Oberwolfach*
(Photo by Gerd Fischer, Düsseldorf)

Mathematics Subject Classification (1991): M00009

ISBN 3-540-57214-7 Springer-Verlag Berlin Heidelberg New York
ISBN 0-387-57214-7 Springer-Verlag New York Berlin Heidelberg

Library of Congress Cataloging-in-Publication Data
Duration and change: fifty years at Oberwolfach/edited by Michael Artin,
Hanspeter Kraft, Reinhold Remmert. p.   cm.
ISBN 3-540-57214-7 (Berlin: acid-free). –
ISBN 0-387-57214-7 (New York: acid-free)
1. Mathematics. 2. Mathematisches Forschungsinstitut Oberwolfach.
I. Artin, Michael. II. Kraft, Hanspeter, 1944–. III. Remmert, Reinhold.
IV. Mathematisches Forschungsinstitut Oberwolfach.
QA7.D87   1994   510–dc20   94-28935   CIP

Typesetting: Reformatting of the authors' input files by Springer-Verlag
and Kurt Mattes, Heidelberg, using a Springer TeX macro package.
Photo composition output: Text & Grafik, B.E.S. GmbH, Heidelberg.
SPIN 10081383   41/3140-5 4 3 2 1 0 Printed on acid-free paper

Preface

The fortieth anniversary of the *Mathematical Research Institute Oberwolfach*, ten years ago, was celebrated by the volume *Perspectives in Mathematics*. This year, for the fiftieth anniversary, we follow suit with the present volume *Duration and Change*. It is a *hommage* to Oberwolfach, to this unique place of mathematical research and conferences, its air of science, which enraptures every visitor – and makes him come again.

What is it that makes the place unique as it is? – The idyllic site, amidst the Black Forest offering magnificent walks? The calm yet intense atmosphere of the house, far from the pressure of everyday business? The very fine, rich library where it is easy to find almost everything one is looking for, and many unexpected things as well? Or is it simply the fact that this place, which welcomes everyone alike, renders this tremendous service to science in such an unassuming and almost perfect manner?

We do not know. But we note that the fascination with "Oberwolfach" has spread all around the globe and has affected colleagues and friends from all parts.

The present volume contains essays from different areas of mathematics like algebra, analysis, arithmetic, geometry, numerical mathematics, and topology. The collection bears witness to the fact that mathematics is still the fundamental language of science: not only the 'Queen of the sciences', but also their servant.

During the past fifty years mathematics has developed and changed enormously. Mathematical thinking has pervaded all parts of society; technology without mathematics is unthinkable. At the same time, mathematicians still find it difficult to communicate their science to the public. Living in a realm with its own peculiar beauty, they are naturally absorbed with the meaning and the inherent value of their discoveries – many of them far from any immediate application. But they are not born communicators (to a larger public) of the marvellous things they contemplate. They may seek and find consolation in Leonardo's word: *Whoever despises the certainty of mathematics plunges into the chaos of thinking.*

During the past decade the Oberwolfach Institute has continued to accompany and influence the evolution of mathematics through its international conferences. Personal encounters of researchers and direct communication have always been essential sources of mathematical research. Innumerable new ideas have originated at Oberwolfach, and many seminal results have been presented for the first time here. The *Lorenzenhof* rests like a solid rock in troubled water, a "monument of unageing intellect". May it remain unchanged in these difficult years. *Ad multos annos!*

Oberwolfach, May 1994   *M. Artin   H. Kraft   R. Remmert*

Vorwort

Vor zehn Jahren erschien zum 40. Geburtstag des Mathematischen Forschungsinstituts Oberwolfach die Festschrift „Perspectives in Mathematics". Ihr stellen wir heuer zum 50. Geburtstag den Band „Duration and Change" an die Seite. Es ist eine „Hommage" an Oberwolfach, an diese einmalige Forschungs- und Tagungsstätte der Mathematiker, deren wissenschaftliche Ausstrahlung jeden Besucher fasziniert und immer wieder anlockt. Was ist es, das diesen Ort so einzigartig macht? Ist es die traumhafte Lage inmitten des Schwarzwaldes, der zu herrlichen Spaziergängen einlädt? Ist es die ruhige, aber dennoch intensive Atmosphäre, frei von der täglichen Hektik? Ist es die vorbildliche und großzügig ausgestattete Bibliothek, in der man beinahe alles findet, auch solche Dinge, die man gar nicht sucht? Oder ist es einfach die Tatsache, daß hier in sehr diskreter, aber beinahe perfekter Weise ein großer Dienst an der Wissenschaft geleistet wird, wo jeder willkommen ist? Wir wissen es nicht! Wir stellen nur fest, daß die Faszination „Oberwolfach" ein weltweites Phänomen ist, das unsere Kollegen und Freunde in der ganzen Welt gleichermaßen in seinen Bann zieht.

Im vorliegenden Buch sind Essays aus verschiedenen Gebieten der Mathematik wie Algebra, Analysis, Geometrie, Numerik, Topologie und Zahlentheorie zusammengestellt. Sie mögen dokumentieren, daß die Mathematik immer noch die ureigene Sprache der Wissenschaft ist: Nicht nur die „Königin", sondern auch ihre „Dienerin".

In den vergangenen fünfzig Jahren hat sich die Mathematik enorm entwickelt und gewandelt. Mathematisches Denken ist heute überall in der Gesellschaft integriert, Technologisierung ohne Mathematik ist undenkbar. Und doch fällt es den Mathematikern nach wie vor schwer, ihre eigentliche Wissenschaft der Öffentlichkeit nahezubringen. Sie leben in einer Welt voll eigenartiger Schönheit und sind durchdrungen vom Sinn und Wert ihrer Erkenntnisse, die oft einer direkten Anwendung fernestehen. Aber es ist ihnen nicht gegeben, die herrlichen Dinge, die sie schauen, einem größeren Publikum mitzuteilen. Sie suchen und finden Trost im LEONARDO-Wort: „Wer die Sicherheit der Mathematik verachtet, stürzt sich in das Chaos der Gedanken."

Auch im letzten Jahrzehnt hat das Institut in Oberwolfach durch seine internationalen Tagungen die Entwicklung der Mathematik stetig begleitet und beeinflußt. Die persönliche Begegnung der Forscher und das direkte Gespräch bildet seit jeher ein wichtiges Fundament der mathematischen Forschung. Unzählige neue Ideen haben ihren Ursprung in Oberwolfach, viele entscheidende Ergebnisse sind hier zum ersten Mal vorgestellt worden. Der Lorenzenhof ist ein „Fels in der Brandung", eine „Festung des Geistes". Möge es auch in diesen schwierigen Jahren so bleiben. Ad multos annos!

Oberwolfach, im Mai 1994   *M. Artin   H. Kraft   R. Remmert*

Contents

Essays

Authors *of the History and of the Essays*

Heinrich Behnke †

Arthur L. Besse   Institut des Hautes Etudes Scientifiques
35, route de Chartres
F-91440 Bures-sur-Yvette, France
e-mail: besse@ihes.fr

Wolfgang Boehm   Angewandte Geometrie und Computergraphik
Technische Universität Braunschweig
Pockelsstraße 14
D-38106 Braunschweig, FRG
e-mail: boehmw@indi.cagd.cs.tu-bs.de

Armand Borel   Institute for Advanced Study
School of Mathematics
Olden Lane
Princeton, NJ 08540, USA
e-mail: borel@math.ias.edu

Henri Cartan   95, Boulevard Jourdan
F-75014 Paris, France

Philippe G. Ciarlet   Analyse Numérique
Tour 55
Université Pierre et Marie Curie
4 Place Jussieu
F-75005 Paris, France

José A. de la Peña   Instituto de Matemáticas
Circuito Exterior
UNAM
México 04510, D. F. México

Simon K. Donaldson   Mathematical Institute
24–29 St. Giles
Oxford OX1 3LB, Great Britain

Peter Gabriel

Mathematisches Institut
Universität Zürich
Rämistraße 74
CH-8001 Zürich, Switzerland

Josef Hoschek

Fachbereich Mathematik
Technische Hochschule Darmstadt
Schloßgartenstraße 7
D-64289 Darmstadt, FRG
e-mail: hoschek@mathematik.th-darmstadt.de

Neal Koblitz

Department of Mathematics
University of Washington
GN-50
Seattle, WA 98195-0001, USA
e-mail: koblitz@math.washington.edu

Matthias Kreck

Fachbereich Mathematik
Johannes-Gutenberg-Universität
Saarstraße 21
D-55122 Mainz, FRG
e-mail: kreck@topologie.mathematik.uni-mainz.de

Jerrold E. Marsden

Department of Mathematics
University of California
Berkeley, CA 94720, USA
e-mail: marsden@math.berkeley.edu

Jürgen Neukirch

Fachbereich Mathematik
Universität Regensburg
Universitätsstraße 31
D-93053 Regensburg, FRG

Thomas Peternell

Fakultät für Mathematik und Physik
Universität Bayreuth
D-95440 Bayreuth, FRG
e-mail: thomas.peternell@uni-bayreuth.d400.de

Georges H. Reeb †

Werner Remmers

Gerhard-Kues-Straße 14
D-49808 Lingen, FRG

Michael Schneider

Fakultät für Mathematik und Physik
Universität Bayreuth
D-95440 Bayreuth, FRG

Wolfgang Schwarz

Mathematisches Seminar
Johann Wolfgang Goethe-Universität
Robert-Mayer-Straße 10
D-60325 Frankfurt am Main, FRG

Joachim Schwermer   Katholische Universität Eichstätt
Mathematisch-Geographische Fakultät
Ostenstraße 26–28
D-85072 Eichstätt, FRG

Hans-Peter Seidel   Institut für Mathematische Maschinen
und Datenverarbeitung
Friedrich-Alexander-Universität Erlangen-Nürnberg
Am Weichselgarten 9
D-91058 Erlangen, FRG
e-mail: seidel@informatik.uni-erlangen.de

Addresses

Stainless steel and brass sculpture (Boy's surface) donated by Daimler-Benz in 1991
(Photo by Gerd Fischer, Düsseldorf). A short description is given on the back of this page.

Stainless steel and brass sculpture (Boy's surface)

On February 15, 1991, a sculpture donated by Daimler-Benz AG was unveiled at the Mathematical Research Institute in Oberwolfach. This sculpture is unusual in a number of ways.

- Not only does the sculpture successfully supplement the architecture of the institute building in artistic terms, it has also been imbued with so much mathematics that the scientific environment of the sculpture could provide enough material for a lecture course lasting several semesters for advanced students of mathematics.
- The technical execution, which was entirely in the hands of the working group SYRKO at the Daimler-Benz plant in Sindelfingen, demanded technical innovation and was only made possible by using the most modern technical methods.

   The design and production of the sculpture made great technical demands: on the one hand, the surface was supposed to be transparent so that the "inner workings" could be studied. On the other hand, it was destined to stand outdoors (and thus had to be weatherproof), which made that transparent materials such as Perspex unsuitable.

   The solution that was finally found satisfied all the requirements: the surface was constructed of a network of 17 + 24 steel strips, which cross each other at right angles. A numerically-guided laser working from computer data was used to cut the strips into shape from 2 mm thick sheet metal and to cut the holes for the rivets, which hold the lattice of steel strips together. A central Möbius strip is marked with brass buttons.

Ministerium für Wissenschaft und Forschung
Baden-Württemberg

In einer von wissenschaftlich-technischen Errungenschaften geprägten Inudstrie-gesellschaft, die Wissen und Erkennen als wichtige Werte betrachtet, steht die Forschung in enger Berührung mit Fragen nach der technologischen, industriellen und gesamten zukünftigen Entwicklung unserer Gesellschaft.

So sehr sich die ihrem programmatischen Selbstverständnis nach wertneutrale Arbeit der Forschung in der Abgeschiedenheit vollzieht, so unübersehbar nehmen ihre Resultate Einfluß auf weite Bereiche unseres Lebens, unseres Alltags, auf unser Weltbild und unsere Zukunft. Es nimmt deshalb nicht Wunder, daß das Thema Wissenschaft und Forschung in wachsendem Maße ins Zentrum der öffentlichen Aufmerksamkeit rückt.

Die Mathematik, das haben wir inzwischen alle gelernt, ist in diesem Zusammenhang von weitaus größerer Bedeutung, als dies noch vor 20 oder 30 Jahren angenommen wurde. Erinnert sei dabei nur an den sogenannten Sputnik-Schock, der als Reaktion auf einen angeblichen technologischen Vorsprung der damaligen Sowjetunion dazu führte, daß in neuer Weise über die Bedeutung der Mathematik und der Physik und über eine entsprechende Erweiterung des Lehrangebotes in den Schulen nachgedacht wurde. Ganze Forschungszweige wären heute ohne die Einbringung mathematischer Erkenntnisse nicht mehr entwicklungsfähig.

Besonders hervorheben möchte ich, daß das Mathematische Forschungsinstitut Oberwolfach auch Herausforderungen nicht ausweicht, die sich durch die Anwendungsmöglichkeiten seiner Forschungsergebnisse ergeben. Die zunehmende Differenzierung und Spezialisierung in allen Fachrichtungen macht es erforderlich, gewonnene Erkenntnisse in einem zweiten Schritt wieder in den Gesamtzusammenhang zu stellen und an die Gesellschaft zurückzuvermitteln. Hierfür erweist sich die Auseinandersetzung mit Methoden und Denkstrukturen aus Nachbardisziplinen als ebenso unabdingbar wie eine breite Diskussion über ökologische, ethische und kulturelle Folgen. Die Orientierungsmaßstäbe traditioneller Gesellschaften, die durch Ausdifferenzierung und Partikulasierung immer weiterer Teilbereiche ins Wanken geraten, können durch den interdisziplinären Diskurs zwar nicht restlos ersetzt werden, doch bietet er in demokratischen Gesellschaften den einzigen Weg der gegenseitigen Verständigung und Koordination.

Forschung im Dialog ist eine Aufgabe, die in Oberwolfach sehr ernst genommen wird. Immer wieder vereint das Institut in wissenschaftlichen Seminaren und Symposien Wissenschaftler, die sich gemeinsam über Forschungsprojekte und die daran anschließenden Problemstellungen verständigen. Das hier praktizierte gemeinsame Nachdenken stellt ein gutes Beispiel auch für die Kooperation mit anderen Wissenschaftszweigen und für diese untereinander dar.

In Zeiten degressiver Finanzierungsmöglichkeiten wächst bekanntlich der Druck der Öffentlichkeit, die Mittelvergabe zu überdenken und zu rechtfertigen, oder dort, wo dies nicht gelingt, auch zu reduzieren. Wir alle werden uns stärker als bisher auf solchen Legitimationsdruck einstellen müssen, und ich verbinde hierfür, wie ich denke, mit Ihnen gemeinsam die Hoffnung, daß uns diese Rechtfertigung gelingt. Die Schwierigkeit jeder Begründung für Forschungsförderung liegt ja auch darin, daß sich ihr Ausbleiben im Gegensatz zu vielen anderen gesellschaftlichen Aktivitäten nicht unmittelbar bemerkbar macht. Der Verlust der Forschung, den Deutschland beispielsweise durch die Vertreibung seiner jüdischen Mitbürger vor und nach 1933 erlebt hat, ist erst viele Jahre später ins Bewußtsein gerückt. Solche Verluste lassen sich nicht im Handstreich wieder ausgleichen. Auch hier hat unsere Geschichte gezeigt, daß es den Lebenszeitraum einer Wissenschaftlergeneration braucht, um wieder Anschluß an den internationalen Standard zu finden, daß Spitzenstellungen, die einmal wahrgenommen worden sind, sich sogar häufig nicht mehr zurückgewissen lassen.

Auch in diesem Zusammenhang ist sich die Landesregierung der herausragenden Bedeutung des international renommierten mathematischen Forschungsinstituts bewußt. Sie wird sich nach Kräften dafür einsetzen, daß das Institut auch in Zeiten einer schwierigeren allgemeinen Haushaltslage seine erfolgreiche Arbeit fortsetzen kann.

Klaus von Trotha

Minister für Wissenschaft und Forschung des Landes Baden-Württemberg

Volkswagen-Stiftung

Oberwolfach, die Vielgegrüßte

Mit keiner Institution ist die Volkswagen-Stiftung über die ganze Dauer ihrer Existenz so beständig verbunden gewesen wie mit dem Mathematischen Forschungsinstitut Oberwolfach. Die Stiftung hat gefördert und gefordert; sie hat Gebäude errichtet, Grund und Boden erworben, Gebäude erweitert und alles schließlich dem Trägerverein geschenkt. Sie hat aber immer wieder auch die Dienste der Mathematiker in Anspruch genommen, sei es als Gutachter und Ratgeber, sei es – in der Form des Trägervereins des Mathematischen Forschungsinstituts Oberwolfach – als Träger für Förderprogramme, die dem mathematischen Nachwuchs oder der internationalen Zusammenarbeit unter Mathematikern gewidmet waren und noch sind. So ist die Beziehung zwischen „Oberwolfach", wie der international gebräuchliche Sammelbegriff für alles ist, was sich personell und inhaltlich mit dem Mathematischen Forschungsinstitut verknüpft, und der Volkswagen-Stiftung in Hannover innig und wechselseitig geworden. Und von daher hat es offenbar nahegelegen, für den Festband, mit dem das 50jährige Bestehen des Instituts gefeiert werden soll, der Volkswagen-Stiftung Gelegenheit zu einem Grußwort zu geben. Die Stiftung entbietet dem Mathematischen Forschungsinstitut Oberwolfach, seinem Trägerverein und der großen Familie der Mathematiker, die in Oberwolfach ihr Zuhause hat, aus Anlaß dieses auch zeitgeschichtlich respektablen Jubiläums gern herzliche Grüße, Dank für alle gute Zusammenarbeit in der Vergangenheit und gute Wünsche für eine gleichermaßen gedeihliche Zukunft.

Rolf Möller

Generalsekretär
der Volkswagen-Stiftung

American Mathematical Society

Greetings to Oberwolfach!

On behalf of the officers and the more than 30,000 members of the American Mathematical Society, it gives me great pleasure to congratulate the Mathematisches Forschungsinstitut Oberwolfach on the occasion of its 50th anniversary. For it is clearly here at Oberwolfach that the flame of mathematics burns brightest, with a tradition of meetings and mathematical communication that is unequaled anywhere else in the world. The fond memories shared by my many colleagues of the idyllic surroundings, superb library, efficient organization and the broad spectrum of wonderful programs have enriched all of our mathematical lives. We can only hope to see that such a treasure will continue to serve mathematicians and mathematics for all future generations as well.

Best regards

Ronald L. Graham

President
American Mathematical Society

Dansk Matematisk Forening

Gruß an Oberwolfach

Oberwolfach! Welche Fülle, welcher Reichtum von Erinnerungen springt nicht hervor, wenn man deinen Namen hört.

Herrliche Herbstwanderungen in Wald und Feld, und eine spannende Wintertour an total eingefrorenen Wasserfällen entlang.

Der alte Lorenzenhof mit seinen winkligen Fluren und großen eckigen Zimmern, wo man mit zuerst fremden, aber nach einer Woche vertrauten Kollegen aus anderen Ländern zusammen wohnte.

Das neue Gebäude, wo man mit wehmutgemischtem Genuß die modernen Bequemlichkeiten des Zimmers in Besitz nahm.

Aber vor allem das Zusammensein mit Mathematikern aus aller Welt.

Vorträge und Diskussionen. Weinabende mit Zwiebelkuchen und Kaiserstühler. Die großen „Namen", die man sich kaum als lebende Personen vorstellen konnte, aber die sich da als lebendige, freundliche Menschen gezeigt haben. Die jungen Kollegen von weit und fern. Einige sah man vielleicht nur ein einziges Mal, andere sind gute Freunde geworden.

Oberwolfach! Wir grüßen dich und wünschen dir alles Gute für die Zukunft.

Deutsche Mathematiker: Wir beneiden Euch – und wir sind dankbar, daß wir mit Euch zusammen dieses Juwel genießen dürfen.

Im Namen des Dansk Matematisk Forening

Ebbe Thue Poulsen

Deutsche Mathematiker-Vereinigung

Es ist mir eine große und ganz besondere Freude, dem Mathematischen Forschungs-institut Oberwolfach und allen seinen Mitarbeiterinnen und Mitarbeitern zum fünf-zigjährigen Bestehen die herzlichsten Glückwünsche der Deutschen Mathematiker-Vereinigung auszusprechen.

Die Geschichte des Institutes ist ungewöhnlich und beeindruckend zugleich: Konzipiert und entstanden in den Wirren der Kriegs- und Nachkriegszeit ist es unter der tatkräftigen Leitung seiner Direktoren – Wilhelm Süß, Hellmuth Kneser, Theodor Schneider und Martin Barner – zu einer Institution herangewachsen, die nicht nur für uns Mathematiker ein Begegnungszentrum von einmaliger Bedeutung ist, sondern die auch Vorbild und Modell für zahlreiche ähnliche Gründungen in Nachbarfächern und anderen Ländern gewesen ist. Mit Sicherheit gibt es keinen Ort auf der Welt, wo nur annähernd so viele Mathematiker gewesen sind und wissen-schaftlich anregende und ertragreiche Diskussionen geführt haben. Oberwolfach ist ja nicht nur ein Tagungszentrum, wo über fertige Ergebnisse berichtet wird, sondern vor allem auch ein Ort der Begegnung, des Gesprächs und der Zusammenarbeit, ein Ort, an dem neue Ideen entstehen.

Die Mathematik ist nicht nur eine der ältesten Wissenschaften, sondern sie ist zugleich auch modern und aktuell. Zu recht wird von ihr im Hinblick auf die komplexen Herausforderungen unserer Zeit gesagt: „Hochtechnologie ist mathe-matische Technologie" oder „Mathematik, der Schlüssel zu den Schlüsseltechno-logien". Um sie zu fördern, benötigt man jedoch keine riesigen Apparate und Geräte, keine Versuchsfelder und großen Meßeinrichtungen. Oberwolfach kommt sogar ohne einen richtigen Computer und fast ohne wissenschaftliches Personal aus. Auch die Administration könnte kaum einfacher sein; weder Sicherheitsvorkehrun-gen, noch Schadstoffordnungen sind notwendig, und auch die Umweltverträglich-keit braucht nicht geprüft zu werden. Müssen die Wissenschaftspolitiker und die Verantwortlichen in Parlament und Regierung nicht geradezu glücklich sein, daß es so einfach und unkompliziert ist, eine Schlüsselwissenschaft so nachhaltig mit verhältnismäßig bescheidenem finanziellen Aufwand zu fördern?

An noch etwas soll bei dieser Gelegenheit erinnert werden, das heute aktuell wie eh und je ist: Vom Tag der Gründung war Oberwolfach ein Symbol für die Zusammenarbeit der Wissenschaftler über nationale und ideologische Grenzen und politische Trennung hinweg. In einer sich neu formierenden politischen Welt gilt es, diese Tradition zu pflegen und zu bewahren.

Die Deutsche Mathematiker-Vereinigung verdankt Oberwolfach unendlich viel. Dafür sei ein herzliches „Dankeschön" gesagt. Wir alle hoffen, daß Oberwolfach bleibt, was es ist: ein Ort wissenschaftlicher Arbeit, an dem es nichts zu verbessern gibt.

Winfried Scharlau

für das Präsidium der DMV

European Mathematical Society

Dear Colleagues,

As President of the European Mathematical Society, I am writing to offer hearty congratulations to the Institute on attaining its 50th anniversary.

From fragile beginnings in the Lorenzenhof, the Institute has achieved world renown and earned the profound gratitude of the mathematical community, the members of whom refer to it, affectionately and simply, as "Oberwolfach". The excellent and carefully constructed programmes, the extensive library und the superb facilities for discourse or meditation have given Oberwolfach a unique position in the annals of mathematics. Mathematicians of all nationalities have benefited greatly from the relaxed atmosphere und from the freedom to engage in their subject. The unobtrusive und highly efficient administration and staff are much appreciated for their skill in maintaining a pleasant ambiance.

In recent years Oberwolfach has generously given of its facilities for several meetings which have ultimately led to the foundation of the European Mathematical Society. The Executive Committee of the EMS is itself much indebted to Oberwolfach for assistance and hospitality.

Oberwolfach has much to be proud of and to celebrate at this anniversary. Over the next fifty years the Institute will undoubtedly continue to occupy a pre-eminent position amongst mathematical research centres. In conveying well-deserved congratulations for the present, I offer the very best wishes of the EMS for the future.

Friedrich Hirzebruch

On behalf of all the members
of the Executive of the
European Mathematical Society

Gesellschaft für Angewandte Mathematik und Mechanik

Die Gesellschaft für Angewandte Mathematik und Mechanik nimmt den 50. Geburtstag des Mathematischen Forschungsinstitutes Oberwolfach zum Anlaß, die Bedeutung dieser Institution nicht nur für die Reine Mathematik, sondern auch für die Angewandte Mathematik und die theoretischen Gebiete der Natur- und Ingenieurwissenschaften zu würdigen. Dieses weltweit einmalige Institut verdankt seinen Ruhm zum einen den ansprechenden und zweckmäßigen Gebäuden und einer guten Bibliothek, aber ebenso dem Geist des Vertrauens (und der Disziplin) und einer Atmosphäre der Ruhe und Konzentration, wie man es anderswo wohl kaum findet. Das eine aufgebaut und das andere auch über turbulente Jahre hinweg bewahrt zu haben, ist vor allem das Verdienst des langjährigen Direktors, Herrn Prof. Dr. Barner. Wir sind ihm besonders dankbar dafür, daß er den Anwendungen der Mathematik seine Unterstützung nie versagt und seinen Einfluß auf das wissenschaftliche Programm in wohlabgewogener Weise ausgeübt hat, und das auch zu einer Zeit, als Anwendungen noch nicht in aller Munde waren.

Die Gesellschaft für Angewandte Mathematik und Mechanik dankt den öffentlichen Geldgebern für ihre großzügige Unterstützung, und sie hofft, daß das Mathematische Forschungsinstitut auch in Zukunft auf diese Unterstützung bauen kann. Sie dankt den Mitarbeitern des Instituts und in besonderem Maße Herrn Barner, der dieses Institut zu einem Teil seines Lebenswerkes gemacht und seinen vielfach bewunderten modus operandi wesentlich geprägt hat.

Reinhard Mennicken

Österreichische Mathematische Gesellschaft

Die Österreichische Mathematische Gesellschaft entbietet dem Mathematischen Forschungsinstitut Oberwolfach die besten Wünsche zu seinem 50. Geburtstag. Es gibt keine vergleichbare Institution, welche in diesen Jahren den Mathematikern aus aller Welt ähnliche Gelegenheiten zu fruchtbarem Gedankenaustausch und zur Diskussion ihrer Arbeiten gewährt hätte. Ohne Zweifel sind die Impulse, die von den Oberwolfacher Tagungen für die klassischen und für neuere Teilgebiete der mathematischen Wissenschaften ausgingen, sehr groß. Alle, die an einer Tagung in Oberwolfach teilnehmen konnten, schätzen dies, die einzigartige Atmosphäre des „alten" wie des „neuen" Lorenzenhofes und die hervorragenden Einrichtungen des Instituts, für welche die Mathematiker von Vertretern anderer Wissenschaften beneidet wurden und werden. Die Einladung zu einer Oberwolfacher Tagung wird daher besonders von jüngeren Mathematikern als Ehre betrachtet.

Die ÖMG gedenkt dankbar des Gründers des Forschungsinstituts, Prof. Süss, und hebt die unschätzbaren Verdienste von Prof. Barner als Institutsdirektor hervor.

Wir dürfen mit dem Ausdruck unserer Hoffnung schließen, daß dem Mathematischen Forschungsinstitut ein zukünftige Wirkung beschieden ist, welche sich würdig den vergangenen Jahren anschließt, und daß österreichische Mathematiker auch weiterhin zu den Gästen dieser ruhmreichen Institution gehören werden, die mit gutem Grund in einem der alten Gästebücher von einem italienischen Mathematiker das „paradiso dei matematici" genannt wird.

Ludwig Reich
Vorsitzender

Russian Academy of Sciences

Dear Colleagues,

On behalf of the Russian Academy of Sciences we have the honour of congratulating the Mathematisches Forschungsinstitut Oberwolfach on the occasion of its 50th anniversary.

During these 50 years this unique institution has become known all over the world. The institute made an important contribution to the development of abstract and applied mathematics and in propagating mathematical knowledge. The thoroughly prepared scientific programmes which took place in Oberwolfach reflect the most important trends in modern mathematics.

For the last 50 years the Oberwolfach Institute played an inestimable role in the scientific collaboration between Western scientists and mathematicians from Russia. Even in the years of the so-called "Cold War", many Russian mathematicians were invited and visited Oberwolfach to enjoy the friendly international scientific atmosphere of the Institute.

We are sure that the Institute will continue its noble activities and will remain one of the most esteemed international mathematical research centres.

We would like to convey the congratulations of the Russian Academy of Sciences to the administration and staff of Mathematisches Forschungsinstitut Oberwolfach for their remarkable work.

Yu. S. Osipov
Academician, President

A. A. Gonchar
Academician, Vice-President

Schweizerische Mathematische Gesellschaft
Société Mathématique Suisse

Herzliche Glückwünsche

zum 50. Geburtstag des Mathematischen Forschungsinstituts Oberwolfach möchte ich Ihnen im Namen der Schweizerischen Mathematischen Gesellschaft zum Ausdruck bringen.

Viele Schweizer Mathematiker haben seit seiner Gründung die Gastfreundschaft dieser internationalen Begegnungsstätte erfahren. Sie denken in Dankbarkeit zurück an interessante Tagungen, wo über die neuesten Ergebnisse ihres Faches berichtet wurde, vor allem aber an anregende Gespräche mit Mathematikern aus aller Welt, die oft zu einer langjährigen Zusammenarbeit und Freundschaft führten. Unvergessen bleiben die Wanderungen durch den schönen Schwarzwald. Die einzigartige Atmosphäre Oberwolfachs mit seinen hervorragenden Einrichtungen, seiner guten Bibliothek, hat diesem Ort der Begegnung unter den Mathematikern weltweites Ansehen verschafft. Es spricht für die Bedeutung von Oberwolfach, daß in den letzten Jahren neu entstandene Forschungsstätten häufig zum Ausdruck bringen, daß sie ähnliche Ziele wie Oberwolfach verfolgen. Wir haben im Laufe des 50jährigen Bestehens von Oberwolfach beobachten können, wie aus dem alten Lorenzenhof, der bei vielen von uns nostalgische Gefühle weckt, das moderne Institut entstanden ist, ohne daß dabei das geistige Klima von Oberwolfach gelitten hat. Das ist vor allem dem unermüdlichen Einsatz des langjährigen Direktors von Oberwolfach, Professor Dr. Martin Barner, zu verdanken. Wir möchten das Land Baden-Württemberg dazu beglückwünschen, daß es dem Mathematischen Forschungsinstitut in Oberwolfach Heimat und Unterstützung gewährt hat.

Wir geben der Hoffnung Ausdruck, daß auch künftigen Generationen von Mathematikern aus aller Welt dieses Forschungsinstitut offen stehen möge.

Harald Holmann

Société Mathématique de France

En témoignage de notre reconnaissance...

La vie des institutions est émaillée de circonstances solennelles; parmi les plus heureuses comptent certainement celles qui permettent de témoigner sa reconnaissance pour l'inspiration donnée. La célébration du cinquantenaire du Forschungsinstitut d'Oberwolfach offre une telle occasion à la Société Mathématique de France. En effet l'activité et le rayonnement de l'institution hors du commun qu'est Oberwolfach ont été tellement exemplaires tout au long de sa vie déja longue d'un demi-siècle que l'existence d'un analogue francais s'est imposé aux mathématiciens français comme une nécessité, et par voie de conséquence comme un modèle à imiter. Aussi le premier rapport qui a ouvert la voie au Centre International de Rencontres Mathématiques de Marseille-Luminy ou CIRM* s'intitule-t-il en 1965, sous la plume de Philippe Courrège et Jacques-Louis Lions, "Pour un Oberwolfach français".

Comme il nous a fallu plus de 15 ans d'une fermeté attentive pour que le projet du CIRM prenne corps sous la forme d'une bastide provençale rénovée, il est facile d'imaginer que la transformation du Schloss original en la cathédrale païenne, fleuron d'une architecture moderne tout entière tournée vers la fonctionnalité qu'est devenu le Centre d'Oberwolfach aujourd'hui, a, elle aussi, requis beaucoup d'opiniâtreté et de soins attentifs et minutieux.

Que nos collègues allemands soient félicités pour le succès de leur entreprise et la justesse de leurs vues. Il est hors de doute qu'Oberwolfach a joué un rôle déterminant dans l'émergence d'une perspective d'intégration mathématique à l'échelle du vieux continent. Henri Cartan, avec la vision prophétique de l'avenir que nous lui connaissons, n'a-t-il pas compté parmi les premiers visiteurs du Centre dans l'immédiat après-guerre? Le flux constant (croissant?) de mathématiciens européens visitant Oberwolfach, et y trouvant l'atmosphère d'échanges informels et prolongés indispensables à une communication profonde, témoigne de la conscience que nous avons tous intériorisée de la réussite de cette alchimie créatrice.

Une autre preuve en est apportée par la tenue à Oberwolfach de nombre des réunions de l'European Mathematical Council, creuset de la Société Mathématique

---

* C'est ainsi que nous le nommons à cause certainement du penchant fâcheux des Français pour les sigles car je n'ai jamais entendu un collègue allemand parler du FIO.

Européenne et lieu de débats et d'espoirs, que l'Histoire récente a transformé en défis.

Pour notre part, nous avons donné dès l'an dernier un tour concret à cette coopération européenne en faisant du CIRM et d'Oberwolfach les partenaires d'un projet d'euro-conférences accepté par la Communauté Européenne, et nous nous sommes sentis honorés par la confiance que nous ont témoignée les responsables d'Oberwolfach en faisant du petit frère CIRM le responsable financier de cette première coopération. Il ne s'agit là, nous en sommes convaincus, que de la première pierre d'un édifice immatériel qui doit nous permettre de transcender l'héritage de l'Histoire pour ouvrir des voies nouvelles et fécondes.

Une tâche lourde, mais ô combien exaltante, attend les nouvelles générations de mathématiciens qui devront continuer cette construction. Nous sommes confiants que ce nouveau défi sera relevé ... pour l'honneur de l'esprit humain.

Jean-Pierre Bourguignon
Pour la Société
Mathématique de France

Wiskundig Genootschap

Aan het Mathematischen Forschungsinstitut te Oberwolfach!

Bij dit 50-jarige jubileum van het Mathematischen Forschungsinstitut te Oberwolfach is het mij een plezier namens het Nederlandse Wiskundig Genootschap enkele woorden te schrijven in lof van dit voortreffelijke onderzoeksinstituut.

Wat zijn de optimale omstandigheden om wiskundig onderzoek te doen, nieuwe resultaten te horen, te bespreken, te presenteren? Worden deze omstandigheden ooit ergens gerealiseerd?

Aan de universiteiten slechts ten dele, omdat ons daar ook andere doelen voor ogen staan, zoals het doorgeven van reeds bekende resultaten aan de volgende generaties, wat eigen eisen stelt, zoals de aanwezigheid van een breed scala van wiskundige specialismen.

Voor intensief onderzoek en de zinvolle communicatie ervan zijn andere voorwaarden noodzakelijk: rust, een goede sfeer, een uitstekende bibliotheek, en niet te vergeten, voldoende gelijk georiënteerde onderzoekers, gekozen niet op geografische gronden, maar onafhankelijk van grenzen, op grond van hun interessen en prestaties.

Al deze goed wiskundig onderzoek bevorderende factoren zijn aanwezig bij de „Tagungen" te Oberwolfach, waar bovendien de zorg van een toegewijde staf en ideale materiële omstandigheden een zeer efficiënt tijdsgebruik mogelijk maken. We kunnen dus bevestigend antwoorden op de tweede vraag hierboven. Ja deze omstandigheden worden soms verwezenlijkt, en in het bijzonder tijdens de Tagungen aan het Mathematischen Forschungsinstitut te Oberwolfach!

Moge het Forschungsinstitut nog een lang en voorspoedig leven beschoren zijn en de komende vijftig jaar even vruchtbaar blijken als de afgelopen halve eeuw!

Namens het Wiskundig Genootschap,

Erik Thomas

Voorzitter

History

Oberwolfach: Lounge in the library building
(Photo by Gerd Fischer, Düsseldorf)

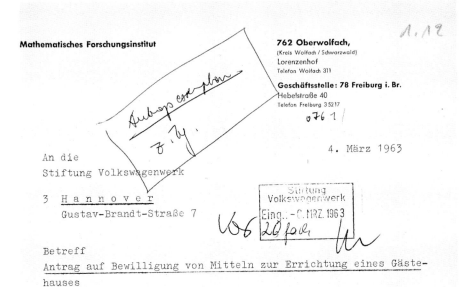

Mathematisches Forschungsinstitut

762 Oberwolfach,
(Kreis Wolfach / Schwarzwald)
Lorenzenhof
Telefon Wolfach 311

Geschäftsstelle: 78 Freiburg i. Br.
Hebelstraße 40
Telefon Freiburg 3 5217

4. März 1963

An die
Stiftung Volkswagenwerk

3 H a n n o v e r
Gustav-Brandt-Straße 7

Stiftung
Volkswagenwerk
Eing. - C. MRZ. 1963

Betreff
Antrag auf Bewilligung von Mitteln zur Errichtung eines Gästehauses

Das Mathematische Forschungsinstitut Oberwolfach bittet die Stiftung Volkswagenwerk um Bereitstellung von Mitteln, die zur Errichtung eines Gästehauses am Mathematischen Forschungsinstitut
Oberwolfach erforderlich sind. Unter Zugrundelegung des ange-
schlossenen Raumprogramms ist nach sorgfältigen Berechnungen durch
das Staatliche Hochbauamt in Offenburg hierfür ein voraussichtli-
cher Aufwand in Höhe von

DM 1.227.000.--

erforderlich. Hinzu kommen gegebenenfalls noch Mittel für den
Erwerb des Geländes und eines Wirtschaftsgebäudes, die den Be-
trag von

DM   250.000.--

keinesfalls überschreiten, vermutlich aber weit darunter liegen
werden.

Eine ausführliche Darlegung der gegenwärtigen Situation des Mathematischen Forschungsinstituts Oberwolfach und eine eingehende
Begründung dieses Antrags sind beigefügt. Das Direktorium des
Mathematischen Forschungsinstituts Oberwolfach und der Wissenschaftliche Beirat der Gesellschaft für mathematische Forschung
e.V. bringen ihre Hoffnung zum Ausdruck, daß die Stiftung Volkswagenwerk den Antrag wohlwollend prüft und sich in der Lage
sieht, ihm zu entsprechen.

- 2 -

Wir versichern -für den Fall einer Bewilligung- ausdrücklich,
die bewilligten Mittel nur zu dem beantragten Zweck zu verwenden,
und werden uns natürlich, was die Abrechnung, die Prüfung des
Rechnungswesens und die Berichterstattung betrifft, nach den Wünschen der Stiftung Volkswagenwerk richten.

(Prof.Dr. Theodor Schneider)
geschäftsführender Direktor
des Mathematischen Forschungsinstituts Oberwolfach

(Prof.Dr. M. Barner)
geschäftsf.Direktor
ab 1.5.1963

*With this letter, accepted as application, more than 30 years of cooperation began between
the Mathematisches Forschungsinstitut Oberwolfach and the Volkswagen-Stiftung (then still
called Stiftung Volkswagenwerk)*

Erinnerungen

Georges H. Reeb

Meine erste Reise nach Oberwolfach fand 1949 statt; mit dem Zug von Straßburg über Kehl, Offenburg, Hausach, Wolfach und dann mit dem Bus nach dem Institut. Kaum 80 Kilometer, aber einige Stunden! Charles Ehresmann, mein „maître", hatte diesen Ausflug geplant; unterwegs wiederholte er zusammenfassend, was er oft erzählte: Wem würden wir begegnen? Wilhelm Süss, dem Hausherrn, „Homme remarquable, aussi par sa tenacité à fonder de solides relations entre deux peuples"; Hellmuth Kneser, Süss' engstem Freund und Helfer; dem Topologen Herbert Seifert „mon ami sincère depuis longtemps" (von Seifert wußte ich seit langem; sein ewiges Buch ‚Topologie' wurde in Clermont-Ferrand eifrig gelesen, als wir Straßburger dort studierten); Emanuel Sperner, auch Topologe; und noch manchen anderen. Nach einigem Zuhören war mir nicht mehr klar, war da Rede von Topologie oder Fußball (Sperners Urberufung)?

Aber beim Eintreffen im Lorenzenhof spürten wir Trauer; auch gleich kam die Nachricht: „Threlfall ist soeben gestorben, Seifert ist höchst betrübt." Süss eilte her, umarmte Ehresmann (vielleicht nur dies einzige mal während einer langen Freundschaft). Das Begräbnis fand statt am nächsten Tag, unter mildem Regen, im Friedhof von Oberwolfach. Still, auch schön, war die Wanderung dorthin. Unvergeßlich ist dieses Begebnis: heimelig war das Haus, das ich später so oft besuchen durfte.

Wie soll ich dies erste Treffen mit Professor Süss beschreiben? Gemüt ist das richtige Wort. Wärme strahlte aus diesem einfachen Mann. Auf den ersten Blick hätte man von einfach auf unbedeutend schließen können; aber jeder weiß es, Süss war alles andere.

Die geplante Zusammenkunft entwickelte sich sehr familiär, im Stile den wir heute noch kennen, aber im alten Schloß. Lebensmittel, fühlte man, waren noch ein Problem, wenn auch nicht wie vor kürzerer Zeit. Das „gouvernement militaire" in Wolfach begrüßte den Besuch aus Frankreich mit Proviant.

Nach den Andeutungen von Ehresmann erhoffte ich ein interessantes Kolloquium; doch meine Erwartung wurde weit übertroffen. Ich kann nicht alle Teilnehmer aufzählen. Großen Eindruck machte G. Bol (mit dem ich später manches mal eng zusammen arbeiten durfte). König, Leichtweiss, viele aus der Schweiz und natürlich Heinz Hopf, damals unermüdlich, so wie er es immer war. Und mit diesen Leuten konnte ich sprechen, ohne Termin, am Abend, am Morgen. Ein Erlebnis!

Das allerwichtigste „événement", das hier festgehalten sein muß, ist das Deutsch-Französische Zusammentreffen wenige Jahre später. Henri Cartan war, von französischer Seite, der Hauptinitiator. Wer über das Schicksal der Familie Cartan weiß, kann daraus das Gewicht und die Bedeutung der Bemühung schätzen.

Da waren mehr als zwanzig junge Leute, darunter Serre (sehr jung, machte auf die deutschen älteren Herren einen verblüffenden Eindruck), R. Thom (den Begriff „Anarchie" fühlte er geeignet, die Struktur des Instituts zu charakterisieren, (selbstverständlich mußte ein „Oberanarch" da sein, um das „efficacement" zu verwalten) [1] B. Charles (er wurde später berühmt als jüngster und sportlichster „Doyen de France"; trotzdem ist er heute ‚nur' Bruder des viel bekannteren Autors der „foire aux cancres"). C. Chabauty, auch treuer Besucher, war diesmal nicht dabei; aber ihm geschah ähnliches: heute ist er besser bekannt als Schwager einer der „Frères Jacques"; J. Braconnier (er lehrte Frau Süss und Kneser den Begriff „indiscipliné"), Riss, ... Bilharz, Gericke, Jehne, Fr. Schlarb, K. Stein, G. Pickert. ... Mathematik stand hoch; selten war ein Seminar so reich an Saat. Über Politik, aktuelle oder vergangene, wurde viel weniger gesprochen, aber trotzdem war da mancher wichtige Austausch. Süss befürwortete – und meines Erachtens hatte er ganz recht –, einen Besuch des Gouverneur militaire (aus Wolfach). Dieser Mann hatte für das Institut öfters lohnende Schritte unternommen. ‚Indiscipline oblige', den jungen Franzosen gefiel das gar nicht, aber Monsieur de Rendinger kam und sein Besuch wurde ein „succès". Nebenbei war auch durch diesen Kanal das Essen um eine Stufe verbessert.

Eine wichtige Persönlichkeit, deren Rolle oft von Süss erwähnt wurde, konnten wir nie sehen; es handelt sich um einen Wissenschaftler, capitaine mit Auftrag ‚contrôler', der sich ganz speziell Süss anhörte und sehr positiv für Oberwolfach arbeitete. Hier sei darüber Zeugnis abgelegt: durch L*** und andere Wissenschaftler, in Bad Ems, wurde da manche Zusammenarbeit begonnen. Damals ging das nicht von selbst; die Zukunft gab diesem Bestreben Recht.

---

[1] Auszug aus dem Gästebuch vom 7.–25.8.1949:

La lecture de ce laius ne nécessite, en principe, aucune connaissance mathématique spéciale ∗; néanmoins, il est destiné à des lecteurs ayant au moins une certaine habitude du climat d' amitié mathématique et polyglotte dont nous avons joui au Lorenzenhof. Il est bien difficile d'analyser l'exquise richesse, des structures qui interviennent dans ce climat; il est encore plus délicat d'ordonner, même partiellement, l'ensemble des faveurs qui nous furent prodiguées par nos hôtes. Pourtant y oserons nous appliquer l'axiome de choix ∗∗ afin d'en distinguer un élément maximal: il nous faut ici remercier Monsieur et Madame Süss de nous avoir permis de donner vie, pour quelques jours, à ce vieux mythe ∗∗∗ si cher à nos coeurs, de l'abbaye de Thélème.

∗ Saint Nicolas, Introduction, premier verset.
∗∗ Saint Nicolas, op. cit, pars prima, lib primus, III, cap. 4.
∗∗∗ F. Rabelais, Opera omnia, passim.

Suivent les signatures Arbault, Serre, Braconnier, Thom, Charles, Perreira Gomez, illisible, ...

Das Wort „laius" (Rede) bedarf eines Kommentars: dieses Wort war derart neu und ungewohnt für Nicht-Frankophone, daß es lange, hochgelehrte Diskussionen auslöste.

Vielleicht kann ich unter diejenigen gezählt werden, die sehr oft in Oberwolfach weilten; auch eine ganze Saison konnte ich im Schloß verbringen und in der Bibliothek schlafen... oder die ganze Nacht lesen. Dies war eine besondere Kur; aber welcher Doktor könnte sie heute noch verschreiben?

Die Gäste, die in dieser Zeit in Oberwolfach verweilten, kann ich natürlich nicht aufzählen, aber einiges Anekdotisches festhalten. Zunächst waren da die vielfachen Gespräche mit Süss; irgendwie hatte ich die Chance, als sein „confident" zu figurieren. Mathematik spielte wohl eine Rolle beim Diskutieren; doch wichtiger war, daß progressiv das große Ziel von Süss klar wurde: unaufhörlich an dem Verständnis zweier Völker zu arbeiten, jeden Tag, trotz der Hindernisse, die auftreten würden. Süss erzählte von den Anfängen des Instituts in trüber Zeit: wie Juden geholfen wurde unter dem Deckmantel numerischen Rechnens; auch die Aufnahme einiger Kriegsgefangenen und von anderen trüben Zeiten, wo Hunger aktuell war und Proviant eine Hauptsorge. Vielleicht war ich nicht der beste Zuhörer, aber eben der anwesende. In diesem Zusammenhang muß G. Hirsch erwähnt werden (auch Topologe), ein belgischer, oft gesehener und beliebter Besucher, sehr im Stile ‚vieille France', obwohl sehr jung. Hirsch verstand Süss und sein Bestreben weit besser und mehr von Herzen als ich; eine gute Lektion!

G. Bol, ein Riese, muß man natürlich neben Süss nennen. Immer bescheiden, immer da, wo es galt einzuspringen, immer fröhlichster Stimmung, und immer mit Süss im Wechselgespräch über Geometrie, so wie diese Kunst von ihnen gepflegt wurde. Wie könnte man alle Mitarbeiter von Süss aufzählen, die im selben Sinne wie Bol das „Institut" geschaffen haben, Stunde für Stunde, Jahr für Jahr!

An pittoresken Besuchern war kein Mangel. Zunächst Professor Behnke aus Münster, ein Stammgast. Er hatte alles miterlebt. Von Politik sprach er gerne, sehr freundlich und von der Seele. Nie werde ich vergessen, wie er eines Morgens, die Zeitung in der Hand schwingend, ausrief: „Denken Sie Süss! Wir (gemeint waren die Deutschen) dürfen wieder mitreden, es wird, wenn auch nur sehr wenig, wieder auf uns gehört, verstehen Sie?: *gehört!*". Den Anlaß habe ich vollständig vergessen. Eisenbahn war sein Hobby (wie dies auch ein Lieblingsthema für O. Haupt war; Haupt, der mit Lothar Hefter bis zum hundertjährigen Geburtstag leben durfte). Behnke belehrte seinen großen Freund Cartan: „Ihr Franzosen kennt betreffend Eisenbahn nur eins: rasen; das ist aber nicht das Wichtige". Seiner Freundschaft mit Cartan war er treu bis zum letzten Besuch zum jubilé in Paris. Dort erwähnte er berührt, was er noch kürzlich erleben durfte, darunter eine kurze Andacht an dem Goethedenkmal in Sesenheim, Friederike. . . .

Wenn wir schon Cartan erwähnen, soll eine typische Anekdote ein Stimmungsbild geben (J. Frenkel war der Zeuge). So anno 1951 kam Cartan nach dem Lorenzenhof mit Auto, gesteuert von einem Pariser Kollegen „Maître de conférence". Zunächst wurde der Mann als (vornehmer) Chauffeur betreut; als klar wurde, er wäre Mathematiker, fand man es selbstverständlich, daß ein Wissenschaftler vom Range Cartans wohl einen außerordentlichen Professor als Chauffeur haben kann.

Ulkig war manchmal die politische Plauderei. Frau Süss erzählte mit Gefühl gemischt mit Heiterkeit, Verblüffung, Staunen, wie an einem Winterabend beim Cheminée B. d'Orgeval, (Schnauzer, Royalist, Adeliger, Rechtshistoriker, Reiter,

Geometer, Winzer und bestimmt Önologe (ganz nebenbei: im Schloss herrschte schon in dieser Zeit eine angemessene Weinkultur. Auch dauerte es nicht sehr lange, bis der Drei-Sterne-Tisch zu Illhäusern besucht wurde.)) die Gäste mit einer ruhigen Frage entsetzte: „Wer bemüht sich um den Hohenzollern-Thron?"

Dieudonné, einen beliebten Gast im Schloß, „hörte" man durchs ganze Haus. Da war oft die Rede von Bourbaki (ein in Oberwolfach sehr begehrtes Thema); wie schön und definitiv klang die ganze Komposition. Cartan hob, sofern es noch möglich war, die Stimmung mit einer Einführung in die (damals entpuppte) Theorie der Kategorien. Neben Mathematik posaunte auch Dieudonné Lektionen über französische Grammatik hinaus, wenn eine eifrige Dame mit bestem Willen sich zu präziös ausdrückte. Bernays war sicher auch eine pittoreske Figur; wie viel Logik habe ich versäumt durch zu flüchtiges Zuhören? Auch von Nevanlinna könnte man erzählen.

Den Lorenzenhof erwähnen, ohne die unermüdliche Leistung von Frau Süss und auch anderer Damen anzuerkennen klingt unmöglich. Nach Tautz war L. Hefter ein Meister der „Damenreden". Sein Talent wäre jetzt notwendig. Leider...

H. Kneser war einer der am meisten angehörten Mathematiker. Neben Tonnen von Geometrie (auch Blaschke sah man) belehrten die Deutschen mit Perlen aus der Algebra, der formalen Logik, mathematischen Ökonomie. Laugwitz, auch Geometer, hatte schon Nichtstandard Mathematik in Sinne (auch Schmieden weilte hier); aber damals habe ich fast nichts verstanden (und Laugwitz durfte nur karg über dieses Thema plaudern). Unter anderem hörte ich zum ersten mal von Lamberts Werk durch den Schweizer A. Burckhardt, der sehr von diesem Elsässer schwärmte. (Hatte doch damals Mulhouse eine Partnerschaft zur Eidgenossenschaft?). Von angewandter Mathematik war nicht viel die Rede. Aber zu dieser Zeit wurde im Institut manche Zusammenarbeit um das Zentralblatt aufgebaut.

Neben Musik und Wanderungen (manche Schwarzwald-Hochzeit konnten wir miterleben) war Philosophie ein Thema (Heidegger ist immerhin ein Mann aus dem Schwarzwald). Die Damen wollten über Sartre hören; sie wußten eher besser als die Lehrer über die Verwandtschaft mit A. Schweitzer. Für mich hatte eine kleine Entdeckung Bedeutung: mit den Bauern auf den Höhen konnte ich mich auf elsässisch leicht und herzlich unterhalten, und es war eine lokale Freude, daß Mathematiker mehr aus dem Norden nur schwierig mitkamen.

Wir sind an den internationalen Charakter der Besuchergemeinschaft des Lorenzenhofes gewöhnt. Doch war in diesen Zeiten ein außerordentlich buntes Bild verschiedenster Nationalitäten zu beobachten: Amerikaner verschiedenster Herkunft (bald ein echter Farmer, bald ein Quäker, bald ein Mann, dessen erste Ankunft im Lorenzenhof in U.S.-Uniform stattfand, bald ein Deutscher, der damals flüchtete oder der Sohn einer, schon 1933, vertriebenen Familie). Engländer waren auch da; dann einer aus Süd-Afrika (dem Manne aus der weiten Landschaft war das enge Zusammenleben unmöglich. Um sich auszutoben, blieb ihm nichts übrig als Krach zu suchen, (den fand er auch bald bei irgend einem Volksfest). Südamerikaner, Russen,...

Die Leute, die sich hier trafen und kreuzten, hatten eines gemeinsam: die Mathematik. Einmalig erscheint die tatsächliche Zusammenarbeit all dieser Mathematiker mit Süss. Es ist wahr, daß man vom Geist im Lorenzenhof das sagen darf, was nach Ostrowsky (auch ein treuer Gast) von den deutschen Universitäten so gegen 1900 galt: „die Institution war eben größer als die Menschen"; nur hier war die Institution improvisiert.

Zwei Mathematiker haben vieles geleistet im Sinne von Süss, vielleicht ganz unbemerkt und spontan aber harmonisch: Ich denke zunächst an Frau und Prof. Ancochea aus Madrid. Plötzlich war nur ein Wort zu hören, sowohl in der Küche wie im Seminar: Charme. Frau Ancochea besaß eben diesen Anmut. Dann Guy Hirsch, den ich vorher erwähnte. Auch Vincensini (schon nahe siebzig), der Korse mit blauen Augen und Tino-Rossi-Stimme, rührte den gesamten Lorenzenhof.

Wie kann man von diesen Erinnerungen Abschied nehmen? Heute ist alles anders: Organisieren, Planieren, Bauen, Haushaltsplan, Beirat! Mehr als genug, um zu vernichten, was von Süss und Mitarbeitern mit Freude und Vertrauen geschaffen wurde. (Zum Beispiel pflegte nicht R. Baer, ein vornehmster Freund des Instituts, sich im Beirat gegen jeglichen Modernismus zu wehren?). Doch es ist klar: der Geist des Hauses besteht, nicht trotzdem, sondern deshalb!

Süss, Behnke, Kneser mit ihren Zeitgenossen sahen (über fast ein Jahrhundert) vieles zusammenbrechen, von dem gedacht wurde, es währe ewiglich. Vieles sahen sie auch wieder aufbauen. ... Vielleicht hat der Lorenzenhof einen Hauch von Ewigkeit?

G. Reeb, geboren 1920 in Saverne, gestorben 1993. Studien in Straßburg. Promotion 1948 mit einer Abhandlung über „Propriétés topologiques des variétés feuilletées". Professor an den Universitäten Grenoble und Straßburg. Ehrendoktor der Universitäten Neuchâtel und Freiburg/Br.

Abschied vom Schloß in Oberwolfach *

Eine Ansprache am 23. Juni 1972 von Heinrich Behnke in Münster

Wenn wir uns nun hier zum letzten Male im Altbau des Mathematischen Forschungsinstitutes Oberwolfach versammelt haben, so ist allen bekannt, welche sachlichen Gründe für den Abbruch des Hauses und seinen Ersatz durch einen streng nach den Anforderungen des Institutslebens gegliederten Neubau im Stil der Betonbauten unserer Tage bestehen. Wenn der Neubau erst steht, wird er sich segensvoll für das Leben im Institut auswirken.

Doch bin ich sicher, daß die ganz überwiegende Mehrzahl der alten Freunde bitter klagen werden, wenn sie zum ersten Mal die Veränderung sehen werden. (Ja, es genügt dazu schon der Vergleich der Bilder vom alten Schloß und der Photomontage für den Neubau.) Ein Stück Romantik und ein Zeugnis jüngster deutscher Geschichte (1905–1972) sind erloschen.

So kann es uns nicht verwehrt sein, dem Abbruch des Schlosses nachzutrauern. Bei vielen von uns hängt auch ein Stück Leben, ja für manchen sogar ein Stück gefährliches Leben, häufig auch ein Teil fachlicher Arbeit mit diesem Hause zusammen. Die so hervorgerufene Dankbarkeit gilt nicht der abstrakten Institution „M. F. Oberwolfach". Im Gegenteil! Alle menschlichen Empfindungen konzentrieren sich in natürlicher Weise auf greifbare Objekte. Bei uns auf dieses Schloß! Manche unter uns haben versucht, durch Erwerb eines charakteristischen Stückes aus dem Schloß bei sich die Erinnerung besonders wach zu halten. Unter allen Wünschen, die mir bekannt geworden sind, scheint mit am treffendsten zu sein: „1 Liter Luft aus der alten Bibliothek". Zugleich wird damit klar, daß man nichts außer Bildern aus der unwiederbringlichen Zeit bewahren kann. Was wir auch tun und sagen, heute abend geht für uns und manche, die heute nicht hier sein können, ein Abschnitt unseres Lebens zu Ende.

Wir haben deshalb zunächst ein Gefühl der Dankbarkeit denjenigen gegenüber zu bezeugen, die uns dieses Schloß als eine für ruhige geistige Arbeit hervorragend geeignete Stätte in schwerster Zeit geschaffen haben.

Diese Gründung eines Forschungsinstitutes konnte wirklich 1944 nicht durch einen einzelnen Kraftakt geschehen. Während der mehrjährigen Planung galt es immer wieder, Gefahren, die von mächtigen politischen Interessen ausgingen, abzuwehren. Schließlich war das Schloß zuletzt in Händen der Führung des NS Lehrerbundes gewesen.

---

* Jber. Deutsch. Math.-Verein. **75** (1973) 51–61.

Aufnahmen: Ottmar Rogg, Oberwolfach/Schwarzwald

Aber am 1. September 1944 konnte das Schloß (vorläufig, d. h. für die Kriegszeit) uneingeschränkt auf Anordnung des Präsidenten des Reichsforschungsrates
von Wilhelm Süss für mathematische Arbeiten übernommen werden. Die Mitarbeiter kamen einzeln. Woher jemand auch immer kam, aus der Heimat oder von
der Front, er wurde überwältigt von der weltabgeschiedenen Stille des Schlosses,
der ungewöhnlichen Bequemlichkeit im Umgang mit der Fachbibliothek, (die meisten brachten zunächst – und wie sich häufig zeigte – umsonst Kisten von Büchern
für ihre Spezialarbeit mit hinauf), die stets vorhandene, aber nie aufgezwungene
Möglichkeit zu Fachgesprächen, die mit sehr wenig Mitteln, aber mit großer Herz-
lichkeit durchgeführten gemeinschaftlichen Abende und die bald hohen Ansprüchen
genügenden musikalischen Darbietungen. Dazu kam noch die Umgebung, der für
Fußgänger einfach unbeschränkt große einsame Wald. Dem Laien scheint es vielleicht in einer Kriegszeit zuviel des Guten zu sein. Aber es gab keine andere Art
der Ablenkung, und so war hier einfach ein Maximum von Möglichkeiten, mathe-
matisch zu arbeiten, geschaffen.

Das Leben im Schloß hat sich sichtbar in den vergangenen 28 Jahren gewandelt. Zuerst überwog natürlich die Arbeit an den mitgebrachten Aufträgen. Aber
auch nachher waren noch lange die Einzelarbeiter in der Mehrzahl, die nicht nur
Wochen, sondern Monate und Jahre blieben und ihre Dissertationen, Habilitations-
schriften oder Bücher schrieben. In den Jahren nach 1948, als alle Älteren unter uns
ihre früheren oder neuen Heimatplätze gefunden hatten, waren in den ungünstigen
Jahreszeiten nur wenige, meistens junge Fachleute oben. Die Tagungen waren auf
wenige Monate verteilt. Im Winter fehlte die Zentralheizung und auch noch die
normale Ernährung. Privatwagen gab es auch in den Jahren nach 1948 noch kaum,
und die öffentlichen Verbindungen waren einfach qualvoll. Aber mit dem Wohlstand wurde alles anders. Man fand bald Geschmack an den Gruppentagungen. Für
immer mehr Spezialfragen fanden sich Gleichgesinnte zusammen. Und es kam die
Zeit, wo für die international stark besuchten, etwa 6 Tage währenden Tagungen
fast alle Plätze beansprucht wurden.

Wurde dann auch am 10. Oktober 1967 mit Eröffnung des neuen Wirtschaftsgebäudes das gesellige Leben auf die neuen Gebäude offiziell verlagert, so konnten
doch damit nicht die Stammgäste aus der überaus gelungenen neuen Bibliothek
(früher Zimmer 1 des Schlosses) verdrängt werden, noch konnte trotz aller Un-
bequemlichkeit die abendliche Geselligkeit für die alten Freunde des Schlosses
aus der vertrauten alten Bibliothek einfach herausgenommen werden. Das dauerte Jahre. Die gemeinsame Erinnerung an Not und Entbehrung vergangener Tage,
wo der Kamin in der Bibliothek uns alle vereinte oder noch einmal eine kleine
Kostbarkeit verteilt wurde, hielt uns zusammen.

Nun müssen wir heute abend zunächst derer gedenken, denen wir den Erwerb
und die Verteidigung dieses Hauses verdanken. Irmgard Süss hat diese schweren
und heiklen Verhandlungen ausführlich geschildert und mit Dokumenten belegt.
(Diese Broschüre ist herausgegeben vom M. F. O. am 10. Oktober 1967.)

Daraus möge aus meiner Sicht ein sehr knapp gefaßter Bericht folgen. – Bei den
Bemühungen, den großen Verlust, den das mathematische Leben durch die grau-
sigen politischen Vorgänge nach 1933 erlitten hatte, wieder wettzumachen, hatte

Wilhelm Süss dem Reichskultusministerium vorgeschlagen, ein mathematisches Forschungsinstitut zu gründen. Dabei ist ihm wenigstens teilweise das Institute for advanced study in Princeton, N.J. Vorbild gewesen. Solche Bemühungen lagen nahe. Zugleich mahnte die große Vergangenheit Göttingens. Das war, nach zahlreichen Vorstufen, von 1900 bis 1933 das große mathematische Zentrum, nach dem die ganze fachliche Welt sich richtete. Was in jener Zeit dort als wichtiges Problem angesehen wurde, war – wenn dann gelöst – als eine der großen wissenschaftlichen Leistungen der Zeit uneingeschränkt anerkannt. Wer sich in Göttingen durch seine Leistungen Respekt verschaffte, gewann ihn auch in der Heimat. So war es selbstverständlich gewesen, daß jeder junge, sich Erfolg versprechende Mathematiker versuchte, zeitweise in Göttingen zu leben. 1933 war diese Stellung Göttingens mit einem Schlage vorbei gewesen. Diese Katastrophe ist wiederholt geschildert.

Wer nun an Göttingen hing und von uns Mathematikern in Deutschland weiterleben wollte oder mußte, dachte an den Wiederaufbau dieser „Mathematiker-Residenz". In den ersten Jahren des braunen Sturms war dieses Bemühen hoffnungslos. Der Wind wehte allen Intellektuellen zu sehr ins Gesicht. Dann begannen langsam die Pläne bei einzelnen verantwortungsbewußten Mathematikern schon Gestalt anzunehmen. Als die schnelle wissenschaftliche Entwicklung in den „Feindstaaten" immer sichtbarer wurde, kam die Zeit, wo Männer, bei denen ein ausgewogenes Verhältnis zwischen Patriotismus und wissenschaftlicher Objektivität niemals aus den Fugen geraten war, trotz der unerträglichen Angriffe der Radikalen nun wieder an hohen Stellen gehört wurden.

Dazu gehörte neben Helmut Hasse unbedingt der Rektor der Universität Freiburg, Wilhelm Süss, der seine Universität – rühmend wurde es hinter vorgehaltener Hand berichtet – aus dem politischen Wirrwarr der damaligen Zeit herausgeholt hatte und der auf ein ruhiges Arbeitsklima an seiner Universität hinweisen konnte. Nicht von ungefähr hatte die oberste Marineleitung (das „OKM"), eine Behörde, die als zurückhaltend gegenüber den „progressiven" Unruhestiftern galt, die Ausbildung der zukünftigen Mediziner nach Freiburg verlegt. Süss gewann das Vertrauen des Ministers Rust[1]. Es gelang ihm bei seinen Bemühungen auch die mathematischen Publikationsorgane wieder in Gang zu bringen sowie die formale und materielle Unterstützung des Reichsforschungsrates zu gewinnen. Natürlich war dies alles kein Passepartout. Es gab genügend Parteiführer vieler Grade, die jede Arbeit von Süss zunichte machen konnten. So erklärte er mir eines Tages – und er tat es sichtlich resigniert – daß das Telefon des Rektors abgehört würde. (Ein bramarbasierender Student hatte in später Abendstunde in einem Lokal davon erzählt, und so hatte ich Süss gefragt.)

Im Laufe der Zeit gelang es Süss, ein Programm für dringende wissenschaftliche Arbeiten aufzustellen und dafür aus Front und Heimat die ihm geeignet erscheinenden Autoren zu gewinnen. Dabei kam er ganz selbstverständlich auf die Idee, ein vom Weltgeschehen möglichst unberührtes Mathematisches Forschungsinstitut zu gründen. Verhandlungen mit Göttingen wegen einer eigenen Berufung nach dort führten erst recht dazu, nachdrücklich die Bitte auszusprechen, ihm ein solches In-

---

[1] Siehe Irmgard Süss, Broschüre M.F.O. vom 10. Oktober 1967.

stitut zu gewähren. Das Ministerium in Karlsruhe wollte ihn in Freiburg halten. So kam es zum ersten Angebot des Lorenzenhofes. Das wiederum war der Fakultät in Göttingen unangenehm. So gingen die Verhandlungen mit Wilhelm Süss hin und her.

Einen gefährlichen Widerstand gegen das Projekt Lorenzenhof leisteten mächtige Parteiinstanzen, die die Parteischule auf dem Lorenzenhof nicht verlieren wollten. Aber im Sommer 1944 ging schließlich für die, welche noch an den Endsieg dachten, die Kriegswichtigkeit vor. (Noch mehr Parteischulung war wirklich nicht möglich!) Für die Parteifunktionäre, die den Zusammenbruch für unvermeidlich hielten und deshalb sich darum bemühten, möglichst viele Deutsche überleben zu lassen (und deren gab es an hohen Stellen glücklicherweise auch einige), waren die Forderungen der radikalen Parteileute, milde gesagt, gefährliches, eigenwilliges Gestammel Besessener.

So konnte Wilhelm Süss zum 1. September 1944 das Institut von Oberwolfach eröffnen. Eine für die damalige Zeit erstaunlich gute Bibliothek war schnell beschafft. Die Verleger in Leipzig eröffneten gern ihre durch den Bombenkrieg höchst gefährdeten Archive für uns. Irmgard Süss hat lebendig geschildert, welche Mühe in den vorangehenden Wochen für die Beschaffung der notwendigen Einrichtungen bis dahin schon erfolgreich aufgebracht war. Die Ernährung aber machte die größte Sorge. Den Pythagoras konnte man den Bauern nicht einmal für ein faules Ei verkaufen. Obwohl das Reisen damals schon recht beschwerlich war (Berlin–Oberwolfach 2 Tage und Nächte und bald darauf 5 und keine Möglichkeit, unterwegs in einem Hotel zu übernachten), kamen bald viele von denen, die Süss für sein Projekt verpflichtet hatte. Zuletzt kam Theodor Schneider direkt von der Front. Seine Anforderung war telefonisch an die Deutsche Bank in Mannheim gegangen. Von da ging sie per Fernschreiben an das Büro Osenbug in Northeim. Wenige Nächte später hatten wir die Bestätigung seiner Abreise. Nun war das Schloß überfüllt. Inzwischen wurde die militärische Lage immer kritischer, und die Verbindung mit der Außenwelt war auf Zufälligkeiten angewiesen.

Die Überrollung durch die Front hat Irmgard Süss geschildert. Wer diese Broschüre kennt, kann begreifen, daß wir trotz verschiedener Interessen und Auffassungen zur Lage eine Schicksalsfamilie waren, die mit unbestimmtem Ziel reiste. Es gab Zeiten, wo wir kaum eine verläßliche Nachricht von draußen hatten. Die ersten Wochen der Besetzung, die sich zunächst wegen der Zerstörung der Brücken nur durch einzelne Boten bemerkbar machte, waren schon wegen der Ungewißheit eine Last. Unser erster Trost waren unsere Ausländer bzw. Halbausländer, Pisot (Elsässer, jetzt Paris), Roger (jetzt Bordeaux), Threlfall (englischer Vater) und Görtler (damals Kanadier). Zweifellos hat jeder geholfen.

Unsere großartigste Hilfe kam von völlig unerwarteter Seite. Unsere Retter in der Not wurden John Todd (der Ehemann von Olga Taussky, beide jetzt Professoren am California Institute for Technology in Pasadena) und G. E. H. Reuter. Sie kamen als englische Stabsoffiziere und zugleich als unsere Freunde. Sie suchten uns persönlich, um die wissenschaftlichen Institute in die Kontrollisten der Militärregierung aufzunehmen. Am Morgen nach ihrer Ankunft war auf der Terrasse ein friedensmäßiger Kaffee gedeckt. Da kam um die Ecke der neue Ortskommandant, ein französischer Hauptmann, um für seine Kompanie das Schloß zu beschlagnahmen. Er verlangte den Hausherrn zu sprechen. Für uns gab es einiges Herzklopfen. Doch nach wenigen Minuten war die Lage durch Colonel John Todd geklärt. Und in den nächsten Tagen gelang es ohne Schwierigkeit, durch die französische Generalität die Ausnahmestellung des Institutes schriftlich bestätigt zu bekommen. Das war die entscheidende Stunde des Institutes.

Der Schreiber dieser Zeilen war in diesen Tagen nicht mehr in Oberwolfach anwesend. Er hatte aber die Gastfreundschaft der Familie Todd noch unmittelbar vor dem Kriege in London genossen, und er wußte, daß er auf der Suchliste des Colonel Todd stand. Es ist aber naheliegend, daß er dieser Stunde auf dem Lorenzenhof schon im Interesse von Wilhelm Süss und des Hauses immer besonders dankbar gedacht hat. Wäre dieser glückliche 9. Juli 1945 nicht gewesen, wäre höchstwahrscheinlich die Zukunft des Forschungsinstitutes verloren gegangen. Der neue finanziell ganz schwache Staat Südbaden hätte bei seinen vielseitigen Verpflichtungen weder die finanzielle noch die moralische Möglichkeit gehabt, neben der Universität Freiburg noch ein mathematisches Forschungsinstitut Oberwolfach zu unterhalten, auch wenn man von allen großen Plänen absah und sich auf das äußerste beschränkte. Der Staatspräsident Wohlleb, der schon wegen Vorkommnissen in der Nazizeit Süss dankbar war, tat das äußerste, was er konnte. Er übernahm die in altem Papiergeld festgesetzte Miete für das Schloß. So blieb vorläufig praktisch Oberwolfach ein Niemandsland. Und während die alten Gäste zu Hause ihre alte Existenz wieder aufzubauen suchten, sah jetzt das Schloß neue Gesichter. Flüchtlinge der verschiedensten Art kamen. Unkontrolliert von der Militärregierung konnte Süss die Ankommenden nicht aufnehmen. Diese mußte natürlich ihre Genehmigung geben. Für einige mußte deshalb an die Nächstenliebe der Dorfbewohner appelliert werden.

Schon am 28. April 1945 fand der erste Kolloquiumsvortrag wieder statt. Die Arbeit im Hause unter der sich langsam ändernden Belegschaft lief weiter. Nur die Not an Lebensmitteln und Heizmaterial nahm noch zu. Dazu kam die Verkehrsnot. Der Eisenbahnverkehr – soweit überhaupt funktionierend – war im wesentlichen nur für die einzelne Zone ausgebildet. Für den Zonenübertritt war eine besondere Erlaubnis erforderlich. Vom Nordwesten mußte man dreimal die französische Zonengrenze überschreiten. Für eine Ausreise ins Ausland brauchte man einen von allen vier Militärregierungen ausgestellten Paß. Der Verfasser dieser Zeilen war im Besitz eines solchen Pseudopasses mit der Nummer 000 10. Hotels zum Übernachten gab es nicht. Unmöglich war es, Lebensmittelkarten an fremden Orten zu benutzen. Das armselige Schuhzeug ließ keine Fußmärsche zu (Verfasser hat auf einem nächtlichen Weg von Oberwolfach nach Hausach seine Schuhsohlen verloren). Aber unerwartet gab es immer wieder Auswege und vor allem menschliche Hilfe. Die wahren Freunde von Oberwolfach blieben nicht aus. Die Lage von Oberwolfach sprach sich bald in den Fachkreisen der ganzen Welt herum. Die große Solidarität der Mathematiker in der ganzen Welt zeigte sich, ja, sie war das Erstaunlichste und Erfreulichste in der Geschichte des Hauses.

Bald kamen die ersten Ausländer. Sie veranlaßten, daß der französische Ortskommandant ein dörfliches Schlachtfest gab. Aus Frankreich kamen Charles Ehresmann, Henri Cartan, Christian Pauc, Gustave Choquet, aus der Schweiz Heinz Hopf, Ostrowski, Hadwiger, Stiefel, aus den Niederlanden Schouten. Und sicher habe ich hier viele vergessen. Die Verbindungen mit den obersten Kommandostellen der Militärbehörden wurden angenehm. Wir konnten mit Hilfe rechnen. Die FIAT-Berichte über die mathematischen Arbeiten in Deutschland 1933–1945 wurden entworfen und beschäftigten viele Schloßbewohner. Bald darauf wurden auch das Archiv der Mathematik und die Studia Mathematica in Oberwolfach gegründet, ebenso die Semesterberichte. Das äußere Leben im Schloß normalisierte sich, wenn auch keineswegs so schnell wie in den Städten des Landes und vor allem nicht wie in der Bizone. Oberwolfach war überdies kein fruchtbares Landstück. Selbst das feinste Holzmehl unseres Nachbarn Krauter konnte nicht zur Ernährung gebraucht werden.

Aber die wissenschaftliche Arbeit und ebenso das Kolloquium nahmen ihren stetigen Verlauf. Bald halfen manchem Care-Pakete aus dem Ausland. So konnte noch in der Notzeit das Haus seinen Forschungszwecken wieder uneingeschränkt zugeführt werden. Immer mehr Mathematiker aus dem In- und Ausland nutzten diese Idylle Schloß für die eigene wissenschaftliche Arbeit und den ungezwungenen Umgang mit Fachgenossen. Man konnte die Bücher ohne Kontrolle in sein Arbeitszimmer mitnehmen und ebenso in heißen Tagen in die lauschigen Lauben und auf die höher gelegenen Waldwiesen. Die Leitung brauchte keine Sorgen um den Bestand der Bibliothek zu haben. Und die vielen ausländischen Besucher brachten auch zu unseren Gunsten ihr Ansehen und ihre Solidarität mit. Selten wird ein wissenschaftliches Unternehmen so von dem Bewußtsein geistiger Gemeinschaft getragen sein wie Oberwolfach in den fünfziger Jahren. Jeder freute sich über die dort möglichen Begegnungen, die doch viel weniger von der Hast und räumlichen Enge gekennzeichnet waren als die Zusammenkünfte auf den Kongressen. Hermann Weyl erklärte, daß er nach dem Kriege zum ersten Male sich im Kreise deutscher Mathematiker wieder restlos wohl gefühlt habe. Nicht aufzählbar sind hier die prominenten Gäste, die alle zu Besuch nach Oberwolfach kamen. Und jeder Gast wurde aufmerksamer und herzlicher behandelt als in jeder noch so guten Unterkunft. Es war die Zeit des Triumphes internationaler Solidarität der Mathematiker.

Und doch war Oberwolfach damals keineswegs so überlaufen wie heute. Auf zahlenmäßig große Zusammenkünfte konnte man sich noch nicht einstellen. Die heutigen vielen Einzelzimmer im Neubau gab es noch nicht. Man mußte vielfach drei und vier Herren in ein Zimmer legen. Und es geschah, daß morgens jemand zum Kaffee kam, als hätte er bittere Zahnschmerzen. Sein Nachbar hatte noch im Bett endlos Witze erzählt, sich dann umgedreht und in gleicher Stärke Lieder ohne Worte gesungen. Da zog man doch lieber die Personalzimmer im Dachgeschoß statt der großen Besuchszimmer im ersten Stock vor. Die drei Fürstenzimmer überließ man sowieso gern der ausländischen Prominenz.

Mit steigendem Wohlstand – etwa in der zweiten Hälfte der fünfziger Jahre beginnend – hatte schon der eine oder der andere sich gescheut, Oberwolfach zu besuchen, wenn er nicht sicher war, ein Einzelzimmer zu bekommen. Auch gab es noch nicht die großen Fonds zum Ersatz der Reisekosten. Wer kam, bezahlte selbst und brachte vielfach einige Kostbarkeiten zur Ernährung für alle mit. Nur da, wo ganz Besonderes an geistiger Kost zu erwarten war, konnte Wilhelm Süss Beiträge oder Stiftungen ausländischer Freunde zur Verfügung stellen. Oberwolfach war wahrlich kein Fürstensitz mehr. Am Essen mußte im allgemeinen mehr als in einer bürgerlichen Familie gespart werden. Aber im Geiste war Oberwolfach elitär.

Es gab Wochen und Monate in den fünfziger Jahren, wo während der langen Winterzeit, aber auch in der ersten Hälfte des Monats August das Haus fast leer stand. Das war im Sommer für die Stammgäste (Hellmuth Kneser, Helmut Gericke, Kuno Fladt, A. Baur-Lübeck, Reinhold Baer, Reeb-Straßburg, dem Unterzeichneten und noch einigen anderen) wohl zu ertragen. Aber das paßte dem Landesministerium in Stuttgart garnicht. Zweimal wurden wir ausgerechnet im August von einem hohen Beamten des Kultusministeriums kontrolliert. Hier drohte ernsthafte Gefahr, daß das neue Land Baden-Württemberg gelegentlich der Neuordnung seiner Finanzen dem Hause die schmale Existenz entzog. Oberwolfach paßte in keinen der üblichen Posten. Forschungsinstitute waren Angelegenheit der Max-Planck-Gesellschaft.

Doch noch einmal fiel ein großer Glanz auf das Schloß, so wie Süss es geplant hatte. Das war am 7. März 1955, dem 60. Geburtstage des Gründers. Ich wurde am späten Vorabend in Hausach von einem jungen französischen Kollegen abgeholt. (Die Abhängigkeit von der Kleinbahn hatte aufgehört. Man fuhr mit gläsernen „Schneewittchen-Särgen" und Autos, die man durch ihren Bauch betrat. Die Jugend machte davon eifrig Gebrauch. Oberwolfach lag jetzt nahe an Offenburg, Stuttgart, Baden-Baden, Straßburg und Basel.)

Die Landschaft lag an jenem Abend unter einer leichten Schneedecke. Und während wir den „Buckel" hinauffuhren (jener Auffahrt zum Schloß, die einer der künstlerisch gelungensten Teile der ganzen Anlage ist) erglänzten alle Fenster des Hauses. Kein Zweifel, das Haus war voll besetzt von allen, die für den nächsten Tag gekommen waren. Da waren Bompiani aus Rom, Hermann und J. F. Weyl aus USA, Nevanlinna aus Finnland, Frostman aus Stockholm, Heinz Hopf und Ostrowski aus der Schweiz, natürlich unsere französischen Freunde, jetzt – soweit aus Straßburg – unsere nächsten Nachbarn, die Gesamtheit der Fachgenossen der Nachbaruniversitäten, natürlich unsere österreichischen Kollegen und viele, deren ich mich jetzt nicht mehr erinnere. Hellmuth Kneser machte die Honneurs, und Ostrowski hielt den Festvortrag. Der Jubilar erschien noch in vollster Gesundheit, und man mußte glauben, daß sich Oberwolfach als ein deutsches Spiegelbild von Princeton entwickeln würde.

Niemand aber ahnte, daß dieses Fest zugleich den Abschied von Wilhelm Süss bedeuten würde. Bald nach diesem Feiertag hörte man, daß Süss sich zur Erholung ganz zurückgezogen habe. Dann kam die Nachricht von seinem schweren Leiden. Am 23. Mai 1958 haben wir ihn in Freiburg begraben müssen. Überall in der Welt, wo Mathematiker beisammen waren, die ein Bewußtsein der Solidarität mit ihren Fachgenossen hatten, und überall, wo man sich ein so schönes mathematisches Leben, wie es Oberwolfach bot, wünschte, trauerte man um ihn.

Zunächst übernahm Hellmuth Kneser die Leitung. Ihn vertrat Gericke. Und dann traten Theodor Schneider und schließlich Martin Barner das Amt an. In den nächsten Jahren wirkten sich die langdauernden Bemühungen um eine wesentliche Verbesserung der finanziellen Grundlagen des Hauses (vor allem von den Herren Weise und Grotemeyer betrieben) aus. Es konnten mehr und mehr Kurse zur Erörterung spezieller Fragen eingerichtet und dazu die ersten Spezialisten der ganzen Welt gebeten werden. Es wurden auch Kurse mit besonderen Ausblicken aus der Mathematik heraus in andere Gebiete aufgezogen, so die glänzend beleumundeten historischen Tagungen, die Tagungen zur theoretischen Physik und zur Volkswirtschaft, die wieder nach besonderen Gesichtspunkten geteilten didaktischen Seminare – noch aus der Zeit, in der Didaktik nicht Gemeingut aller Hochschulen war). Nicht nur die Anzahl der Kurse erreichte bald die maximal höchste Grenze – 50 im Jahr – auch die Teilnehmerzahlen wuchsen entsprechend. Man mußte immer häufiger auf die Gasthäuser im Orte zurückgreifen. So ergaben sich von selbst die Verhandlungen wegen des Neubaus. Die Volkswagenstiftung griff ein.

Und als am 11. Oktober 1967 der Neubau dastand, wirkte das alte Schloß wie aus einem Ankerbaukasten erstellt, den der Großvater schon auf den Enkel vererbt hatte. Und die vielen Ministerwagen parkten am Schloß, wo früher die Reitpferde des Freiherrn an der Longe geführt worden waren und später – in der Zeit des Hungers – der Gemüsegarten des Hauses mit unserem kostbaren „Topinambur" (zu deutsch: „Pferdekartoffeln") gestanden hatte. Am Festabend saßen noch die alten Freunde des Schlosses am vertrauten Kamin der Bibliothek zusammen und gedachten einstiger Zeiten. Die waren unwiederbringlich dahin.

Von jetzt ab gab es Kurse über Kurse, und jeder war überfüllt. Eine schwierige Aufgabe der Leitung wurde es deshalb, die Zahl der Kurse und der Kursteilnehmer beschränkt zu halten (und dabei geht es im Sinne des Instituts mehr um die Beschränkung der Inländer- als der Ausländerzahlen). Aber das Gedränge wird vorläufig an der oberen Grenze des Ertragbaren bleiben. Das liegt an unserer Zeit, an der heutigen riesigen äußeren Ausdehnung des Wissenschaftsbereiches. Natürlich kommt auch in Oberwolfach die Beengung durch den Umbau noch hinzu.

Die jährlichen deutschen Mathematikerkongresse – einst der Höhepunkt mathematischen Lebens in Deutschland – sind durch die vielen deutschen Spezialtagungen gerade in den letzten Jahren in ihrer Bedeutung reduziert. Das liegt wesentlich auch an Oberwolfach. Doch diese ganze Entwicklung paßt nicht in das Konzept von Süss; weder die Beschränkung der deutschen Jahrestagungen noch das starke Anwachsen der Kursteilnehmer.

Für die Leitung von Oberwolfach ist die alte Frage wieder in den Vordergrund getreten: Soll das Institut zweigeteilt werden?

1. Soll ein reines Forschungsinstitut für länger (oder gar dauernd) dort wirkende Fachleute (Modell: Princeton) mit loser Verantwortung für die zweite Aufgabe – wie früher – wieder entstehen? Hier kann man auch fragen: Warum gibt es kein Max-Planck-Institut für Mathematik?)

2. Eine Tagungsstätte für viele Spezialkurse – wie jetzt? Das ist einmal gelegentlich einer Kommissionssitzung der Max-Planck-Gesellschaft ausführlich erörtert worden.

Der Abbruch des Schlosses gibt uns noch einmal Veranlassung, uns das ernsthaft zu überlegen. Wir sind es uns schuldig, uns nicht treiben zu lassen. Zunächst sind wir im Zugzwang. Die Zeit zwingt uns, Oberwolfach vorläufig als Umschlagplatz für die Erörterung von Spezialfragen anzusehen. Aber auch dabei entstehen in natürlicher Weise Eliten. Die wichtigste weitere Aufgabe wäre es, diese zu pflegen. Die Universitäten sind für das wissenschaftliche Leben teilweise schon zu unruhig geworden. Ein Landes-Kultusminister hat schon geäußert: Die Länder bleiben auf die Dauer nicht mehr für die Forschung verantwortlich. Trotzdem muß natürlich gerade die Forschung gepflegt werden. Auf die Vergangenheit projiziert, können wir uns leicht solche mathematischen Schloßbewohner vorstellen: Gauß, Riemann, Dirichlet, Felix Klein, David Hilbert, Carathéodory, Hermann Weyl. Unter den Lebenden gibt es gewiß auch eine Reihe, die dafür geschaffen sind. Sehen wir nach Frankreich hinüber, das mit Oberwolfach ja auch verbunden ist, so wird die Zahl noch größer.

Aber die heutigen großen Kurse, die einer dem anderen folgen, brauchen eine andere Umgebung als das Schloß. Ein reiner Zweckbau ist fällig. Wir können die Sinnhaftigkeit des Abbruches des alten Schlosses, so teuer es uns geworden ist, nicht mehr leugnen.

Allocution du professeur Henri Cartan (Paris)

lors de la cérémonie officielle à Oberwolfach à l'occasion du quarantième anniversaire de la fondation de l'Institut de Recherche Mathématique d'Oberwolfach

Mesdames, Messieurs,

Le professeur Barner m'a demandé de vous dire quelques mots au nom des mathématiciens francais et plus généralement des mathématiciens étrangers qui ont été accueillis dans cet institut depuis bientôt quarante ans. Permettez-moi d'évoquer d'abord quelques souvenirs personnels. Je suis venu ici pour la première fois en 1946 au début du mois de novembre. Il faisait froid dans le vieux château, mais je n'oublierai jamais la chaleur de l'accueil que j'ai recu de la part de Wilhelm Süss et de Madame Süss que j'ai eu le plaisir de pouvoir saluer ici ce matin. C'est lors de cette visite que je revis pour la première fois depuis la guerre mes amis Heinrich Behnke et Hellmuth Kneser. On se chauffait, on se réchauffait autour du grand poêle de faience bleue. Il n'y avait pas grand' chose à manger, mais il y avait un piano. Et il y avait un embryon de bibliothèque mathématique où l'on trouvait entre autres les premiers volumes de Bourbaki.

J'avais été précédé ici par mon collègue strasbourgeois Charles Ehresmann. Nous voulions, sans tarder, reprendre contact avec nos collègues allemands après cette terrible guerre et le professeur Süss faisait beaucoup pour nous y aider. Je lui en garde, ainsi qu'à Madame Süss, une profonde reconnaissance. Je ne saurai dire combien de fois je suis revenu ici, j'ai assisté à la construction progressive des nouveaux bâtiments, à la disparition du vieux château. Peu à peu s'est établie l'organisation modèle si propre au travail que nous connaissons aujourd' hui. Les Francais s'en sont inspirés plus tard pour le création, beaucoup plus modeste il est vrai, du centre de Luminy près de Marsaille.

Je voudrais encore une fois remercier le professeur Martin Barner de m'avoir donné l'occasion d'exprimer ici mes remerciements à tous ceux, et lui en premier, qui ont su développer l'oeuvre de Wilhelm Süss. Et je voudrais aussi souligner l'importance du rôle joué par cet institut pour nos vieux pays d'Europe. Il est essentiel que ce soit sur le sol de l'Europe que puissent se réunir et travailler ensemble les mathématiciens venus du monde entier. Encore une fois, merci.

Henri Cartan

Einweihung der baulichen Erweiterungen
am 26. Mai 1989

Ansprache des Vorsitzenden des Kuratoriums
der Volkswagen-Stiftung, Dr. Werner Remmers

Als ich von Hannover aufgebrochen bin, um mit Ihnen zusammen die neuen Gebäudeteile des Mathematischen Forschungsinstituts Oberwolfach einzuweihen, wußte ich natürlich einiges über das Institut. Aber ich wußte es vom Papier. Jetzt weiß ich, daß man Oberwolfach einmal erleben muß, um es wirklich zu kennen.

Als Wissenschaftsförderer hört und liest man ja gelegentlich, daß da oder dort ein Institut „auf der grünen Wiese" gegründet werden soll. Dann werden gewöhnlich die Argumente pro und contra besonders sorgfältig erwogen. Das Mathematische Forschungsinstitut Oberwolfach liegt nicht nur auf der grünen Wiese, sondern sogar im grünen bzw. im schwarzen Wald. Aber bereits nach kurzem Aufenthalt merkt man: Nicht Enge und Weltferne herrschen hier, sondern Konzentration und Weltoffenheit – was ich übrigens auch in der Architektur des Instituts ausgedrückt sehe.

Ich verstehe sehr gut, daß Oberwolfach eine Verführung für Mathematiker ist. Es ist, wenn ich die Verbindungen zurückverfolge, die zwischen dem Mathematischen Forschungsinstitut Oberwolfach und der Volkswagen-Stiftung bestanden und bestehen, auch eine Verführung für Drittmittelgeber.

Ein anderes Ondit, dem man als Wissenschaftsförderer immer wieder einmal begegnet, lautet: Der Mathematiker braucht zum Arbeiten nichts weiter als ein Blatt Papier und einen Bleistift.

Im Jahre 1962 sollte die Volkswagen-Stiftung – damals hieß sie noch Stiftung Volkswagenwerk und hatte erst wenige Wochen zuvor Ihre Arbeit aufgenommen – einsehen, daß Mathematiker außer Papier und Bleistift und natürlich einer Bibliothek auch noch ein Schwarzwaldtal benötigen.

Nun, im März 1963 bewilligte die Stiftung 1,477 Millionen DM für Oberwolfach. Aber damit hatte es kein Ende. Die gute Tat zeugte fortwährend Nachanträge. Und da auch wir es auf unsere Art und Weise mit Zahlen genau nehmen, darf ich Ihnen die Bewilligungshöhe präzise nennen: 6 719 368 Mark und 76 Pfennige wurden von 1963 bis 1969 für Gästehaus, Tagungs- und Bibliotheksgebäude bewilligt. Für die baulichen Erweiterungen, die einzuweihen wir uns heute hier versammelt haben, hat die Volkswagen-Stiftung im März 1987 noch einmal 2,85 Millionen DM zur Verfügung gestellt und gleichzeitig beschlossen, das Eigentum an Grundstück und Gebäuden des Forschungsinstituts auf die Gesellschaft für mathematische Forschung schenkungsweise zu übertragen.

In diesem Zusammenhang möchte ich auf keinen Fall etwas verschweigen, was die Volkswagen-Stiftung nicht häufig erlebt. Von den bis 1969 bewilligten Mitteln hat das Institut rund 300 000 DM nicht in Anspruch genommen. Sie brauchten nicht ausgezahlt zu werden und konnten anderen wissenschaftlichen Zwecken zugeführt werden.

So gut geht die Rechung diesmal nicht auf, wie ich hörte. Etliche Ausstattungs- und Einrichtungsgegenstände wie auch Dinge, die das Wirken hier angenehm machen sollen, hat das Institut aus eigenen Vereinsmitteln beschafft, weil dafür die Bewilligung der Volkswagen-Stiftung nicht mehr reichte. Professor Barner hat versprochen, deswegen keinen Nachantrag an die Stiftung zu stellen – auch dies ist nicht eben alltäglich im Fördergeschäft.

Zu den besonderen Dingen zählt auch die Hinweis-Stele draußen an der Zufahrt, die von heute an „op ewig" – um es norddeutsch zu sagen – an das fruchtbare Zusammenwirken des Mathematischen Forschungsinstituts und der Volkswagen-Stiftung erinnern soll. Hierfür dürfen wir danken – wie auch Herrn Professor Rossmann und seinem Team, daß es ihm gelungen ist, die Erweiterungen an zwei Baukörpern so zu integrieren, daß sie zwar ihre eigene Sprache sprechen, dennoch aber im Gesamtgefüge der Bauten hier keinen optischen Bruch darstellen.

In der Regel vergibt die Volkswagen-Stiftung ihre Mittel im Rahmen bestimmter Schwerpunkte. Einen eigenen Schwerpunkt im Bereich der Mathematik hat die Stiftung nie gehabt. Sie fördert aber die Mathematik seit jeher auf anderen Wegen, die ihr offenbar angemessener sind.

Mit ihrem Symposienprogramm und ihren Akademie-Stipendien macht die Volkswagen-Stiftung zwei fachoffene Förderangebote. In beiden Bereichen kommt die Mathematik in auffallendem Umfang zu ihrem Recht. Nehmen wir als Beispiel die Akademie-Stipendien, mit denen die Volkswagen-Stiftung besonders qualifizierten und belasteten Professoren deutscher Hochschulen die Möglichkeit gibt, sich während eines Zeitraums von sechs bis zwölf Monaten zugunsten eines speziellen Forschungsvorhabens von den Lehrverpflichtungen befreien zu lassen. Dieses Angebot gilt für alle Bereiche der Wissenschaft und Technik.

Ich will nun nicht in einer alten Streitfrage Partei ergreifen. Aber wenn Sie als Mathematiker sich zu den Geisteswissenschaftlern zählen, war nach unserer Statistik im Jahre 1988 jeder fünfte Akademie-Stipendiat Mathematiker. Noch günstiger sieht es allerdings aus, wenn Sie sich als Naturwissenschaftler sehen, unter denen war dann etwa jeder zweite Stipendiat Mathematiker.

Auf zwei große Förderungen im Bereich der Mathematik aus den letzten Jahren möchte ich ebenfalls kurz hinweisen. Gerade erst im März dieses Jahres hat die Volkswagen-Stiftung 4,5 Millionen DM für die Einrichtung eines „Instituts für experimentelle Mathematik" an der Universität-Gesamthochschule-Essen bewilligt. Das Institut wird sich mit Forschungsaufgaben zu mathematischen Problemen beschäftigen, bei denen der Computer als methodisches Instrument eingesetzt wird.

Im Sommer 1987 hat die Volkswagen-Stiftung mit 3 Millionen DM die Einrichtung eines „Zentrums für praktische Mathematik" unterstützt, das in Zusammenarbeit der Technischen Hochschule Darmstadt und der Universität Kaiserslautern entstanden ist. Dieses Institut dient dem Brückenschlag zwischen der Hochschul-mathematik einerseits und der technisch-naturwissenschaftlichen Welt andererseits.

Stiftungen werden immer wieder gefragt, worin ihr spezieller Nutzen liege, da das Geld für die Wissenschaft doch aus vielen anderen Quellen sprudele. Hier kann die Antwort kurz sein: Stiftungen sind beispielsweise für Institutionen wie das Mathematische Forschungsinstitut Oberwolfach von Nutzen, weil es die sonst wohl nicht gäbe. Wer aber möchte sich Oberwolfach aus der Wirklichkeit wegdenken?

Damit komme ich noch einmal zur gemeinsamen Geschichte des Mathematischen Forschungsinstituts Oberwolfach und der Volkswagen-Stiftung zurück. Bei der Einweihung des Gästehauses am 16. Oktober 1967 hat der damalige Generalsekretär der Stiftung, Gotthard Gambke, den Wunsch und die Hoffnung ausgedrückt, Oberwolfach möge dazu beitragen, daß die Verbindung zwischen den einzelnen Teilgebieten der Mathematik selber, aber auch zwischen der Mathematik und den anderen wissenschaftlichen Disziplinen enger werden möge. Ohne selbst Fachmann zu sein, habe ich doch den Eindruck, daß beides gelungen ist.

So denke ich denn, daß das Mathematische Forschungsinstitut Oberwolfach mit Stolz und die Volkswagen-Stiftung mit Genugtuung auf das Geleistete zurückblicken können, bin aber zugleich überzeugt, daß Herr Professor Barner und seine Mitstreiter auch und gerade anläßlich der Einweihung dieser neuen Gebäude den Blick wie immer nach vorne richten. Für die künftigen Aufgaben wünscht die Volkswagen-Stiftung Ihnen alles Gute!

Die Arbeitsgemeinschaft Oberwolfach

Jürgen Neukirch

Introduktion

Es gibt wohl keinen Mathematiker, der – einmal in Oberwolfach gewesen – dieses Forschungsinstitut nicht als einen Segen für die Mathematik und die Mathematiker priese. Es gibt aber viele, welche jeglicher Tagung dort gleiches Gewicht beimessen. Zu diesen gehöre ich nicht und füge meiner gegenteiligen Überzeugung noch eine weitere hinzu, indem ich ein besonderes Treffen als leuchtende Rechtfertigung der ganzen Institution an die oberste Spitze setze: Die „Arbeitsgemeinschaft Oberwolfach". Die Namen „Kneser-Roquette" trug sie zuerst, bis vor kurzem „Geyer-Harder" und jetzt „Deninger-Schneider"; für die nachfolgenden Zeilen muß sie sich mit dem Kürzel AG begnügen.

Es ist nicht meine Absicht, die Geld oder Tagungszeit gebenden Gremien durch eine nagelfeste Begründung meiner hohen Einordnung für die AG günstig zu stimmen. Ist das Gremium gut besetzt, so wird es die Tagung von selbst nach Kräften fördern, und wird sich im anderen Fall um meine Empfehlungen ebenfalls nicht kümmern. Auch will ich mir nicht zur Aufgabe machen, einen mit Daten, Fakten und Titeln belegten historischen Abriß zu liefern oder einen exakten Bericht über die geleistete Arbeit. Vielmehr möchte ich in diesem Aufsatz meine persönlichen Erinnerungen an eine großartige mathematische Zeit, meine eigenen Beobachtungen und Erlebnisse schildern, von der Hoffnung bewegt, daß sie im Einklang stehen mit dem Eindruck der anderen Mitglieder der Gemeinschaft.

Die Idee

Gegründet wurde die AG im Jahre 1958 von MARTIN KNESER und PETER ROQUETTE. In jener Zeit hatte das französische „Séminaire Bourbaki" begonnen, seinen weltweiten Einfluß auszuüben und war der Anstoß für die beiden deutschen Mathematiker, auch in unserem Lande eine Gemeinschaft zu bilden, die an den üppig sprießenden Entwicklungen mit eigenem Engagement teilnahm. Das französische Vorbild einzelner geschliffener, von Gebiet zu Gebiet springender Vorträge sollte jedoch nicht übernommen werden. Gründlicher sollte es für jeden zugehen. Eine ganze Tagungswoche lang sollte ein ausgewähltes Thema der Gegenstand gemeinsamer Anstrengung sein, ein Thema, das nicht der eigenen Forschung angepaßt sein mußte, die bei der Zusammensetzung der Gemeinschaft ganz unterschiedlich war, sondern das wesentlich für die gegenwärtige und zukünftige mathematische Entwicklung erschien. Jedes Mitglied sollte verpflichtet sein, sich an der Vorbereitung und Ausführung der von einem Tagungsleiter geplanten Vorträge zu beteiligen, die nicht der Souveränität der Experten überlassen werden sollten, sondern dem Einsatz der Verständnis suchenden „Laien". Dieses Prinzip ist bis heute das Grundgesetz der AG geblieben und hat sich als überaus tragfähig erwiesen. Und unter diesem erklärten Zeichen erfuhr dann die erste Tagung vom 8. bis 13. April 1958 ihr Placet durch den folgenden ermunternden Brief des Begründers und damaligen Direktors des Mathematischen Forschungsinstituts Oberwolfach WILHELM SÜSS:

Lieber Martin! *

Deinen Brief erhielt ich gerade vor der Abreise zur Forschungsgemeinschafts-Tagung in Berlin. Herzlichen Dank dafür und zunächst für das Manuskript, das inzwischen schon an die Schriftleitung des Archivs weitergegangen ist. Besonderen Dank aber für Deinen und Euren Plan, dem Lorenzenhof eine schöne Aufgabe zu geben, indem Ihr dort Eure Arbeitsgemeinschaft einrichtet. Rundweg kann ich sagen, daß ich mit allen Deinen Vorschlägen restlos einverstanden bin. Der Termin der Tagung in der Woche nach Ostern dürfte auch passen, und eventuell finde ich auch einen Weg, einen finanziellen Zuschuß an einige der Teilnehmer noch rechtzeitig loszuwerden. Das heißt nämlich, daß ich meine Gelder vor Ende März jeweils ausgegeben haben muß, so daß wir auf einen Weg sinnen müssen, wie dies geschehen kann, ohne den Tagungstermin vorzulegen, was ich nicht gern täte, weil für März schon allerhand Pläne bestehen. Vielleicht können wir das in Kürze einmal miteinander bereden. Zum Beispiel könnte man die in Frage kommenden Teilnehmer aus Eurem Kreis als Mitläufer einer vorhergehenden Tagung finanziell führen, was nur vorher verabredet werden müßte. Am besten läßt Du mich die entsprechenden Teilnehmer wissen, sobald Du sie kennst. Dein Plan entspricht durchaus gewissen Vorstellungen, die ich mir schon immer über die Verwendung des Lorenzenhofes gemacht habe. Möge er gelingen und andere zur Nachahmung verlocken.

Das Thema dieser ersten Tagung lautete „Stellenringe und Schnittmultiplizitäten in der algebraischen Geometrie" und die Teilnehmer hießen GASCHÜTZ, HABICHT, HUPPERT, JACOBS, JEHNE, KNESER, KNOBLOCH, LEPTIN, RÖHRL, ROQUETTE. Die vorwärtstreibende Kraft und Initiative ROQUETTES wird von KNESER besonders hervorgehoben.

Die Thematik

Die Tagung fand von Anfang an im Frühjahr und im Herbst eines jeden Jahres statt. Über die Vielfalt, die Verschiedenartigkeit und das Niveau der Themen mag der folgende Auszug sprechen (in Klammern die Tagungsleiter).

„Indexsatz von Atiyah-Singer" (Harder-Hirzebruch, 1970)
„Singularitäten komplexer Hyperflächen" (Th. Bröcker, Herbst 71)
„Modulfunktionen" (Helling, Frühjahr 73)

---

* Die Anrede erklärt sich daraus, daß die Familien Süss und Kneser aus gemeinsamer Zeit in Greifswald befreundet waren.

„Invariantentheorie" (Geyer, Frühjahr 75)
„Monodromie" (Brieskorn-Lamotke, Herbst 75)
„Himmelsmechanik" (Rüßmann-Zehnder, Herbst 76)
„Periodenabbildung" (Steenbrink, Herbst 77)
„Die Weil-Vermutung" (Freitag-Kiehl, Herbst 78)
„Iwasawa-Theorie" (Neukirch-Tamme, Herbst 79)
„Quillen's K-Theorie" (Waldhausen, Frühjahr 80)
„Die Vermutungen von Hodge und Tate" (Langlands-Rapoport, Frühjahr 81)
„Relativitätstheorie" (Karcher-Schmidt, Herbst 81)
„Das Eisensteinideal" (Coates-Frey, Herbst 82)
„Transzendenztheorie" (Wüstholz, Frühjahr 84)
„Die Beilinson-Vermutung" (Schneider-Rapoport, Frühjahr 86)
„Yang-Mills-Theorie" (Arbeiten von Donaldson) (Lübke - van de Ven,
    Herbst 86)
„Integrable Hamiltonsche Systeme" (Knörrer-Trubowitz, Frühjahr 87)
„Arakelov-Theorie auf Arithmetischen Flächen" (Forster, Herbst 87)
„Degeneration von Hodge-Strukturen" (Esnault-Steenbrink-Viehweg,
    Frühjahr 88)
„Brownsche Bewegung auf Mannigfaltigkeiten" (Deninger-Thalmaier,
    Herbst 88)
„Tamagawazahl und Selbergsche Spurformel" (Clozel-Rapoport, Herbst 89)
„Connections between Mathematical Physics (Quantum groups, von Neumann
    Algebras) and the Theory of Knots" (Turaev, Frühjahr 91)
„Logarithmic Algebraic Geometry" (W. Bauer-Illusie, Herbst 92).

Wie kann man das alles lernen, und welchen Sinn hat es, das alles zu lernen?
An dieser Frage will ich nicht vorbeigehen, ganz besonders im Hinblick auf junge
Mathematiker, die vor der Entscheidung stehen, sich der Arbeitsgemeinschaft zu-
zugesellen oder nicht. Alle Mathematiker leben insgeheim mit der Furcht oder der
Gewißheit, zu dumm für die Mathematik zu sein – mit Ausnahme der Dummen.
Daraus entsteht eine Schutz verheißende Überzeugung, die ebenso verbreitet wie
falsch ist. Sie äußert sich etwa so: „Wenn ich in der Mathematik zu Leistung
und Erfolg kommen will, so muß ich meine ganze Energie auf mein gewähltes
Forschungsgebiet wenden und darf mich nicht dadurch verzetteln, daß ich meinen
Verstand mit anderem verstopfe und meine beschränkte Zeit auf das Studium frem-
der Gebiete lenke, die für meine eigene Forschung nichts liefern können". Oft geht
damit uneingestanden die latente Angst einher, es könnte sich herausstellen, daß
man der Aufgabe nicht gewachsen ist, sich die großen Theorien und Ergebnisse
mit ihren komplexen Verzahnungen und Vertiefungen zu eigen zu machen, so daß
man dann vor sich selbst und den anderen einigermaßen schwächlich dasteht.
    Diese Einstellung ist, wie gesagt, ein Irrtum. Ich glaube, alle ständigen Mitglie-
der der AG werden bezeugen können, daß die Erarbeitung der genannten Gebiete
zwar Mut, Überwindung und nicht geringe Anstrengung erforderten. Im Rückblick
aber stellte sich stets heraus, daß es doch gar nicht so schwer gewesen war, wie
man geglaubt hatte, und daß gerade die *gemeinsame* Arbeit, das Zusammenstecken

verschieden denkender Köpfe, manches zunächst unmöglich Scheinende durchaus möglich machte. Durch diese Erfahrung, ein paar mal gemacht, verliert sich die Furcht, und der Einstieg in fremde mathematische Welten wird nur immer leichter. Was aufgeschrieben ist, läßt sich auch verstehen – von wenigen horrenden Beispielen abgesehen, vor denen man rechtzeitig gewarnt wird. Es versteht sich von selbst, daß man vieles wieder vergißt, vielleicht sogar das meiste. Aber die Übung schafft Vertrauen.

Über den anderen Irrglauben, die Beschäftigung mit fremden Gebieten sei eine Belastung, die den Kopf nicht freihält für die eigene Forschung, ist mir der in einem Disput gemachte lapidare Ausruf des AG-Mitglieds UWE JANNSEN im Gedächtnis geblieben: „Man kann gar nicht genug lernen!" Von der Wahrheit dieses einfachen Ausspruchs werde ich am Ende dieses Aufsatzes Zeugnis geben. Die Beschäftigung mit der Mathematik in ihrer ganzen Breite schafft auch die ganze Freude an ihr, stärkt die eigenen Kräfte, öffnet den Blick für das Denkbare und ermöglicht viele für die eigene Arbeit fruchtbare Gespräche mit verständigen Mitstreitern. Dies ist doch die Erkenntnis unseres Jahrhunderts, daß die Mathematik nicht in verschiedene Einzeldisziplinen zerfällt, sondern eine Einheit bildet.

Von den vielen Tagungen, an denen ich teilnahm, sind mir drei als ganz besondere Erlebnisse unvergeßlich geblieben, die „Weil-Vermutung", das „Eisensteinideal" und die „Beilinson-Vermutung". Wie ein Lauffeuer war seinerzeit die Nachricht vom DELIGNEschen Beweis der Weil-Vermutung um die Erde gegangen: Die Nullstellen ρ der Zetafunktion $\zeta_X(s)$ einer glatten projektiven Varietät $X/\mathbb{F}_p$ der Dimension d haben den Realteil $Re(\rho) = \frac{1}{2}, \frac{3}{2}, \ldots, \frac{2d-1}{2}$, und die Polstellen den Realteil $Re(\rho) = 0, 1, \ldots, d$. Dies zu begreifen mußte und sollte unsere nächste Aufgabe sein! Mehr noch als das Resultat entzückte uns der geometrische Beweis, der so klar den Grund für das Gesetz aufdeckte. Dabei mußten wir zunächst lernen, was es mit der Etalkohomologie auf sich hatte, ein weites und für die meisten neues Feld. Wie wichtig sollte diese Kenntnis für uns werden, und wie sehr hat sich die Mühe ausgezahlt. „Grothendieck, Grothendieck" – überall wurde der Name mit Bewunderung genannt. Aber allem die Krone setzte der Delignesche Beweis auf. Die vielen verschiedenen, ineinandergreifenden Gedanken, die das glänzende Resultat wie ein Gebäude umschlossen, lösten eine allgemeine Begeisterung aus. Diese erlebnisreiche Woche wurde durch eine von Teilnehmern der Tagung aufgeführte Oper abgeschlossen, wohl die erste in Oberwolfach – doch davon später.

Ähnlich ging es mit dem „Eisensteinideal". Hinter dem abstrakten Titel verbarg sich das konkrete MAZURsche Resultat, daß jede elliptische Kurve über $\mathbb{Q}$ höchstens sechzehn rationale Torsionspunkte hat, daß also jede Gleichung $y^2 = 4x^3 - ax - b$, $a, b \in \mathbb{Q}, a^2 - 27b^3 \neq 0$, höchstens sechzehn rationale Lösungen endlicher Ordnung besitzt (die Ogg-Vermutung). Auch hier bestand das eigentliche Erlebnis darin, daß es die modernen, abstrakten Theorien waren, wie die Theorie der abelschen Schemata und der quasi-endlichen Gruppenschemata, die die elementare Wahrheit zutage förderten. Die hochfliegenden theoretischen Ausschweifungen wurden aber nicht nur durch die konkreten Resultate, sondern auch durch die erdverbundene Austragung des „Eisenstein-Cups" an den Boden gepflockt – doch davon später.

Aus den vielen schönen Tagungen der AG hebt sich noch eine weitere heraus: Die „Beilinson-Vermutung". Es war die Begegnung mit einer faszinierenden mathematischen Vision, die Werte der Zetafunktionen an ganzzahligen Stellen in umfassender Allgemeinheit betreffend, die jedoch uns allen durch ihre übergroße, in diesem Maße nie dagewesene Schwierigkeit in Erinnerung geblieben ist. Diese Schwierigkeit ergab sich durch den Umstand, daß sich die Tagung auf eine Abhandlung des genialen russischen Mathematikers A.A. BEILINSON stützte, in der für die Resultate die Beweisideen, nicht aber die Beweise selbst gegeben wurden. Letztere mußten also von den Vortragenden ausgegraben werden. Dies wurde zu einer überaus schmerzlichen Mühe, weil für die außerordentlich tiefliegenden Beweise alle mathematischen Register gezogen werden mußten. Das Ergebnis dieser Mühen wurde aber schließlich zu einer runden Genugtuung für die AG und wurde in einem Buch „Beilinson's Conjectures on Special Values of L-Functions" von M. RAPOPORT, N. SCHAPPACHER und P. SCHNEIDER herausgegeben. Diese Tagung ist für mich die größte Leistung der AG gewesen, auf die sie mit allem Recht stolz sein kann. Beilinson selbst soll dazu gesagt haben: „I never could have done that."

Von den vielen Begegnungen mit Befremdlichkeiten wie den Quantenfeldern, den Kristallen, den Strings, den schwarzen Löchern, den Martingalen etc. und dem Staunen darüber muß ich schweigen. Der hin und wieder geäußerte Verdacht einer Einseitigkeit der Tagungen traf immer ins Leere. Zwei Tagungen aber will ich extra nicht erwähnen, weil sie mir mißfallen haben. Die eine wegen des enttäuschenden mathematischen Gehalts einer Theorie, die in der nicht-mathematischen Öffentlichkeit laut als die moderne Mathematik verkauft wurde, die andere, weil einige unerfahrene Mathematiker mit ihren Vorträgen den Erfahrenen zeigen wollten, wie erfahren sie seien, ein Versuch, der dem Geist der Tagung entgegenstand und denn auch gottlob daneben ging. Ich sage dies nur, um nicht mit uneingeschränktem Lob das Bild zu verfälschen.

Die Vorträge

Die AG ist stets eine junge Gemeinschaft gewesen. Dies ist nur natürlich, denn es ist vor allem die Jugend, die nach Wissen und Kenntnis drängt, während dieses Drängen bei der reiferen Generation im Verlaufe verfällt. Es muß der letzteren aber auch zugutegehalten werden, daß sie sich um der selbstlosen Förderung des jugendlichen Schwungs willen rechtzeitig von Leitung und Maßgabe zurückgezogen hat. Eine Bereicherung war stets die kleine, aber wesentliche Schar der holländischen Mitglieder. In jüngster Zeit sind die ostdeutschen hinzugekommen. Die AG war nie eine geschlossene Gesellschaft. Es wurde jeder aufgenommen, wenn er nur den Willen bekundete, nicht als „Trittbrettfahrer" dabei zu sein, sondern tatkräftig mitzuarbeiten – und wenn es der Platz erlaubte. Auf diese Weise war die Gemeinschaft stets im Wandel, doch sorgte andererseits ein harter Kern für Beständigkeit.

Die Vorträge wurden nach Inhalt und Reihenfolge von den gewählten Tagungsleitern zusammengestellt, die somit das jeweilige Gebiet wenigstens halbwegs kennen mußten. Sie schickten ihre Entwürfe an die verschiedenen mathematischen Institute und mußten aus den eingehenden Vortragswünschen die endgültigen Vortragenden bestimmen. Bei diesem Gerangel haben wir Regensburger uns einer äußerst geschickten Strategie bedient, indem wir vorwiegend die letzten Vorträge beantragten. Diese waren nicht so begehrt wie die ersten, zwangen uns, die ganze Sache zu verstehen, bescherten uns die Aufgabe, die interessantesten Resultate vorzustellen, und wurden uns dadurch zum größten Gewinn. Da wir überdies häufig die größte Fraktion bildeten, und da bekannt war, daß wir unseren Auftrag stets ordentlich erfüllten, so wurde uns auch eine ausreichende Zahl von Vorträgen zuteil. Auf diese Weise wurde denn manche Tagung am Donnerstag und Freitag durch die „Regensburger Walze" zu Ende gebracht. Ich sollte mich schämen, meine eigene Fakultät unter den trefflichen anderen so hervorzuheben. Da aber in diesem Aufsatz von Objektivität ohnehin keine Rede ist, will ich mir auch dies erlauben. Es ging nicht ohne alle Kritik für uns ab. So wurde uns halb scherzend, halb ernsthaft vorgeworfen, unlauteren Wettbewerb zu betreiben, indem wir uns schon vorher zu Hause in Seminaren mit dem Stoff auseinandersetzten, und es daher kein Wunder sei, wenn unsere Vorträge so perfekt ausfielen. Dieser Vorwurf war allerdings nicht schwer zu ertragen. Man muß jedoch gerechterweise hervorheben, wieviel leichter wir es im Vergleich zu denen hatten, die sich zu Hause allein mit dem umfangreichen, unbekannten Stoff auseinandersetzen mußten, und die insofern einen größeren Respekt verdienten.

Aus dieser Schilderung geht schon hervor, wie groß der Ehrgeiz bei allen war, einen schönen und verständlichen Vortrag zu halten. Auf diese Weise wurde das Zuhören zu einer fesselnden Zeit des Lernens. Die Zahl der täglichen Vorträge war weise beschränkt: 2 am Vormittag, 2 am Nachmittag. Dies war einerseits der Konzentrationsfähigkeit angepaßt und gab andererseits Zeit für die vielen mathematischen Diskussionen. Überall und bis in die Nacht hinein sah man die Freunde beim gemeinsamen Grübeln und Diskutieren. Eine bittere Notwendigkeit oft, denn selbst wenn man glaubte, seinen Vortrag in der Tasche zu haben, konnte es passieren, daß einem gesagt wurde, man habe die Sache ganz falsch verstanden, und müsse die Dinge so und so sehen. Das bedeutete dann, in fiebriger Nervosität von Neuem anzufangen. Wenn man spät abends die Gesellschaft beim Wein zusammen sah, konnte man erkennen, wer schon vorgetragen hatte und wer nicht. Die einen bei lachendem „Prost", die anderen blaß und verhärmt. Nicht jeder Vortrag war gelungen zu nennen, doch im ganzen war der Erfolg der Bemühungen der Sprecher zu bewundern. Ein besonderes Verdienst haben sich die Tagungsleiter erworben, die all ihre Umsicht und Klugheit bei der Aufstellung und Einteilung der Vorträge eingesetzt haben.

Die Wahl

Die Festlegung der Tagungsthemen war demokratischer Wahl unterworfen. Diese Wahl fand donnerstags 20°° (früher mittwochs) im großen Vortragsraum statt und wurde in der Regel zu einem dramatischen Ereignis. Schon beim Mittagessen war überall die Frage zu hören „Haben Sie schon eine Idee, was man vorschlagen könnte?" Der Wunsch, sich auf das Kommende einzustellen und womöglich sogar mit eigenem Vorschlag aufzuwarten, war spürbar. Ein Glockenzeichen berief die abendliche Versammlung pünktlich ein. Das Wahlgeschehen konnte jedoch nicht gleich beginnen. Eiliges Heraus- und Hineinhuschen mußte zunächst für die Verpflegung mit Saft, Wein und Bier sorgen, und mancher glaubte sogar, einen verbotenen Aschenbecher im Versteck gesehen zu haben. Ich selbst, sonst immer nur links am lichten Fenster sitzend, wählte hier stets den rechtesten Vordersitz, von dem aus ich den besten Überblick über die Gesellschaft hatte und sie am wirkungsvollsten beeinflussen konnte. Der Raum war abgedunkelt, nur der vordere Teil wie eine Bühne erhellt, das Holzgetäfel drückte sanft die Gemütlichkeit in das Dämmerlicht, die Gesellschaft war schließlich vollzählig versammelt, erwartungsvoll, doppelt schöner als am Tage, Ruhe trat ein, es konnte losgehen.

Mit gewaltiger Kreide stand GÜNTER HARDER an der Tafel, der anerkannte und vielgeschätzte König der Gemeinschaft, um die Vortragsvorschläge aufzunehmen. Der fleißige Schwamm lag griffbereit, um sie wieder wegzuwischen. Zunächst zögerlich, dann im Accellerando durch die Luft fliegend wurden die Themen von hinten nach vorn gerufen und von Harder an der Tafel notiert und numeriert. Jedes Thema mußte mit der Nennung der möglichen Tagungsleiter einhergehen, damit die Realisierung gesichert war. Wenn die Zahl der Vorschläge auf die Zwanzig zuging, wurde ein Ende gemacht.

Da lag nun schier die ganze Mathematik vor uns, und wir mußten finden, wo wir schließlich an Land gehen sollten. Zunächst waren die Vorschlagenden aufgefordert, eine kurze Erläuterung ihres Themas zu geben, Inhalt, Bedeutung und Aktualität zu beschreiben. Dies allein war spannend genug. Nach der anstrengenden, alle Konzentration hinnehmenden Mühe um das Verständnis der Lemmata, Definitionen, Beweise in den Tagesvorträgen, wurde nun im Plauderton die mathematische Entwicklung der Zeit wie ein weicher Teppich ausgebreitet, und es durfte sich das beschwerte Mathematikerherz an den einfachen, erzählenden Worten erlaben. Nicht immer wurde klar, was gemeint war, doch was machte es, Klarheit war der Geist des Tages, nicht der Nacht. Auch konnte es passieren, daß der Vorschlagende von seinem Thema nicht mehr wußte als nur den klingenden Titel. Dann geschah es oft, daß Harder mit seinem ausgebreiteten Wissen andeutete, was es mit der Sache auf sich hatte. Im ganzen aber war dieser Teil von einer erstaunlichen Verständigkeit bestimmt und wurde dadurch zur höchst interessanten Unterhaltung.

Nach diesen Erläuterungen und abermaligem, neue Nahrung schaffenden Raus- und Reinhuschen war dann der erste Wahlgang dran. Dieser war von weiser, alle politische Wucht zurücksetzender Milde getragen, indem ein jeder so viele Stimmen hatte, wie er wollte, und für alles stimmen durfte, was ihm gefiel. Immerhin zeigte dieser erste Gang schon Tendenzen und raffte ein, zwei, drei Vorschläge frühzeitig dahin. Beim zweiten Wahlgang durfte jeder mit drei Stimmen hantieren. Hier wurde nun schon durch Flüstern und Tuscheln die zunächst heimliche Wahlbeeinflussung bemerkbar. Für die Themen, die bleiben sollten, wurde eine Punktzahl festgesetzt, alles was darunterlag, fiel dem fleißigen Schwamm zum Opfer. Beim nächsten und vorletzten Wahlgang standen jedem zwei Stimmen zur Verfügung und die Unruhe wuchs. Mit beschwörenden Appells wurde die Qualität *dieser* Themen hervorgehoben, mit Witzen und Gelächter sollten *jene* disqualifiziert werden. Hin- und herwogende Argumente brachen die Temperamente auf, die Szene steigerte sich zum Kampf. Freude und Enttäuschung über den Ausgang dieses Wahlgangs wurden nicht mehr zurückgehalten und die Leidenschaft wuchs ihrem Höhepunkt entgegen. Nur noch zwei Themen standen an der Tafel, und nur eine Stimme war zu vergeben – der Augenblick der Entscheidung war gekommen. Der Ausgang war denkbar ungewiß, denn keiner konnte ahnen, welchem Thema sich die bisherigen Verlierer zuwenden würden. Noch einmal wurden flammende Plädoyers gehalten, Egoismen und die Parole „Gut ist, was mir frommt" gegeißelt, die Berücksichtigung der zur Tagung nicht gekommenen Mathematiker gefordert und der Blick auf das Ganze beschworen. Schließlich brachte Harder die Gemüter mit dem ihm eigenen trockenen Humor zur Ruhe und es wurde abgestimmt, gezählt, zur Verhinderung von Wahlfälschung nochmal gezählt und der Strich gezogen: Das Thema für die nächste Tagung war gewählt!

Der Kampf war vorüber und machte wieder versöhnlichen Scherzen Platz. Die Gesellschaft verließ den Raum, doch hielt die Erregung noch eine Weile an. Bei freigegebenem Pfeifen- und Zigarettenrauch – natürlich nicht für alle Geister das beruhigende Labsal – wurde das Ergebnis kommentiert und begutachtet, zog das Wenn und das Aber seine Kreise, kehrte aber schließlich doch eine allgemeine Zufriedenheit über das so heftig errungene Resultat ein. In der Tat konnte man fast immer über die Wahl sehr zufrieden sein. Denn gewöhnlich ist die Demokratie die Trägerin des Mittelmaßes, hier brachte sie stets das Vortreffliche nach oben.

Vergnügungen

Im Verbund mit dem Heiteren wird der Ernst zum besten Förderer der Arbeit, und so müssen auch die Vergnügungen dieser Tagung erwähnt werden.

Am Mittwoch fand die obligate Wanderung statt, die allerdings für mich durch die völlig falsche Einstellung der Gesellschaft zu dieser Sache ein recht getrübtes Vergnügen wurde. Durch lange Erfahrung gewitzt, pflegte ich dem allgemeinen Aufbruch eine Viertelstunde vorauszugehen, um den mühseligen Aufstieg zu bewältigen, den der entsetzlich steile und lange Hang verlangte. Erschöpft und keuchend oben angekommen, suchte ich entkleidet meinen schweißgebadeten Körper von der Sonne trocknen zu lassen, träufelte das kühlende Wasser des Rieselbaches auf mein Haupt und trachtete, das bebende Herz und den fliegenden Atem zu beruhigen. Daraus wurde nichts, denn schon stürmte er heran, der Troß, Oberstürmer Harder an der Spitze, mit schleudernden Armen und stampfenden Beinen stürmte er heran, maß mich kurz mit beleidigendem süffisanten Lächeln, hinterließ mir ein beißendes Spottwort, stürmte weiter und ward meinen Augen entschwunden. Hastig mich ankleidend rannte ich hinterher und suchte, wenigstens nicht verloren zu gehen, doch konnte mein Blick meistens nur noch die Hacken des um die Ecke eilenden letzten Kollegen erhaschen. Das war nicht meine Vorstellung einer Wanderung. In der menschlichen Begegnung, im kontemplativen Austausch, im erbaulichen Ergehen der schönen Natur hätte ich den Sinn gesehen, die ja auch nicht hetzt – wer hätte jemals einen rennenden Baum, einen keuchenden Vogel, eine vor Anstrengung schwitzende Wiese gesehen? Diese Wanderung hieß zu Recht die „Ochsentour". Aber auch das kühle Bier im Gasthof „Zum Ochsen" wurde mir nicht zur ruhigen Freude, denn kaum stand es vor mir, so sauste die Gesellschaft schon wieder los, ihrem Vergnügen des Rückmarsches entgegen, und saß mit rotem Kopf, aber wie aus dem Ei gepellt munter tafelnd beim Abendbrot, wenn ich mich müde und geschlagen ins Haus schleppte. Das war also nichts, wie der mitfühlende Leser zugeben wird.

Der sportähnlichen Betätigungen gab es viele. Im Schachspiel hatte jemand wie ich, der in seiner Jugend immerhin manche Partie von Aljechin, Lasker, Capablanca auswendig kannte, nicht den Hauch einer Chance gegen die Großmeister wie G. Harder, W. Bauer etc. Vom Tischtennis schien jeder etwas zu verstehen, während am Billardtisch die Stümperei wahre Feste feierte. Hier gab es nur eine bestaunte Ausnahme, nämlich JOSEF STEENBRINK, jedoch konnte er schließlich alles, er sang, spielte Billard, Klavier, Tischtennis, Theater, Fußball – von seinen mathematischen Künsten ganz zu schweigen. A propos Fußball muß ich eine denkwürdige Begebenheit doch etwas ausführlicher erzählen.

Bei einem übermütigen Zusammensein der Regensburger Mitglieder nach der Wahl des Themas „Das Eisensteinideal" flüsterte uns der Weingeist ein, die übrigen, also den „Rest der Welt" bei der nächsten Tagung zu einem Fußballspiel herauszufordern. Die Antwort hieß „Topp!", und so trafen denn am 14.9.1982, 14°° die beiden Kontrahenten im Stadion zu Kirche aufeinander:

1.FC RM (Regensburger Mathematiker) – Kickers RdW (Rest der Welt).

Der 1. FC, in blau-weißem Gala-Dress und exakter Reihe, führte unter dem Kommando ihres Coaches K. WINGBERG vor, was Disziplin war: Linkes Bein hoch, Recken nach rechts, Strecken nach links, dies hieß „Aufwärmen". In Wahrheit sollte das Muskelspiel den in lottriger Alltagskleidung unordentlich und staunend herumstehenden Gegner in Panik versetzen und seine Verteidigungsbereitschaft lähmen. Die Aufstellung sah so aus:

Schulze-Pillot

Gekeler   Berndt

Neukirch

Lorenz   Steenbrink   Herrlich

Gerritzen   Schülting   Harder   Laska

Jannsen   Wingberg   Schneider

Diekert   Knebusch   Köhnen

Tamme

Silhol

Die Linienrichter waren W.D. GEYER und N. SCHAPPACHER, der Schiedsrichter GERHARD FREY. Letzterer hielt mir, in dem er offenbar den gefährlichsten Spieler zu sehen meinte, schon vor dem Spiel drohend die rote Karte vor die Nase in Form eines

dünnen Buches mit dem Titel „Graphentheorie". Von den beiden Mannschaftska-
pitänen Harder und Neukirch wurden die Vereinswimpel ausgetauscht (echte, mit
Troddeln), dann kam der Anpfiff. Ich hatte kraft meiner Autorität in äußerst klu-
ger Voraussicht für die Ungültigkeit der Abseitsregel gesorgt. Dies brachte für den
Gegner keinen Vorteil, der wie ein Hornissenschwarm auf unser Tor zustürmte,
aber von Verteidiger Tamme im Griff gehalten wurde, der hinten wie ein Berserker
wütete, die brennende Zigarette mal links, mal rechts in der Hand. Mir aber schuf
der listige Schachzug die Möglichkeit zu schneidigen, das Spiel belebenden Kon-
tern. Merkwürdigerweise kann ich mich nicht mehr an den Ausgang des Spieles
erinnern. Ich erinnere mich nur noch, daß ich von der nächsten Tagung – gern hätte
ich das Spiel in seiner Einzigkeit zur Legende werden lassen, doch gehörte von
da an der Fußball zum Tagungsprogramm – ich erinnere mich, daß ich in Tokyo
eine Postkarte erhielt mit der in großen Ziffern gemalten einzigen Information **5 : 1**
und den Unterschriften meiner Mannen. Eine tückische Karte, sollte sie mir doch
eigentlich nur andeuten, wie gut es flutschte, wenn ich nicht dabei war.

Viel Kulturelles gab es am freitäglichen Weinabend. Da war der obligate, von
W.D. GEYER komponierte und einstudierte Kanon auf einen die Tagung kennzeich-
nenden Text. Dargeboten wurden instrumentale Kammermusik, Loewe-Balladen,
Gesangsduette und Terzette von Dvorak und Schumann, vierstimmige Chorgesänge
etc. Auf der Tagung „Iwasawa-Theorie" erlebte man sogar eine Uraufführung. Sie
war von den Tagungsleitern und ihrer Assistentin PILAR BAYER eigens für diese
Tagung komponiert worden und hieß

„Triosonate A-dur"

Auftraten die weltberühmten Künstler Fred Emat, Giacomo Lehmann und Mer-
cedes Zapalata, in Frack und Abendkleid gehüllt, alle drei Violinvirtuosen und
Pianisten. Die Komposition hatte durchaus mathematischen Bezug, indem jeder
Künstler in zyklischer Vertauschung für den von ihm komponierten Satz am Flügel
saß. Wer nun meinte, daß Emat und Zapalata eine Geige zum letzten Mal als
Fünfjährige in der Hand gehabt hatten und Lehmann noch nie, dem wurde der
schnöde Verdacht durch die raffinierte Komposition zerschmettert. Beim einleiten-
den „Grave" etwa wurde der schwere Klaviersatz von eindrucksvollem Ostinato
der im forte angerissenen leeren G-Saiten der Geigen begleitet, während im letzten
Satz „Allegro virtuoso" die echte Virtuosität der Pianistin (das einzige Pfund, mit
dem wir wuchern konnten) durch rasendes Terzentremolo der Geigen zum Furioso
gesteigert wurde. Die Begeisterung des Publikums war komplett.

Bei der Tagung „Weil-Vermutung" wurde sogar eine ganze Oper zur Aufführung
gebracht:

„Ein Abend im Gesangsverein zu Bummelsdorf"

Es handelte sich um den tragischen Konflikt des Dirigenten mit der Unzu-
verlässigkeit seiner Sänger, mit hinderlichen Beerdigungen, Alkoholismus und Tur-
nerleidenschaften – eine gemarterte Seele, die zum Schluß noch in einer alles beim
Alten belassenden Versöhnung in den gemeinsamen Jubelgesang einstimmen muß.

Das Orchester wurde von Pilar Bayer markiert und die Opernsänger hießen Neukirch, Steenbrink, Mayer, Knebusch, Tamme.

Genug der lustigen Geschichten! Nichts wäre falscher als der Eindruck, es handele sich um eine Gesellschaft vergnügungssüchtiger Hallodris. Die Heiterkeiten geben Zeugnis von der optimistischen Haltung und der menschlichen Zugeneigtheit, in der manch enge und bleibende Freundschaft entstanden ist. Die mathematische Arbeit ist durch diese Atmosphäre nur immer gefördert worden, und nicht nur für die Glockengießer gilt das Schillersche Wort

„Wenn gute Reden sie begleiten, dann fließt die Arbeit munter fort".

Auswirkung

Die große Tradition der Mathematik in unserem Land hat durch die Nazizeit und den Krieg einen erheblichen Einbruch erlitten, insbesondere durch den Exodus der vielen hervorragenden jüdischen Mathematiker. Wohl kamen nach der Katastrophe einige junge, ausgezeichnete Persönlichkeiten wieder herauf, jedoch mußten sie ganz von vorn beginnen, das einst so fruchtbare Feld zu bestellen. Auf vielen Gebieten mußte man die Überlegenheit der Franzosen, Amerikaner, Russen, Japaner anerkennen. Heute sieht man die Mathematik in Deutschland wieder in ihrer alten Stellung und in internationalem Ansehen. Dazu hat die AG nicht wenig beigetragen; der sie prägende Wille, an den modernen Entwicklungen mit eigener Arbeit teilzunehmen, hat seinen Weg nach oben gefunden.

Die auf Breite zielende Aktivität der AG in Oberwolfach hat sich in vielen Oberseminaren in diesem Geiste fortgesetzt. Die jungen Mathematiker der Universitäten Bonn, Köln, Wuppertal, Münster etwa haben sich zu einer weiteren Arbeitsgemeinschaft zusammengefügt und scheuen die ständigen Reisen an die wechselnden Orte ihres Treffens nicht, um ihr Wissen zu erweitern. Eine ähnliche Gemeinschaft bilden die Heidelberger, Mannheimer, Karlsruher und Straßburger AG-Mitglieder. Der geweitete Blick hat auch auf viele andere Tagungen in Oberwolfach eine nach vorn treibende Auswirkung gehabt. Die auf den AG-Tagungen erworbenen Einsichten gaben Anlaß zu weiteren wichtigen Spezialtagungen.

Aber nicht nur Kenntnis und Wissen hat das aufs Lernen gerichtete Bemühen der AG gebracht, es war auch von großer Wirkung auf die Forschung. Nicht zu zählen sind die wissenschaftlichen Arbeiten, die diesem Studium entsprungen sind, und kaum zu überschätzen ist ihre Auswirkung auf den Fortgang der Mathematik. Wichtige Bücher waren der direkte Ausfluß der Tagungen, wie etwa FREITAG-KIEHL „Etale Kohomologie und Weil-Vermutung" und das schon erwähnte Buch RAPOPORT-SCHAPPACHER-SCHNEIDER „Beilinson's Conjectures on Special Values of L-Functions", ein echtes Gemeinschaftswerk, das zu einer Grundlage moderner Forschung wurde und überall zitiert wird.

Ein direktes Ergebnis gemeinsamer Tagungsarbeit war der (von C. KLINGENBERG und D. RAMAKRISHNAN vervollständigte) Beweis von HARDER-LANGLANDS-RAPOPORT der verwegenen Tate-Vermutung für die Hilbert-Blumenthal-Flächen. Die entscheidenden Ideen zu diesem bedeutenden Theorem wurden nach dem Vortrag von

HARDER auf der „Hodge-Tate-Tagung" von den drei Mathematikern, ich muß es zugeben, zwar nicht auf der „Ochsentour", aber doch auf der Wanderung nach St. Roman gefunden.

Ein anderes herausragendes Resultat, ebenfalls das direkte Ergebnis einer AG-Tagung, war der Beweis der Beilinson-Vermutung für elliptische Kurven mit komplexer Multiplikation von C. DENINGER. Das Aufsehen erregende Ergebnis war die Folge des Vortrags von DENINGER und WINGBERG auf der „Beilinson-Tagung", in dem die Vermutung nur für das Argument $n = 0$ bewiesen wurde. Angesichts des großen Aufwandes allein für diesen Spezialfall konnte eine Ausdehnung auf alle n eigentlich nur ein hoffnungsloses Wunschdenken sein, und es erschienen mir Deningers daran anknüpfenden Spekulationen als realitätsferne Träumerei. So kam ich aus dem Staunen nicht heraus, als ich sehen mußte, wie alles, was nicht zusammenpaßte, sich schließlich doch der Phantasie und dem Willen dieses Mathematikers fügen mußte.

Überaus verblüffend war der Fall MARKUS ROST. Der junge Topologe wurde von den Regensburger AG-Mitgliedern überredet, einmal mitzukommen auf eine AG-Tagung. Das Thema war die Algebraische K-Theorie und hatte mit seinem Arbeitsgebiet nichts zu tun. Im Zentrum stand der berühmte Satz von MERKURJEV-SUSLIN

$$K_2(F)/nK_2(F) \cong H^2(F, \mu_n^{\otimes 2}).$$

Rasch hatte sich Rost in den tiefliegenden Beweis eingearbeitet. Heraus kam er mit einem völlig überraschenden, von ganz neuen Ideen lebenden Beweis der Isomorphie

$$K_3(F)/2K_3(F) \cong H^3(F, \mu_2^{\otimes 3}),$$

ein Ergebnis, das um die Welt sofort die Runde machte (aber, wie sich herausstellte, gleichzeitig von Merkurjev und Suslin bewiesen wurde).

Durch all diese Leistungen hat die AG eine große internationale Beachtung gefunden und hat sich auf diese Weise um die Institution „Mathematisches Forschungsinstitut Oberwolfach" verdient gemacht. Eine beeindruckende Zahl von Professoren ist aus ihr hervorgegangen. Mein Gedächtnis und mein Überblick läßt mich nur einige nennen:

P. Bayer, S. Böcherer, C. Deninger, V. Diekert, H. Ésnault, G. Faltings, E.-U. Gekeler, F. Grunewald, F. Herrlich, U. Jannsen, E. Kani, I. Kersten, N. Klingen, H. Knörrer, G. Martens, B.H. Matzat, E. Nart, H. Opolka, M. Rapoport, J. Rohlfs, N. Schappacher, C.-G. Schmidt, P. Schneider, A.J. Scholl, R. Schulze-Pillot, J. Schwermer, W. Singhof, P. Slodowy, U. Stuhler, E. Viehweg, R. Weissauer, K. Wingberg, J. Wolfart, G. Wüstholz, u.a.

Den ständigen AG-Mitgliedern CHRISTOPHER DENINGER, GÜNTER HARDER, MICHAEL RAPOPORT, PETER SCHNEIDER wurde der „Leibniz-Preis" der Deutschen Forschungsgemeinschaft verliehen und das frühere AG-Mitglied GERD FALTINGS ist Träger der „Fields-Medaille".

Es sind dies, ich will es abschließend noch einmal sagen, meine persönlichen Erinnerungen und meine subjektiven Eindrücke von der „Arbeitsgemeinschaft Oberwolfach", die nicht mit Beckmesser-Augen betrachtet werden möchten. Manches

Wichtige werde ich ausgelassen haben, anderes mag zu sehr herausgehoben erscheinen. Ich hoffe aber doch, ein im Ganzen stimmiges Bild gegeben zu haben. Die Tagung liegt erneut in den Händen umsichtiger und verantwortungsvoller Leiter. Sie setzt ihre gute, sich ständig erneuernde Wirkung auf das mathematische Leben in unserem Lande fort und verdient es, mit allen Kräften gefördert zu werden.

Der Oberwolfachpreis und seine Preisträger

Die Gesellschaft für mathematische Forschung e.V. hat in ihrer Sitzung am 17. Februar 1990 beschlossen, alle zwei Jahre einen „Förderpreis des Mathematischen Forschungsinstituts Oberwolfach" zu vergeben. Preisträger können frei bis zu drei Monaten im Mathematischen Forschungsinstitut Oberwolfach verbringen. Der Preis wird für herausragende Leistungen an europäische Mathematiker(innen) bis zum Höchstalter von 35 Jahren vergeben.

Der Preisträger 1991 war Herr Peter B. Kronheimer (Oxford). Im Jahre 1993 waren die Herren Jörg Brüdern (Göttingen) und Jens Franke (Bonn) die Preisträger. Die Laudationes auf Kronheimer bzw. Brüdern bzw. Franke wurden von Simon Donaldson (Oxford) bzw. Wolfgang Schwarz (Frankfurt) bzw. Joachim Schwermer (Eichstätt) gehalten.

The Work of Peter B. Kronheimer

Simon Donaldson

Kronheimer's work in this period was mainly concerned with the subject of "hyperkahler geometry". Recall that a *Kahler manifold* is a Riemannian manifold with a complex structure I on its tangent spaces which is preserved by the parallel transport of the Levi-Civita connection: equivalently, it is a manifold whose holonomy group reduces to the unitary group. A *hyperkahler manifold* is a Riemannian manifold with a triple of complex structures I, J, K, satisfying the algebraic identities of the quaternions, each of which yields a Kahler structure. Equivalently, it is a manifold whose holonomy group reduces to the compact symplectic group. A hyperkahler structure gives a natural kind of "quaternionic manifold" and they are also very interesting from the point of view of Riemannian geometry: in any dimension they have vanishing Ricci tensor, and in dimension 4 they can be characterised, locally, as metrics which are simultaneously Ricci-flat and "self-dual".

Kronheimer's first principal achievement involved the construction of "Gravitational Instantons". As a consequence of a theorem of Cheeger and Gromoll, any complete, Ricci-flat Riemannian manifold which is "asymptotically Euclidean" is in fact flat Euclidean space. This is a natural result from the point of view of relativity, where one can think of the Ricci tensor as the source term for the metric, regarded as a gravitational field. However, it was realised by Gibbons and Hawking, in the context of Quantum Gravity, that there are non-trivial *asymptotically locally Euclidean*, Ricci-flat metrics. These are non-compact spaces which are asymptotic at infinity to a cone $\mathbf{R}^4/\Gamma$, where Γ is a finite subgroup of $SO(4)$. In the case when the metric is also self-dual, so the manifold is hyperkahler, they are known as Gravitational Instantons and, as the terminology suggests, they are analogous to the well-known Instanton solutions of the Yang-Mills equations. In this case the group Γ lies in one factor in the decomposition $Spin(4) = SU(2) \times SU(2)$, and so (since $SU(2)$ is locally isomorphic to $SO(3)$), is essentially one of the finite symmetry groups of 3-space, studied since antiquity. Hitchin obtained further examples of these Gravitational Instantons, using the Penrose twistor space construction which encodes a self-dual 4-manifold in an associated 3 dimensional complex manifold. In Hitchin's examples the twistor spaces were obtained from the simultaneous resolution of the complex surface singularity $\mathbf{C}^2/\Gamma$ which, as the Kleinian or simple singularities, had been extensively studied by algebraic geometers. On the basis of this, Hitchin conjectured that Gravitational Instantons should exist for every finite subgroup $T \subset SU(2)$. It was this conjecture which Kronheimer proved, inventing an elegant general construction which gave indeed a complete classification of all Gravitational Instantons [1, 2].

The setting for Kronheimer's approach was the "hyperkahler quotient" construction which had been introduced shortly before, and which in turn has its roots in symplectic geometry. If a group G acts symplectically on a symplectic manifold X there is, under some general hypotheses, a "moment map" μ from X to the dual of the Lie algebra of G. This generalises the notion of a Hamiltonian for flows. The "symplectic quotient" is the space $\mu^{-1}(0)/G$ and this has a natural induced symplectic form. A Kahler manifold is symplectic and one can show that in this case the quotient is again Kahler in a natural way. Now if X is hyperkahler it has three different symplectic structures and thus three moment maps μ_I, μ_J, μ_K and it was observed by Hitchin, Karlhede, Lindstrom and Roceck that the quotient of the common zero set $\mu_I^{-1}(0) \cap \mu_J^{-1}(0) \cap \mu_K^{-1}(0)$ has a induced hyperkahler structure. This can be used to generate many interesting examples, even when the original space X is flat.

In his construction Kronheimer begins with the vector space $Q \otimes EndR$, where R is the regular representation of a finite group $\Gamma \subset SU(2)$, and Q is the given 2-dimensional representation. This is naturally a quaternionic vector space and so has a flat hyperkahler structure: the same is true of the subspace X of Γ-invariant elements. Then Kronheimer considers the quotient of X by the group G of unitary automorphisms of R which commute with the Γ action. Now in a case when the Lie algebra has a non-trivial centre the moment maps are not quite unique, one can change them by adding constant central elements. Equivalently, if one fixes the maps one can make a quotient using the preimages $\mu_I^{-1}(\xi_I)$ etc., where ξ_I, ξ_J, ξ_K are central elements. In Kronheimer's setting, with the natural normalisation, the hyperkahler quotient obtained from $\xi_I = \xi_J = \xi_K = 0$ is the singular space $\mathbb{C}^2/\Gamma$, with the flat cone metric. For generic parameters ξ_I, ξ_J, ξ_K the hyperkahler quotient is smooth and gives a Gravitational Instanton with the symmetry group Γ at infinity. Kronheimer then showed, via a sort of "Torelli Theorem" that this gives all the Gravitational Instantons, with the parameters ξ_I, ξ_J, ξ_K appearing instrinsically as the periods of the three symplectic forms, in the cohomology of the quotient. This description is bound up with the "McKay correspondence" between the representations of the finite group Γ and the 2-dimensional cohomology of the resolution of the Kleinian singularity, which in turn can be interpreted in terms of the Dynkin diagram of a simply-laced Lie group associated to Γ.

We now turn to the other major project of Kronheimer during this period: his work on the hyperkahler geometry of co-adjoint orbits. Again, this brings original and deep ideas to a topic which fits tidily into a complex tapestry of existing mathematics. Recall first that the co-adjoint orbits of any Lie Group, in the dual of the Lie algebra, have canonical Konstant-Kirillov symplectic structures. If the group is complex one gets complex symplectic orbits, while if the group is compact then it is well-known that the orbits are Kahler, Kronheimer's work provides, from one point of view, a common generalisation of these two cases: for complex semi-simple groups he found hyperkahler structures on the co-adjoint orbits, so that if $\omega_I, \omega_J, \omega_K$ are the three symplectic forms, the combination $\omega_J + i\omega_K$ is the holomorphic Kostant–Kirillov form while the form ω_I is analogous to the Kahler metric in the compact case. In fact Kronheimer treated to extreme cases: in [4] the generic case of "regular orbits" G/T, where G is complex semi-simple and T is a maximal torus, and in [3] the other extreme of the nilpotent orbits (whose closure contains zero). Actually the main thrust of [3] is not towards the hyperkahler metric but rather the elucidation of a famous theorem of Brieskorn which asserts that the singularity in the nilpotent variety of a complex semi-simple group G, transverse to the subregular orbit, has the form $\mathbf{C}^2/\Gamma$, for $\Gamma \subset SU(2)$. Kronheimer obtained a new proof by finding two descriptions of this space which show that it is hyperkahler but in one of which there is an obvious action of $SU(2)$ by isometries, mixing the three complex structures. Brieskorn's theorem is an immediate consequence of this and becomes more transparent from the hyperkahler viewpoint, in which the different complex structures have the same status.

Kronheimer's approach to finding these hyperkahler metrics again involved the quotient construction, but now in an infinite dimensional version. Atiyah and Bott had observed that one could view the natural Kahler metric on a moduli space of flat bundles over a Riemann surface as an example of the Kahler quotient, in which one starts with the flat space of all connections and then carries out the quotient procedure with respect to the action of the infinite-dimensional gauge group. (This point of view has more recently become extremely important in the geometric approach to Conformal Field Theory.) These ideas have a 4-dimensional counterpart, in which one starts with a hyperkahler 4-manifold Z, then any moduli space M of Yang-Mills instantons, or anti-self-dual connections, over Z has a hyperkahler structure, which can be regarded as the hyperkahler quotient of the space of connections by the gauge group. This applies in particular when Z is flat space $\mathbf{R}^4$, and also to various reductions of the problem in which one imposes symmetries on the connections. The case which is relevant to Kronheimer's work is when one looks at connections over $\mathbf{R}^4$ (or a portion thereof) which are preserved by translation in three orthogonal directions. The instanton equations then reduce to a system of ODE, *Nahm's equations*

$$\frac{dT_i}{dt} = \left[T_j, T_k \right] .$$

Here we have three variables T_1, T_2, T_3 taking values in the Lie algebra of the given structure group, and the indices i, j, k run over cyclic permutations of $1, 2, 3$. The essence of Kronheimer's method is to show that, for suitable "boundary conditions", the moduli space M of solutions of Nahm's equations can be identified with a co-adjoint orbit of the complexified group. On the other hand the general hyperkahler reduction mechanism above gives a hyperkahler structure on M.

References

1. P. B. Kronheimer: The construction of ALE spaces as hyperkahler quotients. J. Diff. Geom. **29**, 665–685 (1989)
2. P. B. Kronheimer: A Torelli-type theorem for gravitational instantons. J. Diff. Geom. **29**, 685–699 (1989)
3. P. B. Kronheimer: Instantons and the geometry of the nilpotent variety. J. Diff. Geom. **32**, 473–490 (1990)
4. P. B. Kronheimer: A Hyper Kahlerian structure on co-adjoint orbits of a semi-simple complex Lie group. J. Math. Soc. (London) **42**, 193–208 (1990)

Zur Verleihung des Preises an Jörg Brüdern

Wolfgang Schwarz[1]

Die *Vorlesungen über die Entwicklung der Mathematik im 19. Jahrhundert* beginnt Felix Klein mit den Worten: „... liegt es ... auf der Hand, daß in einer solchen Übersicht der kulturbildenden Faktoren ... unsere Wissenschaft nicht fehlen darf. Vielmehr muß versucht werden, auch der Mathematik die Stellung einzuräumen, die ihr als einer der ältesten und edelsten Betätigungen des menschlichen Geistes und als einer der richtunggebenden Kräfte in seiner Entwicklung gebührt, die sie aber im Bewußtsein der Gebildeten, wenigstens in Deutschland, leider nur selten einnimmt."

Kürzer drücken David Hilbert und S. Cohn-Vossen diese Gedanken im Vorwort ihres Buches *Anschauliche Geometrie* 1932 aus: [Denn] „im allgemeinen erfreut sich die Mathematik, wenn auch ihre Bedeutung anerkannt wird, keiner Beliebtheit. Das liegt an der verbreiteten Vorstellung, als sei die Mathematik eine Fortsetzung oder Steigerung der Rechenkunst."

Auf eine provokative Frage, was Mathematiker noch zu tun hätten, da Computer doch viel schneller rechnen könnten, kam meine den Sachverhalt verkürzende Antwort: Die Mathematiker rechnen nicht, sie *denken*. Damit war gemeint, daß einen Mathematiker folgende Fähigkeiten auszeichnen sollten:[2]

- Phantasie und Einfallsreichtum,
- die Fähigkeit zur Beschränkung auf das Wesentliche,
- Abstraktionsfähigkeit,
- die Fähigkeit zur Formalisierung und zum Ausarbeiten von Lösestrategien,
- die Fähigkeit zur Heuristik,
- Flexibilität, Einarbeitungsfähigkeit in unterschiedliche Problembereiche.

---

[1] Diese Laudatio basiert auf dem am 13. Februar 1993 gehaltenen Vortrag; sie ergänzt den in den *Mitteilungen der DMV*, Heft 2, 1993, S. 29–30, abgedruckten Bericht *Förderpreis des Mathematischen Forschungsinstitutes Oberwolfach*. Dort werden technische Details gegeben, hier wird Wert auf allgemeine Gesichtspunkte gelegt. Die beiden Darstellungen sind weitgehend disjunkt.

[2] Man vgl. hierzu auch die Antrittsvorlesungen *Das Selbstverständnis von Mathematikern und das Image der Mathematik* von H. Behr, Frankfurt 1976, und *Mathematik als kreatives Schaffen* von W. Roggenkamp, Stuttgart 1975. Nach diesem ist ein Mathematiker ein „Problemlöser und Theorienschöpfer".

Diese Auffassung vom Wesen der Mathematik scheint nicht Allgemeingut zu sein. Karl Jaspers' Einsichten, die in seinem Buche *Nikolaus Cusanus*[3] geäußert werden,[4] können als Hinweis betrachtet werden, daß sich das Bild der Mathematik in der Öffentlichkeit zu ändern beginnt.[5] O. H. Peitgen, bekannt durch reizvolle Computergraphiken zur Theorie der Dynamischen Systeme, stellte 1990 fest: „Die Mathematik, die vielen in der Schule oft nur eine Plage war und ein Buch mit sieben Siegeln geblieben ist, hat heute, wenn auch von der breiten Öffentlichkeit weithin unbemerkt, bereits viele Bereiche unseres Lebens erobert. Die explosive Vermehrung natur- und ingenieurwissenschaftlichen Wissens geht Hand in Hand mit einer Mathematisierung aller Wissenschaften … .“

Vielleicht *auch* mit der Absicht, die überfällige Korrektur des öffentlichen Mathematikbildes zu beschleunigen, haben hochherzige Mathematiker, wie Otto und Edith Haupt oder Alexander Ostrowski, *mathematische Preise* gestiftet, und dasselbe haben für die Belange der Mathematik zuständige Organisationen getan, wie die Deutsche Mathematiker-Vereinigung mit ihrer Georg-Cantor-Medaille oder die *Gesellschaft für Mathematische Forschung e.V.* mit dem *Förderpreis* des Mathematischen Forschungsinstitutes Oberwolfach, der am 13.2.1993 zum zweiten Male vergeben wurde.

Man könnte einwenden, daß erfolgreiche Wissenschaftler eine gesonderte Auszeichnung gar nicht nötig hätten. Darauf wies schon Friedrich Schiller in seiner *Akademischen Antrittsrede zu Jena*, 26. Mai 1789, hin: „… der philosophische Geist findet in seinem Gegenstand, in seinem Fleiße selbst, Reiz und Belohnung. Wie viel begeisterter kann er sein Werk angreiffen, wieviel lebendiger wird sein Eifer, wieviel ausdauernder sein Muth und seine Thätigkeit seyn, da bey ihm die Arbeit sich durch die Arbeit verjüngt. … " Trotzdem wird dieser Preis den Lau-

---

[3] Unter IV. Die Wahrheit in der Mathematik, Die Beziehungen von Mathematik und Philosophie.

[4] Der gewaltige Eindruck der Mathematik auf die Philosophen beruht auf immer wieder erstaunlichen Tatsachen. Es sind dies

- die Evidenz der vom Geiste konstruierten mathematischen Verhältnisse,
- der Zauber der ‚Schönheit‘ mathematischer Erkenntnisse, in denen eine verborgene Zweckmäßigkeit zu walten scheint,
- der Unterschied trivialer Richtigkeiten von wesentlichen mathematischen Einsichten, der gleichgültigen von den faszinierenden Ergebnissen,
- die Fähigkeit unseres Geistes, aus von ihm selbst gesetzten Voraussetzungen Operationen vollziehen zu können, die ins Grenzenlose neue Herrlichkeiten zu entdecken erlauben,
- die Eignung der Mathematik zur Erkenntnis der Natur, soweit sie nach Maß, Zahl und Gewicht geordnet ist,
- dieses gar nicht Selbstverständliche einer (teilweisen) Koinzidenz der mathematischen Erkenntnisse mit Naturerkenntnissen, die in beiden gefundenen Ordnungen und Gesetzmäßigkeiten.

[5] In den Vereinigten Staaten sieht es besser aus. Im Vorwort der "State of the Art Reviews", USA 1986, heißt es: "Mathematics has become, even more directly, the language and the foundation of science, technology, and social organization. Indeed, it is a fundamental driving force in the worldwide progress that is altering the economic, political, and social balance among nations. … "

reaten ein nochmaliger Ansporn sein und ihre Forschungstätigkeit – auch durch den vorgesehenen Aufenthalt in der anregenden Oberwolfacher Atmosphäre mit einer Bibliothek, die ihresgleichen selten findet – beflügeln.[6] [7]

Die Diplomarbeit, von J. S. Patterson betreut, brachte Herrn Brüdern einen Preis „für besondere studentische Leistungen" des Niedersächsischen Ministers für Wissenschaft und Kunst, und war für die Wahl der Arbeitsrichtung von Herrn Brüdern entscheidend, denn *The child is father of the man* (Wordsworth).

Das Promotionsstudium erfolgte im märchenhaften Oxford mit gotischen Colleges, in denen ein Student ein stilvolles Zuhause findet und eine Prägung auf seine Universität erfährt. Herr Brüdern arbeitete mit Roger Heath-Brown und Bryan John Birch.[8] Mit *Iterationsmethoden in der additiven Zahlentheorie* promoviert er im April 1988; Doktorvater war J. S. Patterson. Die Habilitationsschrift (1991) *Sieves, the Circle Method, and Waring's Problem for Cubes* ist richtungweisend: Die Kreismethode wird mit dem „Kleinen Sieb" und dem „Großen Sieb" kombiniert.

In den 1986 erschienenen *State of The Art Reviews – Mathematical Sciences: A Unifying and Dynamic Resource* des National Research Council heißt es: "Difficult problems that have been unsolved for many years are now being solved with amazing frequency – a strong confirmation of the vitality of mathematics." Zur Richtigkeit dieser Aussage hat Herr Brüdern beigetragen. Sein anwendungsfernes Arbeitsgebiet ist ein kulturgeschichtlich klassischer Problembereich. Herr Brüdern befaßt sich mit dem nach Edward Waring, 1734–1798, Magdalen College, Cambridge, benannten Problem: Ist die Gleichung

$$N = x_1^k + \ldots + x_s^k$$

---

[6] Ein Abriß des *Lebenslauf*s von Herrn Brüdern wurde in den DMV–Mitteilungen gegeben. – Das Studium der Mathematik mit den Nebenfächern Theoretische Physik und Astronomie erfolgte in Göttingen, wobei die Regelstudienzeit lässig unterboten wurde.

[7] Georg Christoph Lichtenberg, vor 251 Jahren bei Darmstadt geboren, seit 1790 a.o. Professor in Göttingen, ein scharfsichtiger Beobachter, dem nach J. W. von Goethe „eine ganze Welt von Wissen und Verhältnissen zu Gebote stand, um sie wie Karten zu mischen und nach Belieben schalkhaft auszuspielen", beschreibt Göttingen folgendermaßen:

Seitdem mein Kutscher und mein Schicksal,	Dies geistliche Schlaraffenländchen.
Mich, Teuerster, aus Deinem Blick stahl,	Liebst Du die gare Wahrheit, heißt es,
Leb' ich in diesem Vaterstädtchen	So öffne hier das Maul des Geistes,
Von hoher Weisheit in Traktätchen,	Nur aufgesperrt, mein lieber Sohn,
Berühmt in allerlei Bedeutung	Das andre gibt sich selber schon.
Durch Würste, Bibliothek und Zeitung,	Hier trieft der Honig der Erkenntnis
Durch Professoren und schlechtes Wetter	Und dort die Sahne vom Verständnis.
Und breite Stein' und Wochenblätter.	Kommt, Jünglinge, die ihr gebessert
Du kennst zwar schon aus einem Bändchen	Sein wollt, und trinkt sie ungewässert!

Die Lichtenberg-Zitate sind dem von Paul Requadt herausgegebenen Bande G. C. Lichtenberg, *Aphorismen, Briefe, Schriften*, Alfred Kröner Verlag 1939, entnommen.

[8] "International competition and competitiveness are concepts that traditionally are alien to the study of mathematics, in fact, cooperation among individuals of different nations is a vital part of the pursuit of solutions to mathematical problems. Mathematics is intrinsically international, with its own language cutting across barriers of geography and culture .... ." (State of the Art Reviews).

bei gegebenem N in ganzen Zahlen $x_1, \ldots, x_s$ lösbar? Welches ist die kleinste Zahl $s = G(k)$, so daß jedes hinreichend große N in dieser Gestalt darstellbar ist? Wie groß ist die Anzahl der Lösungen?

Das Problem läßt sich *abwandeln*, man kann die Monome x^k durch Polynome ersetzen, man kann nach der Lösung solcher Gleichungen in Primzahlen oder Fast-Primzahlen fragen, man kann entsprechende Kongruenzen betrachten, und man kann die Lösbarkeit diophantischer Ungleichungen untersuchen. Alle diese Frage-stellungen haben ihre Bedeutung; manchmal ist ein Lösungsweg vorhersehbar, oft ist aber eine Lösungsmethode nicht in Sicht und muß, wie im Falle der Fastprim-zahlen, durch Herrn Brüdern erst geschaffen werden.

Man kann aber auch deutliche *Verschärfungen* vorliegender Ergebnisse anstre-ben – hier ist z. B. die obere Abschätzung der Minimalanzahl $G(k)$ der notwendigen Summanden interessant. Solche Verschärfungen erfordern i.a. methodisch wenig-stens eine neue Idee.

Nach David Hilbert, 1909, bescherte die zweite Lösung des Waring-Problems durch G. H. Hardy und J. E. Littlewood der additiv-analytischen Zahlentheorie eine effektive Methode, die "Circle Method", um deren Verfeinerung und Verschärfung sich Herr Brüdern, auf den Schultern von Riesen stehend, verdient gemacht hat. Mit der Urform der Hardy-Littlewoodschen Methode ließen sich freilich die besonderen Erfolge von Herrn Brüdern, nicht erzielen. Aber Vaughan, Wooley und Brüdern haben nach einer längeren Stagnationsperiode wegweisende Verbesserungen bei der Abschätzung der Zahl $G(k)$ erzielt. Methodisch wichtige Neuerungen, die es Herrn Brüdern ermöglichten, über bisher vorliegende Ergebnisse hinauszukommen, sind z.B.:

- Multiplikative Bedingungen an die Variablen können die Lösungsanzahl von Hilfssystemen diophantischer Gleichungen „klein" halten. Der Umgang mit sol-chen Bedingungen erfordert Ideenreichtum und technische Versiertheit.
- Die phantasievolle Kombination der "Circle Method" mit dem „Kleinen Sieb" und dem „Großen Sieb" erlaubt es, Ergebnisse zu erhalten, die außerhalb des Anwendungsbereiches jeder einzelnen dieser Methoden liegen.
- Die Einführung des „Vektorsiebes" (gemeinsam mit E. Fouvry) stellt eine noch wenig ausgeschöpfte Neuerung dar, die es z.B. erlaubt, die Lösbarkeit der Glei-chung $N = x_1^2 + \ldots x_4^2$ in *Fastprimzahlen* $x_1, \ldots, x_4$ zu beweisen.

Einige Ergebnisse von Herrn Brüdern sind in Kurzform in den *Mitteilungen der DMV* zusammengestellt worden und werden hier nicht reproduziert.

Die Preisträger sollten der Mathematik treu bleiben, denn „Das hohe Alter mancher Mathematiker (Fontanelle, Euler, Leibniz) könnte eine Folge sein der Be-trachtung ihrer selbst, des Subjektivischen bei den Körpern .... So könnte die Mathematik zur Verlängerung des Lebens beitragen," und[9] „... keiner der Alten, der als groß erfunden wurde, ging schwierige Probleme anders als durch mathe-

---

[9] Nikolaus Cusanus, „... et nemo antiquorum, qui magnus habitus est, res difficiles alia similitudine quam mathematica aggressus est. Ita ut Boethius, ille Romanorum littera-tissimus, assereret neminem divinorum scientiam, qui penitus in mathematicis exercitio careret, attinger posse."

matische Vergleiche an. So versichert der gelehrte Römer Boethius, daß jemand, der in der Mathematik völlig ungeübt sei, das Wissen vom Göttlichen nicht erreichen könne". Auf Grund der nunmehr für Nachwuchswissenschaftler günstigeren Altersstruktur der Hochschullehrerschaft werden sie optimistischer in die Zukunft sehen können – im Gegensatz zu den letzten zwanzig Jahren, in denen eine ganze Generation von Nachwuchswissenschaftlern trotz ausgezeichneter Leistungen nahezu keine Berufungchance hatte. Nochmals sei aus den *State of the Art Reviews* zitiert:

With the necessary support, mathematics will continue to flourish, to attract excellent minds, and in the coming years to produce much essential new mathematics on an international basis. The role of any particular country in this development is hard to predict. Mathematical leadership will depend on many factors, primary among them the support that individual nations give to basic sciences in general and mathematical research in particular.

Dieselbe Idee wird auch in Heinz Maier-Leibnitz, *Der geteilte Plato*, 1981[10] angesprochen:

Aber die größte Aufgabe liegt darin, unsere Hochschulen wieder für die besten Wissenschaftler attraktiv und fruchtbar zu machen. Hier, bei den Besten, scheint mir der Hebel zu sein, wo man ansetzen muß. Von der Breite der Personenzahl, der Ausstattung her sind die Hochschulen nicht notleidend. Eine Hilfe an der Spitze, mehr Freiheit, mehr Förderung derer, die mehr leisten können und dazu bereit sind, das ist mit geringen Mitteln möglich. Aber man muß es wollen!

Der Förderpreis des Mathematischen Forschungsinstitutes Oberwolfach trägt zu diesem Ziele bei.

---

[10] Die zitierte Aussage über die Ausstattung ist freilich im Jahre 1993 nicht mehr zutreffend, wie auch aus der HRK-Broschüre *Konzept zur Entwicklung der Hochschulen in Deutschland*, 1992, hervorgeht.

Laudatio auf Jens Franke

Joachim Schwermer

Der Bitte, einige Worte über Herrn Dr. Jens Franke zu sagen, folge ich sehr gerne. Als ich ihn während meines Aufenthaltes am Institute for Advanced Study in Princeton, USA, im Herbst 1990 kennenlernte und begann, mit ihm zu diskutieren und zusammenzuarbeiten, wurde mir sehr rasch klar, daß er ein junger Mathematiker mit außergewöhnlichen Talenten ist und gleichzeitig ein ungewöhnlich breites und tiefes Wissen besitzt.

Herr Jens Franke wurde am 29. Juni 1964 in Gera geboren, besuchte dort die Oberschule ‚Rudolf Scheffel‘ (1971–1979) und legte dann an der Erweiterten Oberschule ‚Otto Grotewohl‘ (1979–1983) im Jahre 1983 das Abitur ab. Während der letzten fünf Jahre seiner Schulzeit erhielt er zusätzlichen mathematischen Unterricht durch Herrn Dr. T. Runst, Mitarbeiter am Mathematischen Institut der Friedrich-Schiller-Universität Jena. Dort nahm er dann auch das Studium der Mathematik auf, schloß es schon 1985 nach zwei Studienjahren mit dem Grad des Diplom-Mathematikers ab. 1986 wurde er mit einer Dissertation [2] zu einem Thema aus der mathematischen Analysis (betreut durch H. Triebel) zum Dr. rer. nat. promoviert. Schon während der Schulzeit studierte Herr Franke, neben Büchern zur Analysis, Topologie, auch Werke zur Zahlentheorie, durch die er mit den klassischen Ergebnissen von Gauss, Hilbert, Minkowski, Hecke, Hasse vertraut wurde. Dies führte während des Studiums dazu, daß er die anstehende Kursvorlesung Algebra des Dozenten K. Haberland mit der Auflage erlassen bekam, das bekannte Kompendium zur algebraischen Zahlentheorie von Cassels-Fröhlich durchzuarbeiten. Das auf Anraten der Kollegen Triebel und Haberland von Herrn Franke 1986 aufgenommene Zusatzstudium in Moskau bot ihm dann reichlich Gelegenheit, in Richtung Zahlentheorie und arithmetisch-algebraische Geometrie weiterzuarbeiten.

Herr Franke war von einem tiefliegenden Problem in der Analysis auf arithmetisch definierten lokalsymmetrischen Räumen angezogen, etwa den Räumen, die man als Quotienten aus der Aktion der speziellen linearen Gruppe ganzzahliger $(n \times n)$-Matrizen der Determinante 1 auf dem Raum der positiv definiten symmetrischen reellen $(n \times n)$-Matrizen erhält. In Fortsetzung von Arbeiten von A. Weil, Y. Matsushima–S. Murakami, M. S. Raghunathan, G. Harder, A. Borel und anderen vermutete man seit langem, daß die Kohomologie eines solchen Raumes (d.h. die Kohomologie einer arithmetisch definierten Gruppe Γ in einer algebraischen Gruppe G) vollständig durch die Kohomologie desjenigen Komplexes von Differentialformen auf dem zugehörigen lokal symmetrischen Raum

$\Gamma \backslash X (X = G(\mathbb{R})/K$, K maximal kompakte Untergruppe der Gruppe $G(\mathbb{R})$ der reellen Punkte von G) beschrieben werden kann, die automorph sind. Diese Eigenschaft bedeutet, daß die Formen in einem endlich dimensionalen Raum stabil unter $z(\mathfrak{g})$, der Algebra der biinvarianten Differentialoperatoren auf $G(\mathbb{R})$, enthalten und von moderatem Wachstum „im Unendlichen" des Quotienten $\Gamma \backslash X$ sind.

Ist der arithmetische Quotient $\Gamma \backslash X$ kompakt, so ist dies ein Ergebnis, das sich als eine Verfeinerung der klassischen Hodge-Theorie gewinnen läßt. Wenn der Raum jedoch nicht kompakt ist, so sollte man diese Frage als eine vergröbernde Erweiterung der Hodge-Theorie auf solche arithmetische Quotienten sehen, für die im einfachsten Fall der Laplace-Operator ein kontinuierliches Spektrum besitzt. Die Struktur des kontinuierlichen Spektrums einer arithmetischen Gruppe wurde 1964 druch R. P. Langlands aufgeklärt und vollständig durch Eisensteinreihen beschrieben. G. Harder hatte diese Theorie schon 1971 ausgenutzt, um im Falle von Gruppen von k-Rang 1, etwa SL_2 über einem algebraischen Zahlkörper, die Kohomologie der arithmetischen Quotienten $\Gamma \backslash X$ ‚im Unendlichen' durch Eisensteinreihen oder Residuen solcher zu beschreiben. Obwohl Harders Intention (und seine Resultate) auf wesentlich feinere Beschreibungen und insbesondere arithmetische Anwendungen abzielte, waren diese Ergebnisse ein starkes, unterstützendes Moment für die angesprochene Vermutung. So lag es für Franke nahe, sich intensiv mit den Arbeiten von R. P. Langlands zur Theorie der Eisensteinreihen auseinanderzusetzen. Die Nicht-Kompaktheit des Quotienten macht es erforderlich, eine sehr sorgfältige Analyse des asymptotischen Verhaltens der Kohomologieklassen in der Nähe der verschiedenen Randkomponenten unterschiedlicher Kodimension durchzuführen. Eine Fülle von kombinatorischen und analytischen Schwierigkeiten ist hier zu überwinden; an der Lösung des Problems und der damit zusammenhängenden Frage über die Beziehung zwischen L^2-Kohomologie und Schnittkohomologie im Falle arithmetischer Varietäten, arbeiteten auch andere Mathematiker.

Neben der Beschäftigung mit diesem Fragenkreis war Herr Franke in Moskau bis zum Jahre 1988 regelmäßiger Teilnehmer der Seminare von I. M. Gelfand, Yu. I. Manin und A. Parshin. Dies half ihm über manch toten Punkt bei der Arbeit in der Theorie der automorphen Formen hinweg, zumal er gemeinsam mit Yu. I. Manin und Y. Tschinkel die Theorie der Eisensteinreihen fruchtbringend zur Lösung eines Problems in der Diophantischen Geometrie einsetzen konnte. Diese Resultate [5] bestärkten Franke, weiter über Anwendungen der Analysis nachzudenken. Gleichzeitig wurde durch A. Parshin Frankes Interesse an dem durch P. Deligne formulierten Problem über eine verfeinerte Version in funktorieller Form des Riemann-Rochschen Satzes geweckt. Hierzu hat dann Herr Franke, jetzt tätig als Mitarbeiter am Karl-Weierstraß-Institut in Berlin, eine Serie von Arbeiten [6–8] geschrieben, über deren Ergebnisse er auch 1988 bei der Mathematischen Arbeitstagung in Bonn in einem Vortrag berichtete. Schon hier wurde sehr deutlich, daß eine der Stärken von Herrn Franke in der abstrakten, konzeptionellen Mathematik liegt. Auf Grund seiner hochentwickelten Auffassungsgabe kann er sehr komplexe und abstrakte mathematische Zusammenhänge schnell und sicher erfassen. Es ist ein mathematischer Arbeitsstil, der an große Vorbilder erinnert, Dinge und Probleme dadurch klar zu sehen, daß sie im richtigen allgemeinen Kontext betrachtet werden. Herr Franke betont selbst, daß ihm explizite Fragen, Beispiele, Spezialfälle oder konkrete Rechnungen weniger liegen. Aber man sollte ihm dabei nicht ganz trauen, da er dies ohne Zweifel auch beherrscht.

Auf Anregung von A. Borel, der ebenfalls Teilnehmer der Arbeitstagung 1988 in Bonn war, ging Herr Franke 1989 an die School of Mathematics am Institute for Advanced Study in Princeton, USA. Wegen der Ungewißheit über die weitere politische Entwicklung in der ehemaligen DDR verlängerte er seinen Aufenthalt um ein Jahr bis zum Frühjahr 1991. Während dieser Zeit gelang es Herrn Franke, das schon angesprochene Problem aus der Kohomologietheorie arithmetischer Gruppen zu lösen. Er zeigte, daß die Kohomologie einer durch Kongruenzen definierten arithmetischen Untergruppe Γ einer reduktiven algebraischen Gruppe G, definiert über einem algebraischen Zahlkörper, durch die Kohomologie des Raumes der automorphen Differentialformen auf $\Gamma \backslash X$ bestimmt ist, i.e., $H^*(\Gamma \backslash X, E) = H^*(\Omega_{\text{automorph}}(\Gamma \backslash X), E)$, wobei E ein durch eine irreduzible endlichdimensionale komplexe Darstellung von $G(\mathbb{R})$ gegebenes Koeffizientensystem bezeichnet. In sechs Vorträgen berichtete er im Herbst 1990 am Institute for Advanced Study über seine Ergebnisse. Frankes Beweis [9] dieser Vermutung zeigt ein sehr tiefgehendes Verständnis der Theorie der automorphen Formen und zeichnet sich durch eine wirkungsvolle Kombination von Methoden der Analysis mit solchen der kohomologischen Algebra aus. Für Gruppen höheren k-Ranges als Eins ist die Geometrie „im Unendlichen" von $\Gamma \backslash X$ sehr kompliziert. Um diese Schwierigkeiten zu meistern, ist eine Filtrierung auf gewissen Funktionenräumen eingeführt, die nicht nur die geometrischen Gegebenheiten (etwa die Kodimension des Randstratums) berücksichtigt, sondern auch nach der Wachstumsgeschwindigkeit unterscheidet. Dieses Ergebnis und die dabei benutzten Methoden haben in der Fachwelt große Beachtung gefunden.

Unter Ausnutzung der Filtrierung ist im Anschluß gezeigt worden [10], daß die Kohomologie von Γ eine natürliche Summenzerlegung besitzt

$$H^*(\Gamma \backslash X, E) = H^*_{\text{cusp}}(\Gamma \backslash X, E) \oplus \bigoplus_{\{P\} \in \mathcal{C}} \bigoplus_{\Phi_{\{P\}}} Eis_{\{P\}, \psi} \, ,$$

in der der Summand $H^*_{\text{cusp}}(\Gamma \backslash X, E)$ von den cuspidalen harmonischen E-wertigen Differentialformen auf $\Gamma \backslash X$ aufgespannt wird, und in der die weitere Zerlegung durch die Klassen $\mathcal{C}$ assoziierter parabolischer $\mathbb{Q}$-Untergruppen von G parametrisiert wird. Die zweite Summe erstreckt sich dabei über eine geeignete Menge $\Phi_{\{P\}}$ von Klassen assoziierter irreduzibler cuspidaler Darstellungen der Levi-Komponenten der Elemente von $\{P\}$, denen Eisensteinsche Kohomologieklassen zugeordnet sind. Es sei bemerkt, daß dieses Ergebnis keine Beschreibung der inneren Struktur der mit transzendenten Methoden definierten Räume $Eis_{\{P\}, \psi}$ beinhaltet.

Neben den bisher angesprochenen Fragenkreisen hat Herr Franke auch über Probleme der stabilen Homotopietheorie, genauer, über die algebraische Konstruktion der elliptischen Kohomologie gearbeitet [12]. Sein klarer Blick für die zugrundeliegenden konzeptionellen Strukturen kam auch hier wieder zum Tragen.

Herr Franke ist ein sehr bescheidener und zurückhaltender Mensch. Dies wird ihm gelegentlich als Verschlossenheit ausgelegt. Doch besitzt er, bei näherem Kennenlernen, eine heitere Freundlichkeit. In den vielen mathematischen und persönlichen Gesprächen, die ich mit ihm seit unserem Kennenlernen 1990 hatte, habe ich immer wieder bemerken können, mit welcher Liebe er an der Mathematik hängt und wie er von ihr begeistert ist. Und ich bin davon überzeugt, daß es ihm auch Freude macht, diese Begeisterung auf Studenten der Universität Bonn, an der er seit Sommer 1992 als Professor arbeitet, weiterzugeben. So wie ich ihn als stets gut vorbereiteten Vortragenden erlebt habe, der auf hohem Niveau auch komplizierte Sachverhalte klar darstellen konnte, so wird Herr Franke auch das Interesse der Studenten finden. Der ihm verliehene Förderpreis des Mathematischen Forschungsinstituts Oberwolfach gibt ihm den Raum, seine Forschungsarbeiten fortzuführen, in der Stille dieses Instituts, verbunden mit der Möglichkeit zum Gespräch in einem internationalen Rahmen.

Schriftenverzeichnis von Jens Franke

[1] On the spaces F_{pq}^p of Triebel-Lizorkin type. Pointwise multipliers and spaces on domains. Mathematische Nachrichten **125**, 29–68 (1986)

[2] Elliptische Randwertprobleme in Besov-Triebel-Lizorkin-Räumen. Dissertation, Jena 1985

[3] On the admissibility of function spaces of type B_{pq}^p, F_{pq}^p, and boundary value problems for non-linear partial differential equations (mit T. Runst). Analysis Mathematica **13**, 3–17 (1987)

[4] Non-linear perturbations of linear non-invertible boundary value problems in function spaces B_{pq}^p and F_{pq}^p (mit T. Runst). Czechoslovak Mathematical Journal **38**, 623–641 (1988)

[5] Rational points of bounded height on Fano varieties (mit Yu. I. Manin und Y. Tschinkel). Inventiones mathematicae **95**, 421–435 (1989)

[6] Chow categories. Compositio Mathematica **76**, 101–162 (1990)

[7] Chern Functors. In: Arithmetic Algebraic Geometry (ed. G. van der Geer et al.). Progress in Mathematics 89, Boston Basel 1991

[8] Riemann-Roch in functorial form. Vorabdruck 1989

[9] Harmonic analysis in weighted L_2-spaces. Vorabdruck 1991

[10] Decomposition of spaces of automorphic forms and rationality properties of automorphic cohomology classes of GL_n (mit J. Schwermer). Vorabdruck 1992

[11] A topological model for some summand in the Eisenstein cohomology of congruence subgroups. Vorabdruck 1992

[12] On the construction of elliptic cohomology. Mathematische Nachrichten **158**, 43–65 (1992)

[13] Cohomology of S-arithmetic subgroups in the number field case (mit D. Blasius und F. Grunewald). Inventiones mathematicae **116**, 75–93 (1994)

Essays

One of the many lectures held at Oberwolfach
(Photo by A. Einzmann, Dortmund; Volkswagen-Stiftung, Hannover)

Some Trends in Riemannian Geometry

Arthur L. Besse[*]

G	H	P	A
C	M	F	L
S	J	P	B
G	B	P	B

The sessions "Differentialgeometrie im Großen", initiated by Professor Wilhelm Klingenberg and Professor Shiing Shen Chern, have brought to Oberwolfach a very international crowd of geometers for many years. We begin this review by presenting some highlights of talks given there in the last ten years. In the next paragraphs, we selected four lines of development of the field that we tried to put in perspective, namely, Isospectrality, Dynamical Geometry, Deforming metrics, and Moduli spaces of geometric objects. The influence of other areas of Mathematics, and also of Physics, will often be perceivable.

1. Some Highlights from the Most Recent Meetings "Differentialgeometrie im Großen"

Constant Mean Curvature Tori in Euclidean Space. It had been conjectured for a long time (H. Hopf 1950) that such tori could not exist. This was disproved in 1984 by Harry Wente, who gave an analytic existence proof, a beautiful contribution from non-linear Partial Differential Equations to Geometry. Abresch [1] modified the parameters in Wente's construction until, using a computer, he could draw pictures. He then noticed that the curvature lines of the surfaces "looked" planar. This allowed him to find explicit solutions, in terms of elliptic functions, that he presented at the 1985 meeting. This initiated a nice theory of complete integrability for certain PDEs on the 2-torus. All solutions can be found, either by solving ODEs (line followed by Ulrich Pinkall and the Berlin school as reported by Ian Sterling at the 1989 meeting), or by algebro-geometric means (Nigel Hitchin).

Isoparametric Submanifolds. Isoparametric hypersurfaces were defined and studied by Elie Cartan in 1938. This notion was extended by Chuu Lian Terng in 1985 to submanifolds. A submanifold is called *isoparametric* if its normal bundle is flat

* Je suis redevable à nombre d'institutions pour l'aide qu'elles m'ont apportée au cours des ans. Je voudrais particulièrement remercier le Centre National de la Recherche Scientifique, le Centre de Mathématiques de l'Ecole Polytechnique, l'Institut Elie Cartan, l'Institut Fourier, l'Université Paris-Sud, Centre d'Orsay, et, last but not least, la Direction Générale XII de la Communauté Européenne pour le soutien apporté au programme G.A.D.G.E.T.

and the second fundamental form (with respect to any parallel normal vector field) has constant eigenvalues. Codimension one orbits of isometric group actions are examples of isoparametric hypersurfaces, but there are a few other examples of special interest. Thorbergsson [95] explained at the 1989 meeting that these exotic examples cannot exist for higher codimensions: *an isoparametric submanifold in the sphere with codimension at least 2 must be a principal orbit of the isotropy representation of a symmetric space*. Elaborating on earlier work by Wu Yi Hsiang, Richard S. Palais and Chuu Lian Terng, he attaches a topological Tits building (incidence geometry) to an isoparametric submanifold. In his proof (which relies on deep work by Jacques Tits) emerges an unexpected parallel between the theory of isoparametric hypersurfaces and the rigidity theory of locally symmetric spaces.

Manifolds with Positive Curvature Operator. The beautiful theorem proved by Morio Micallef and David G. Moore, presented at the 1989 meeting, is well in the spirit of the theme "topological implications of curvature assumptions". Precisely, on a compact manifold, if one complexifies the tangent bundle, one can extend the Riemannian metric as a complex bilinear form. It has then isotropic vectors. Now, *if the curvature operator* (i.e., the Riemann curvature tensor viewed, using its skew symmetries, as an endomorphism on the bundle of complexified exterior 2-forms) *is positive on totally isotropic 2-planes, then the manifold is a homotopy sphere.* The proof uses the now central theory of harmonic maps. Roughly speaking, if some homotopy group is non zero, there exists a harmonic map from the 2-sphere into the manifold with small index. On the contrary, the curvature assumptions allow one to show that any such map must have a large index, hence a contradiction.

Infinitely Many Closed Geodesics on the 2-Sphere. This was a rare and nice moment when V. Bangert explained us at the 1991 meeting that, by combining their geometric and dynamic results, he and J. Franks could give a positive answer to a 60 years old conjecture, namely *any metric on the sphere S^2 has infinitely many closed geodesics*. This long story started in 1917 when G. Birkhoff proved the existence of at least one simple closed geodesic. Then the celebrated theorem of Ljusternik and Shnirelmann stated that, on a surface of genus 0, there are at least three simple closed geodesics.

To get some flavor of the clever proof given by V. Bangert and J. Franks, choose a simple closed geodesic, and call it the equator. Any other geodesic can either be trapped in one "hemisphere" or crosses again and again that equator. In the first case V. Bangert proves that this produces infinitely many closed geodesics unless the equator has a very special "bad" feature, but in that case he can find a better equator. If all the other geodesics are not trapped, the first recrossing position and angle define an area preserving mapping of the annulus. But we don't know that the rims have opposite rotations. Here, comes J. Franks who proves that area preserving maps of the annulus still have infinitely many periodic points or no periodic points at all. But this last possibility would imply that the equator is the only simple closed geodesic contradicting the result by Ljusternik and Shnirelmann!

A Global Overview. From these excerpts, it should be clear that varying the points of view is indeed one of the salient features of this series of meetings. This is to testify that the wind of diversity that Professor Shiing Shen Chern and Professor Wilhelm Klingenberg brought to Oberwolfach in this area is still blowing. Diversity is one of the richness of this field, at the confluent of many different areas of Mathematics. Among the main trends, one can notice the very substantial contributions coming from Analysis and Dynamical Systems.

2. Isospectrality

By *Spectral Geometry*, one usually understands the part of Riemannian Geometry which deals with the interplay between the geometry of a Riemannian manifold (M, g) and the *spectrum* of (M, g), i.e., the spectrum of its Laplace-Beltrami operator Δ acting on functions (or, more generally, the spectra of all natural operators constructed from the Riemannian metric).

A challenging problem in spectral geometry, since the early sixties, has been the *isospectral problem*, namely the problem of deciding whether two compact Riemannian manifolds which have the same spectrum are necessarily isometric. In case the manifolds have a boundary (and we consider either Dirichlet or Neumann condition on the boundary), this problem has been rephrased by Mark Kac as *"Can one hear the shape of a drum?"* During the last decade, progress towards a better understanding of the isospectral problem were made in several directions.

Another interesting aspect, which has also been developed during the last decade, is the interplay between spectral geometry and graph theory, leading to a better understanding of *small eigenvalues*, and of the problem of *prescribing* (finitely many) *eigenvalues* (with multiplicities).

2.1 The Isospectral Problem: Basic Results

Until 1983, only sporadic examples of isospectral non-isometric manifolds were known, namely pairs of flat tori (J. W. Milnor, 1964), pairs of hyperbolic manifolds (M. F. Vigneras, 1980), and pairs of lens spaces (A. Ikeda, 1980). As the 1980 examples show, the spectrum does not in general determine the topology (not even up to homotopy) in dimension larger than 2. On the other hand, V. Guillemin and D. Kazhdan (1979) proved that a closed Riemannian surface with negative (variable) curvature is spectrally rigid (this was slightly generalized by M. Min-Oo). For references concerning the isospectral problem up to 1989, we refer to the report [8].

Special Nilmanifolds. The *first striking breakthrough* on the isospectral problem was published in 1984 by C. Gordon and E. Wilson [41]. They indeed gave a general method for constructing *isospectral deformations* on compact nilmanifolds. Let G denote a connected, simply-connected nilpotent Lie group and assume it admits a discrete co-compact subgroup Γ.

Definition 2.1. Let φ be an automorphism of G with the following property: for any element $a \in G$, the element $\varphi(a)$ is conjugate to a (by an element which a priori depends on a). Such automorphisms are called *almost inner automorphisms*.

The almost-inner automorphisms of G form a subgroup $AIA(G)$ of the group $\text{Aut}(G)$ which obviously contains the group of inner automorphisms. As shown in [41], there are many instances in which the inner automorphisms form a proper subgroup of $AIA(G)$. The main result in [41] can be stated as follows.

Theorem 2.2. *Let G and Γ be as above and let g be any left-invariant Riemannian metric on G. Assume there is a continuous one-parameter family φ_t of almost-inner automorphisms of G that are not inner. The metrics $\varphi_t^* g$ project down to metrics $\underline{g}_t$ on $M = \Gamma\backslash G$ and the manifolds $(M, \underline{g}_t)$ are pairwise isospectral.*

In order to use this theorem for the isospectral problem, one needs to show that the manifolds under consideration are not isometric (inner automorphisms lead to isometric manifolds as is easily seen). One can do so, in the case at hand, by either looking at small values of t as in [29] or by carefully studying the isometry groups of compact nilmanifolds [41]. Another possibility is to look for Riemannian invariants which do change with t (see [30]). It has also been shown, in the case of a 2-step nilpotent group G, that the deformations described above (slightly generalized) are the only isospectral deformations of left-invariant metrics on G [80].

A Connection to Group Theory. The *second striking breakthrough* was made by T. Sunada. In his paper [87], published in 1985, he gives a general method for constructing pairs of isospectral Riemannian manifolds. The main idea comes from number theory. Let $(G; H, K)$ be a triple of finite groups with H and K subgroups of G.

Definition 2.3. The subgroups H and K are said to be *almost conjugate* subgroups in G if there exists a bijection φ from H onto K such that the image $\varphi(h)$ of any element $h \in H$ is conjugate to h in the group G (by an element which a priori depends on h).

Let us state T. Sunada's main result in a simple setting.

Theorem 2.4. *Let $(G; H, K)$ be a triple of finite groups, with H and K almost conjugate in G. Let M be a closed manifold on which the group G acts without fixed points. Lift any metric g_0 on $G\backslash M$ to a metric g on M, and then project g down to g_H (resp. g_K) on $H\backslash M$ (resp. $K\backslash M$). The Riemannian manifolds $(H\backslash M, g_H)$ and $(K\backslash M, g_K)$ are isospectral.*

Our formulation of T. Sunada's theorem aims to point out the analogy with Theorem 2.2 (compare Definitions 2.1 and 2.3). As above, if the groups H and K

are conjugate in G, the manifolds $(H \backslash M, g_H)$ and $(K \backslash M, g_K)$ are isometric. One needs to choose triples such that H and K are almost-conjugate, non-conjugate subgroups in G.

Example 2.5 (*a typical one*). One instance of a triple $(G; H, K)$ such that the subgroups are almost conjugate, non-conjugate in G is the triple in which G is $\mathrm{Sl}(3, \mathbb{Z}_2)$, H is the subgroup of matrices whose first column $(1, 0, 0)^t$, and K the subgroup of matrices whose first line is $(1, 0, 0)$ (see [18]). This particular example has been used for the first time by R. Brooks to construct isospectral hyperbolic surfaces of genus 4 (see [21]).

Planar Isospectral Domains. The *third striking result* was obtained by C. Gordon, D. Webb and S. Wolpert in June 1991. Back in 1987, P. Buser noticed that to the triple given in Example 5, in which $G = \mathrm{Sl}(3, \mathbb{Z}_2)$, correspond planar graphs (Cayley graphs for the action of G on the coset spaces). This observation led him to constructing (with paper and scissors) non-isometric, flat surfaces with boundary, embedded in $\mathbb{R}^3$, which are isospectral (with respect to both Dirichlet and Neumann conditions). These surfaces are diffeomorphic to planar domains. C. Gordon, D. Webb and S. Wolpert observed that P. Buser's flat isospectral surfaces admit a hyperplane of symmetry in $\mathbb{R}^3$. This was the first step towards the following result.

Theorem 2.5 [42]. *The Euclidean domains shown below are non-isometric and isospectral (for both Dirichlet and Neumann conditions).*

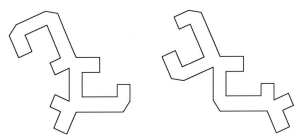

Fig. 1. Isospectral domains in the plane

2.2 The Isospectral Problem: Further Results

Taking Some Distance. The methods introduced by C. Gordon, E. Wilson and T. Sunada were subsequently developed. They were put into a unified and more general framework by D. DeTurck and C. Gordon [29] and later by P. Bérard [9]. P. Buser greatly contributed to the understanding of the combinatorial aspect of T. Sunada's method. He showed how to construct examples of isospectral manifolds by gluing isometric building blocks according to the combinatorics given by the Cayley graphs of the action of the group G on the coset spaces $H \backslash G$ and

$K \backslash G$. This method led to the construction of isospectral non-isometric surfaces with curvature -1 and any genus greater than or equal to 4 (see [21] for more details). In 1977, S. Wolpert [98] had shown that the set $\mathcal{V}_\gamma$ of hyperbolic surfaces of genus γ which admit an isospectral companion is contained in an analytic set of positive codimension in the corresponding Teichmüller space. Using T. Sunada's theorem, one can show that the set $\mathcal{V}_\gamma$ has positive dimension for any $\gamma \geq 4$. It is still an open problem to decide whether $\mathcal{V}_\gamma \neq \emptyset$ for $\gamma = 2$ or $\gamma = 3$.

In the 2-dimensional case, one has the following general result due to B. Osgood, R. Phillips and P. Sarnak [79].

Theorem 2.6. *Any set of isospectral closed surfaces is compact in the C^∞ topology of metrics.*

All the examples of isospectral manifolds constructed by the above mentioned methods share at least two important properties. First the manifolds under consideration are locally isometric; second they are isospectral with respect to any natural differential operator (e.g., the Hodge-de Rham Laplacians on p-forms). Examples of non locally isometric isospectral manifolds have recently been given by C. Gordon; examples of manifolds that are isospectral on functions, but not on differential forms have been given by C. Gordon and by A. Ikeda [58].

2.3 Spectral Geometry Versus Graphs: Small Eigenvalues

Since H. McKean's classical paper [73], much interest has been devoted to small eigenvalues of hyperbolic surfaces. By small eigenvalues, we mean the positive eigenvalues which are less that $\frac{1}{4}$, if any. Recall that 0 is always in the spectrum of a finite volume manifold and that $\frac{1}{4}$ is the lower bound of the spectrum of the Laplacian on the hyperbolic plane. That there exist instances of small eigenvalues has been proved by B. Randol back in 1974, using the Selberg trace formula.

Theorem 2.7. *Let M be a closed hyperbolic surface. One can find n-fold coverings M_n of M such that the first positive eigenvalue $\lambda_1(M_n)$ tends to 0 as n tends to ∞.*

Let Γ_0 denote the fundamental group of the surface M, and let $S = \{\gamma_1, \ldots, \gamma_p\}$ denote a set of generators of Γ_0 (not necessarily minimal). Let Γ_n denote the fundamental groups of the manifolds M_n. One may now consider the coset spaces $\mathcal{G}_n = \Gamma_0/\Gamma_n$, and the Cayley graph $\mathcal{C}_n$ associated to the action of Γ_0 on the left on $\mathcal{G}_n$ and to the set S. Let $\lambda_1(\mathcal{C}_n)$ be the least positive eigenvalue of the graph $\mathcal{C}_n$ for the usual combinatorial Laplacian. The following result has been proved by R. Brooks [17], using Cheeger's constante, and generalized by M. Burger [19], and T. Sunada [88].

Theorem 2.8. *Let M_n be a family of finite coverings of M_0 and $\mathcal{C}_n$ the corresponding graphs. Then $\lambda_1(M_n)$ goes to 0 as n goes to ∞ if and only if $\lambda_1(\mathcal{C}_n)$ goes to 0 as n goes to ∞.*

2.4 Spectral Geometry Versus Graphs:
Prescribing Finitely Many Eigenvalues

Let M be a Riemannian surface and let u be a non-zero eigenfunction of the Laplacian on M. It is a well-known fact that u cannot vanish at infinite order. A simple local analysis shows that the nodal set $u^{-1}(0)$ of the function u consists of finitely many regular arcs which meet at points where both du and u vanish; at such points, the arcs form an equi-angular system. It follows that to each eigenfunction of the Laplacian on M one may associate an embedded graph in M (the edges are the regular arcs in the nodal set and the vertices the points at which at least three arcs meet). One may use this graph to prove the following theorem.

Theorem 2.9. *Let (M, g) be a closed Riemannian surface with Euler characteristic $\chi(M)$ and let $0 = \lambda_0 < \lambda_1 \leq \lambda_2 \leq \lambda_3 \leq \ldots$ be the spectrum of the Laplace-Beltrami operator acting on functions. Let m_k denote the multiplicity of the kth eigenvalue λ_k. One has*

1) *if $M = \mathbf{S}^2$, then $m_k \leq 2k + 1$;*
2) *if $M = \mathbf{P}^2$, then $m_k \leq 2k + 3$;*
3) *if $M = \mathbf{T}^2$, then $m_k \leq 2k + 4$;*
4) *if $M = \mathbf{K}^2$, then $m_k \leq 2k + 3$;*
5) *if $M = M_\chi$, then $m_k \leq 2(k + 2 - \chi(M)) + 1$.*

The first results in this direction were obtained by S. Y. Cheng: quadratic bound with respect to the Euler characteristic; sharp bound for $k = 1$ in Case (1). They were then improved by G. Besson: linear bound in the Euler characteristic; sharp bound for $k = 1$ in Cases (2-3); by Y. Colin de Verdière: sharp bound for $k = 1$ in Case (4); by N. Nadirashvili who gave the bounds which appear in the preceding theorem, for $k > 1$, by making use of the graph mentioned above. We refer to [12] and [77] for more details and references. A consequence of Theorem 2.9 is that there is an obstruction to prescribing eigenvalues (even finitely many) in dimension 2. It is conjectured by Y. Colin de Verdière that the maximum multiplicity of the first positive eigenvalue on a surface M is given by $\mathrm{Chr}(M) - 1$, where $\mathrm{Chr}(M)$ is the chromatic number of the surface (see [26] for more details and references). The following striking result [27] shows that the situation is very different in higher dimensions.

Theorem 2.10. *Let M be an n-dimensional manifold, with $n \geq 3$, and let $0 = \lambda_0 < \lambda_1 \leq \lambda_2 \leq \ldots \leq \lambda_N$ be any finite sequence (with multiplicities). Then there exists a Riemannian metric whose N first eigenvalues are the given sequence.*

3. Dynamical Geometry

Riemannian manifolds, one of the geometers' favourite objects of study, generate a natural dynamical system, the *geodesic flow*. This is the one parameter group of diffeomorphisms of the unit tangent bundle obtained when pushing at unit speed a unit tangent vector along the geodesic it determines. Dynamical geometry is this part of Riemannian geometry which is concerned with the dynamical properties of the geodesic flow, e.g., the long time behaviour of geodesics.

3.1 Going Back Into History

The subject can be traced back, at least, to 1898, when Jacques Hadamard investigated geodesics on surfaces in Euclidean 3-space which have everywhere negative curvature [48].

On such a (necessarily noncompact) surface, geodesics emanating from a point x can be divided into two types: those which are trapped in an end, and those which do not tend to infinity. Hadamard observed that this dichotomy is independent of the particular point x. This remark defines what is now known as the "ideal boundary" of the surface.

As pointed out by Hadamard himself, the dynamics of geodesics on negatively curved surfaces is rich and interesting. Its properties are described nowadays in terms of *ergodicity* and *entropy*, two concepts that have their origin in statistical Physics. Most of the present report will be devoted to this aspect of the theory.

A second source for our subject is Blaschke's problem about manifolds all of whose geodesics are closed, see for example [10].

The round sphere has the property that all its geodesics are closed with the same length. W. Blaschke asked whether this property, which is indeed a dynamical property of the geodesic flow, is characteristic of the round metric. It turns out not to be the case: there indeed exist many metrics of revolution on the 2-sphere (the so-called Zoll metrics) whose geodesic flow is the same as that of the round metric. However, the same problem has a positive answer for the projective plane (the quotient of the 2-sphere by the antipodal involution): the canonical metric on the projective plane is indeed characterized by its geodesic flow.

Blaschke's problem is a special case of the inverse problem for geodesics: *to what extent is a Riemannian metric determined by its geodesic flow?* We shall see that this question is related to the famous isospectral problem for the Laplace-Beltrami operator. When not included, references can be found in the report [81].

3.2 Negative Curvature: Basic Properties
of the Geodesic Flow, Basic Examples

Chaos and Stability. There are many ways of expressing the fact that orbits of a dynamical system visit the whole space. Increasing degrees of disorder are represented by *topological transitivity*, i.e., the existence of dense orbits and *ergodicity*, i.e., the fact that no measurable function can be invariant unless it is almost everywhere constant.

What is more remote from the idea of disorder than the Riemann surface attached to a Fuchsian group, such as Sl(2, ℤ)? Still, in 1934, H. Hedlund showed that the geodesic flow on such a (finite area, constant curvature) surface is ergodic.

Hedlund's result has an algebraic generalization which leads to a very important class of examples. A locally symmetric space of noncompact type is of the form $\Gamma \setminus G / K$ where K is a maximal compact subgroup in the noncompact semisimple Lie group G, and Γ a discrete torsion free subgroup of G. The geodesic flow is a one parameter subgroup of G. It is ergodic if and only if the symmetric space G/K has strictly negative curvature, i.e. G has rank one.

On the other hand, Eberhard Hopf succeeded in 1939 to generalize Hedlund's result to surfaces with variable curvature. Part of the Lie theoretic picture (the Iwasawa decomposition) survives in variable curvature: the *stable* and *unstable* foliations. Nevertheless, the generalization to higher dimensions was achieved only in 1962 by D. V. Anosov. In this work, D. V. Anosov singled out the important class of diffeomorphisms and flows that bears his name, and he proved that such flows are *structurally stable*. This means that the orbit structure does not change under a small perturbation: there is a homeomorphism that takes orbits of the perturbed flow to orbits of the initial flow. For the geodesic flow on a compact negatively curved manifold, even more is true: a geodesic in the universal cover is determined by its endpoints on the *ideal boundary*, and the ideal boundary can be constructed from the fundamental group only. In other words, the orbit structure is entirely determined by the fundamental group. In particular, many concepts and problems can be expressed just in terms of the combinatorics of the fundamental group, a recurrent theme of this study.

Entropy. In the 1950's, A. N. Kolmogorov and Ya. Sinai invented a numerical measurement of disorder: the entropy of a transformation. The definitions we give here are due to R. Bowen and A. Katok. Let T be a map of a compact metric space X to itself. A family of distances d_n^T is defined on X by $d_n^T(x, y) = \max_{0 \le k \le n}\{d(T^k x, T^k y)\}$.

Definition 3.1 (Kolmogorov-Sinai). Let $N_n^T(\varepsilon)$ be the minimum number of ε-balls needed to cover X in the metric d_n^T. The *topological entropy* of T is

$$h_{\text{top}}(T) = \lim_{\varepsilon \to 0} \lim_{n \to +\infty} \frac{1}{n} \log N_n^T(\varepsilon) .$$

If an invariant measure on X is given, then, for each $\delta > 0$, one defines the number $N_n^T(\varepsilon, \delta)$ to be the minimum number of ε-balls needed to cover some subset in X whose complement has measure less than δ. Then the *metric entropy* of T is

$$h_{\text{met}}(T) = \lim_{\delta \to 0} \lim_{\varepsilon \to 0} \lim_{n \to +\infty} \frac{1}{n} \log N_n^T(\varepsilon, \delta) .$$

The topological (resp. metric) entropy of the geodesic flow is the topological entropy (resp. metric entropy with respect to the Liouville measure) of the time one map. It does not change when taking coverings and scales by t^{-1} if the distance is scaled by t. Metric entropy is never greater than topological entropy.

A wonderful feature of the topological entropy is the fact it has many different interpretations, at least for negatively curved manifolds.

1. Topological entropy is the exponent of exponential growth of the volume of balls in the universal covering space (A. Manning, G. A. Margulis).

Theorem 3.2 [72]. *For each point x,*

$$\text{vol } B(x, R) \sim c(x)\, e^{h_{\text{top}} R} \quad as \quad R \to +\infty. \tag{1}$$

The function $c(x)$ is poorly understood.

2. Topological entropy can be extracted from the lengths of periodic geodesics. Again the curvature is assumed to be negative.

Theorem 3.3 (Bowen-Margulis). *Let $N(t)$ be the number of closed geodesics of length less than t. Then,*

$$N(t) \sim \frac{e^{h_{\text{top}} t}}{h_{\text{top}} t}. \tag{2}$$

3. Topological entropy is the Hausdorff dimension of the ideal boundary of the universal covering space, equipped with a distance associated to the geometry of horospheres [49].

4. Topological entropy has an analytic counterpart: the bottom of the L^2-spectrum of the Laplace-Beltrami operator on the universal covering space, which we denote by λ_0. One has in general

$$\lambda_0 \leq \tfrac{1}{4} h_{\text{top}}^2. \tag{3}$$

Does a reverse inequality $\lambda_0 \geq \text{const.}\, h_{\text{top}}^2$ hold for negatively curved manifolds? The equality $\lambda_0 = \tfrac{1}{4} h_{\text{top}}^2$ is characteristic of asymptotically harmonic manifolds, F. Ledrappier [66]. A metric is said to be *asymptotically harmonic* if the mean curvature of all horospheres is constant.

Although topological entropy is not an invariant of the fundamental group only, it has purely algebraic counterparts which lead to interesting comparison problems.

When topological entropy is not zero, it can be used to normalize a Riemannian metric (i.e., adjust a constant scaling factor so that $h_{\text{top}} = 1$), and the range of variation of other more traditional Riemannian invariants gives rise to appealing problems.

For example, define the *entropic volume* of a compact manifold M to be the infimum of volumes of metrics on M with topological entropy 1. When is this finite? non zero? Is the infimum achieved? For which metrics?

Conjecture 3.4. *If a compact manifold M carries a locally symmetric metric g, then, among all metrics on M with topological entropy* 1, *g has minimum volume.*

In 2 dimensions, the entropic volume is known.

Theorem 3.5 [13], [13]. *On a compact surface, among all Riemannian metrics with* $h_{\text{top}} = 1$, *constant curvature metrics have minimum volume.*

In higher dimensions, a local minimum property is established by G. Besson, G. Courtois and S. Gallot in [14]. Note that the question makes sense as well for Finsler metrics, and, in this extended framework, it is open even in dimension 2.

In some sense, Gromov's *simplicial volume* is a kind of homotopy invariant entropic volume. Indeed, it bounds entropic volume from below, but this bound does not seem to be sharp in any dimension. G. Besson, G. Courtois and S. Gallot have modified Gromov's definition to make it sharp, at least in 2 dimensions [13].

The *entropic diameter* of a manifold M is the least diameter of a Riemannian metric on M with topological entropy 1. Here is an easy way to construct a homotopy invariant entropic diameter. Let Γ be a finitely generated group with exponential growth. Each finite symmetric generating set S determines a *word metric* on Γ. The volume (number of elements) of balls grows exponentially with exponent h_S. The least value of h_S when S varies is our algebraic entropy. When Γ is the fundamental group of a compact manifold M, it (unsharply) bounds the entropic diameter from below. Question: are there homotopy invariants that give optimal lower bounds for the entropic diameter?

The corresponding question for the *entropic maximal sectional curvature* has received a complete answer. Let $K(g)$ denote the maximum of the sectional curvatures for a metric g. Among all metrics with topological entropy 1, a locally symmetric metric achieves the minimum of K (P. Pansu and U. Hamenstädt, [81]).

The question of equality of the two entropies is again a nice problem.

Conjecture 3.6 (A. Katok). *On negatively curved Riemannian manifolds, equality of topological and metric entropies is characteristic of locally symmetric metrics.*

This was solved in 2 dimensions by A. Katok [60].

Since there is a formula for metric entropy in terms of the mean curvature of horospheres, a first step would be to show that equality of entropies implies that all horospheres have the same constant mean curvature, hence the metric is aymptotically harmonic. The second step is fairly well understood. F. Ledrappier [66] and G. Knieper [63] have independently shown that compact asymptotically harmonic manifolds of negative curvature have constant curvature if the dimension equals 3. In all dimensions, their geodesic flow is the same as that of a locally symmetric space [36].

In spite of the existence of formulae for the variation of entropy under a deformation of a metric (or of a flow) due to A. Katok, G. Knieper, H. Weiss and M. Pollicott, entropy does not seem to be exactly computable on examples. Here is a specific challenge. We describe a construction called *symplectic change of Hamiltonian*. Remember that the geodesic flow can be viewed as a Hamiltonian flow on the cotangent bundle T^*M. Transport the Hamiltonian by a symplectomorphism φ of T^*M. If φ is exact, the new Hamiltonian flow is conjugate to the original one. In general, the new flow differs from the original one by a smooth change of parameter depending only on a cohomology class in $H^1(M, \mathbb{R})$. The metric entropy does not change. What is the new topological entropy?

3.3 Characterizations of Locally Symmetric Geodesic Flows

These are in some sense the only negatively curved geodesic flows one can put one's hands on. Most compact examples have an arithmetic origin, and their mere existence is nontrivial. Hence, it is particularly striking that they can be distinguished among all flows by simple dynamical features.

The key feature is the so-called *Anosov property*. Let (φ_t) denote a flow on some space. For a point x, the strong stable leaf through x is the set $W^{ss}(x)$ of points y such that the distance between $\varphi_t x$ and $\varphi_t y$ tends to zero as t tends to $+\infty$. Reverse time to obtain the unstable leaf $W^{su}(x)$. By definition, for an Anosov flow (e.g., a negatively curved geodesic flow), W^{ss}, W^{su} and the orbits form three complementary transverse foliations.

Although each individual leaf is smooth, the foliations are not much more than Hölder continuous, a fact which caused E. Hopf and D.V. Anosov a lot of trouble. In fact, they are probably of class C^2 only for locally symmetric geodesic flows (and their obvious modifications) and suspensions.

Y. Benoist, P. Foulon and F. Labourie have the following result in this direction (after partial results by E. Ghys, M. Kanai, R. Feres and A. Katok surveyed in [81]).

Theorem 3.7 [7]. *If an Anosov flow has smooth stable and unstable foliations, and if it leaves a smooth contact form invariant, then it is smoothly conjugate to a locally symmetric geodesic flow, up to a finite covering or a symplectic change of Hamiltonian.*

In the special case of a negatively curved geodesic flow, this says that the ideal boundary is not smooth in general, i.e., the fundamental group does not act smoothly on the ideal boundary. A general problem is to classify smooth actions of fundamental groups of negatively curved manifolds. A recent result of E. Ghys [40] asserts that all smooth actions of a surface group on the circle with maximal Euler number arise from a metric of constant curvature on the surface.

3.4 The Inverse Problem

When we say that two Riemannian manifolds M_1 and M_2 have the same geodesic flow, we mean that there exists a smooth diffeomorphism F between unit tangent bundles that takes one flow to the other, i.e., $F \circ \varphi_t^1 = \varphi_t^2 \circ F$. Sometimes the map F is merely a homeomorphism and one speaks of a C^0-conjugacy.

All negatively curved geodesic flows on a given manifold are orbit equivalent, i.e., there exists a homeomorphism between unit tangent bundles that takes orbits to orbits (M. Gromov). Thus the preservation of the parameter along orbits is essential, and the lengths of closed orbits are often the crucial data.

We now define the *marked length spectrum*. Remember that each free homotopy class of curves in a compact manifold contains at least one closed geodesic. The minimum length of such a geodesic is called the *length* of the corresponding conjugacy class in the fundamental group. The *length spectrum* of a Riemannian manifold is the collection of all these numbers, a subset of the real line. Two Riemannian manifolds M_1 and M_2 have the same marked length spectrum if there is a length preserving isomorphism between fundamental groups. In negative curvature, having the same marked length spectrum is equivalent to C^0-conjugacy of geodesic flows (Livšič).

Conjecture 3.8. *If two negatively curved Riemannian manifolds have the same marked length spectrum, then they are isometric.*

In 2 dimensions, the conjecture has been solved independently by C. Croke and J. P. Otal. Later, C. Croke, A. Fathi and J. Feldman (cf. [81]) have weakened the assumptions, leading to the following result.

Theorem 3.9. *If two metrics on a compact surface have the same marked length spectrum, and if one of them has nonpositive curvature, then they are isometric.*

In higher dimensions, the only existing results are local. Indeed, a small deformation of a sufficiently negatively curved metric that does not change the length spectrum has to be isometric (V. Guillemin and D. Kazhdan, M. Min Oo).

A less ambitious question is to determine which naive Riemannian invariants can be read from the geodesic flow. The simplest one is the total volume, which is a C^1-conjugacy invariant. Hence one can ask: can the volume be extracted directly from the length spectrum?

3.5 Analysis and the Geodesic Flow

There seems to be a rich and subtle interaction between the Laplace-Beltrami operator and the geodesic flow.

First instances of such interaction are the Poisson formula for flat tori and the Selberg trace formula for constant curvature surfaces. They establish a reversible relation between the spectrum of the Laplace-Beltrami operator and the length spectrum, one set determining the other and vice-versa.

A weaker relation persists for general Riemannian manifolds: the lengths of closed geodesics can be heard from the singularities of the wave kernel (in fact in very degenerate cases – which cannot occur if curvature is negative – some lengths might become inaudible). This follows from results by Y. Colin de Verdière, J. J. Duistermaat and V. Guillemin.

84 Arthur L. Besse

Theorem 3.10 [25, 34]. *For metrics of negative curvature, the asymptotic behaviour of the Laplace-Beltrami spectrum determines the length spectrum.*

Examples (see the report [8]) show that the marked length spectrum cannot be recovered. Is there a more direct way to extract, for example, the topological entropy from the Laplace-Beltrami spectrum?

The spectrum of the Laplace-Beltrami operator on differential forms contains even more information, like holonomies around closed geodesics. The wave kernel method gives little insight into the converse question: to what extent does the length spectrum – if need be augmented with extra information like linearized Poincaré maps or holonomies – determine the Laplace-Beltrami spectrum? Note that another Laplace-Beltrami spectral invariant, the η-invariant, is expressible in terms of closed geodesics.

Brownian motion provides another link between the Laplace-Beltrami operator and the geodesic flow. Naively speaking, a Brownian orbit moves along geodesics but keeps forgetting its direction. In negative curvature, Brownian orbits on the universal cover converge to points on the ideal boundary as was shown by Y. Kifer and J. J. Prat [83].

This is reflected in the behaviour of harmonic functions on the universal cover. Extremal positive harmonic functions correspond to points on the ideal boundary, and bounded harmonic functions are expressible in terms of their boundary values on the ideal boundary. In particular, the Martin and Poisson boundaries coincide with the ideal boundary. As a consequence, harmonic measures can be viewed as measures on unit tangent spheres. Again a nice problem is to decide when the harmonic and Lebesgue measure classes coincide.

Since the ideal boundary can be constructed directly from the fundamental group, one can ask whether analogous constructions exist for the Martin and Poisson boundaries. This amounts to comparing Brownian motion and random walks on the fundamental group.

3.6 Beyond Negative Curvature

Many general questions do not have a satisfactory answer yet. We just propose some that we feel belong to the circle of ideas which has matured recently.

Question 3.11. i) *Which Riemannian manifolds have a completely integrable geodesic flow?*

ii) *For which Riemannian manifolds is the geodesic flow of Anosov type (see [35])?*

iii) *Does every manifold carry a metric with ergodic geodesic flow?*

Note that the question of which nonpositively curved geodesic flows are ergodic is close to be but not quite solved. Apart from these, just a few examples are known (see [20]).

Question 3.12. *Classify Riemannian manifolds whose geodesic flow is the same as that of a locally symmetric space.*

This ambitious problem contains as subproblems two famous conjectures.

Conjecture 3.13 (W. Blaschke). *Canonical metrics of the projective spaces (real, complex, quaternionic and octonionic) are characterized by the property that their geodesic flow is periodic.*

The case of real projective spaces has been settled affirmatively by Leon Green, Marcel Berger and Jerry L. Kazdan. See [10] for more information on the remaining open cases and related questions.

Conjecture 3.14 (E. Hopf). *A metric without conjugate points on a torus is flat.*

Leon Green [43] has solved the conjecture in 2 dimensions. Since the property of having no conjugate points can be expressed in terms of the geodesic flow only, this is indeed a result on the inverse problem. Another encouraging result has been announced recently by C. Croke and B. Kleiner: *any metric which possesses a parallel vector field is determined by its geodesic flow.*

How general is the failure of the geodesic flow to determine a metric, or in other words, is the geodesic flow of a Riemannian metric flexible? In the Finsler framework the geodesic flow is extremely flexible: a small exact symplectic change of Hamiltonian does not change the flow but changes the Finsler metric. The Zoll examples, which can be grafted onto any given Riemannian manifold, show that every manifold carries metrics which can be deformed without changing the geodesic flow. This geodesic flexibility seems to have little to do with complete integrability or ergodicity.

3.7 From Geodesics to Submanifolds

A higher dimensional analogue to the geodesic flow is the *Weyl chamber flow.* Given a locally symmetric space M of rank r, this is the natural action of $\mathbb{R}^r$ on the bundle of totally geodesic maps of $\mathbb{R}^r$ into M. However such an example seems very specific, and in particular is not deformable [61].

Another possible generalisation, which makes sense for arbitrary Riemannian manifolds, is obtained by weakening the condition totally geodesic and replacing it by harmonic. Geodesics can also be characterized as (locally) energy minimizing maps from $\mathbb{R}$ into a manifold. It is well known since the work of J. Eells and J. H. Sampson, that their higher dimensional analogues, harmonic maps, behave remarkably well in the context of negative curvature. The same can be said of the other analogue, namely minimal surfaces. Classically people have been interested in harmonic mappings with a compact source, or compact minimal surfaces, which play the role of closed geodesics.

Of course one topic of great interest, if we stick to this analogy with the geodesic flow, would be the study of the collective behaviour of these maps or surfaces. In [47], M. Gromov introduces the higher dimensional analogue of the unit tangent bundle, namely the space of pointed harmonic maps, or of pointed minimal surfaces. Now, the dynamics is represented by a foliation. He also proved an analogue of structural stability in a specific situation. Namely, given a compact hyperbolic manifold, consider the foliation of the Grassmannian bundle of p-planes by totally geodesic planes. Then for a nearby metric M. Gromov constructs a topologically equivalent foliation whose leaves are lifts of minimal submanifolds. M. Gromov has applied this method to extend Margulis' superrigidity theorem to some rank one symmetric spaces.

The 2-dimensional case is of special interest. In this situation, changing the conformal type of the leaves is the analogue of reparametrisation for a flow. D. Sullivan has recently introduced the concept of the Teichmüller space of a foliation by surfaces, that is the space of all conformal structures up to the natural equivalence. For instance a natural question is now to understand the Teichmüller space of the grassmannian bundle of 2-planes (foliated by the lift of totally geodesic planes) of a hyperbolic manifold and its relation to the change of metrics. More specifically, thanks to the result of M. Gromov alluded to before, a nearby metric will define a new point in the Teichmüller space of the foliation (which we may think as an analogue of the length spectrum); is this point different from the original one and does it characterize the metric?

4. Deforming Metrics

In the seminal paper [Ha1], R. Hamilton not only solved the deep question in Riemannian geometry whether positive Ricci curvature was characteristic of the (smooth) 3-sphere, a statement that the author considered a main step towards the solution of the Poincaré conjecture. He also provided the geometers with a powerful tool to investigate topological and/or algebraic structures through the existence of metrics with specific curvature properties: the Ricci flow.

Some Background. The idea of getting a metric with "better" properties by properly deforming it is natural enough, and led to some results as early as the mid 70's (see [15]. But this early Riemannian deformation theory was infinitesimal in nature. It only involved properties of the first two jets of the curvature of a given metric.

In this direction, to view the Ricci curvature as a preferred vector field on the space of metrics on a given manifold had been considered since a long time, Indeed, David Hilbert already derived the first variation of the total scalar curvature to get the Ricci curvature (more precisely the Einstein tensor field: $z_g = \mathrm{Ric}_g - \frac{1}{2} \mathrm{Scal}_g \, g$) as its gradient.

In the 60's, H. Yamabe asked whether, in a prescribed conformal class, there always exist metrics with constant scalar curvature. This led to a method which can be viewed as an ancestor of Hamilton's work. Although it required more work than anticipated by Yamabe, this metric can (more or less) be obtained by integrating the gradient of the total scalar curvature, a functional which, when restricted to a conformal class, admits precisely metrics with constant scalar curvature as critical points.

Some years later, the fundamental paper of J.E. Eells and J.H. Sampson on harmonic maps must have been a direct source of inspiration to Hamilton. In it, they establish the existence of a harmonic map in a given homotopy class of maps between Riemannian manifolds provided the target manifold has non-positive curvature. They do so by finding minima of the *energy* functional $E(\varphi) = \frac{1}{2} \int_M |d\varphi|^2$ along gradient lines of E.

4.1 The Ricci Curvature Field Is Integrable

Local Existence of the Ricci Flow. Then, why not try to get Einstein metrics by integrating the gradient field of the total scalar curvature functional, whose critical points on the space of metrics with normalized volume are precisely Einstein metrics. The space of metrics, a cone in the vector space $C^\infty(M, S^2 T^* M)$, is an infinite-dimensional (Fréchet) manifold. Hence, the Cauchy-Lipschitz theorem does not apply to prove (even local) integrability. When switching to some Hölder or Sobolev spaces, the difficulty becomes that none of the geometric vector fields we would like to consider, like $\varphi \in C^\infty(M_1, M_2) \mapsto \Delta_g \varphi C^\infty(M_1, \varphi^* T M_2)$ or $g \in \mathcal{M} \mapsto \mathrm{Ric}_g \in C^\infty(M, S^2 T^* M)$ are Lipschitz. This loss of regularity makes integrability more delicate.

In the case of harmonic maps, the equation $\Delta \varphi = d^* d\varphi$, though non-linear (indeed, the codifferential operator d^* involves the φ-pulled-back connection) is an elliptic semi-linear operator in diagonal form, for which it is not too difficult to derive local existence of the flow from classical parabolic theory.

As for the more geometric Ricci flow, a straightforward computation of the symbol of the operator $g \mapsto z_g = \mathrm{Ric}_g - \frac{1}{2} \mathrm{Scal}_g\, g$ shows it is neither non-positive nor non-negative, hence exhibits *backward heat propagation directions*, which should lead to nonintegrability. Therefore, proving local integrability of the gradient of the total scalar curvature functional is a difficult problem. Nevertheless, the previous computation tells us that some combinations of the type $\mathrm{Ric}_g - \alpha\, \mathrm{Scal}_g\, g$ have non-negative symbols, i.e., are (degenerate) elliptic operators. Although the classical parabolic theory proves insufficient there, the local existence question can be solved on compact manifolds via Nash-Moser theory.

Notice that the degeneracy of the operator $g \mapsto \mathrm{Ric}_g - \alpha\, \mathrm{Scal}_g\, g$ follows from its invariance under the action of the group of diffeomorphisms of M. As initially noted by D. deTurck, this gauge symmetry can be broken by adding an *extrinsic* potential to the operator – explicit enough to enable us to go back to an integral curve of the geometric vector field we are interested in – and show local integrability using arguments developed by analysts in the 60's when studying parabolic systems (see [28], and, for a systematic treatment, [68] Chap. 1). At this stage, one can remark that the operator $g \mapsto \text{Ric}_g -\alpha \text{Scal}_g \, g$ is *not* in diagonal form. This means we cannot content ourselves neither with the scalar theory nor with the semi-linear one – in contrast to the harmonic map case – since the higher order term is itself non linear.

Later on, local integrability has been extended by W.-X. Shi [85] to complete manifolds with well-behaved metrics at infinity (e.g., admitting uniform bounds on the sectional curvature). That some a priori taming is needed for a uniform local solution is clear: just consider properly chosen revolution metrics on $\mathbb{R} \times S^k$, $k \geq 2$. Shi proved his result by solving the associated Dirichlet problems on larger and larger compact domains exhausting the manifold, and proving uniform convergence of the Dirichlet solutions to the wanted local flow, at least on any compact set.

To close this paragraph on the local integrability of the Ricci flow, let us remark that M. Min-Oo and E. Ruh [74] succeeded in finding for this flow a variational status it lost when passing from $g \mapsto \text{Ric}_g -\alpha \text{Scal}_g \, g$, for $\alpha \in -(\frac{1}{2(n-1)}, \infty)$ to the Ricci flow. They observed that the Ricci flow is induced by the gradient flow of the Yang-Mills functional $D \mapsto \int_M |F^D|^2$ on the space of Cartan connections of hyperbolic type. This makes more natural what might have earlier been considered a lucky "tour de force".

Global Existence. In any case, local existence is not enough to get any deep geometric results. What remains is to characterize the *maximal* integral curve of the Ricci tensor field through a given metric. If the maximal existence interval is not the whole real line, one can show that some sectional curvature blows up in finite time. This result has been established initially by R. Hamilton for the Ricci flow, and it somewhat surprisingly extends to all fields $g \mapsto \text{Ric}_g -\alpha \text{Scal}_g \, g$, for which local existence can be proved ($\alpha > -\frac{1}{2(n-1)}$) as studied in [68] Chap. 2.

This is the first stage where the (parabolic) maximum principle – a fundamental tool in the theory of metric deformations as the following lines should make it clear – shows up. At least for $\alpha = 0$, i.e., for the Ricci flow itself, the estimates can be proved precisely by checking that the norms of the iterated covariant derivatives of the curvature satisfy parabolic inequalities. From them, uniform a priori bounds on any (finite) interval can be inferred via the maximum principle.

4.2 First Geometric Results

A General Scheme. Up to now we have not made any geometrical assumption on the manifold, so that the Ricci flow appears as a universal tool which can be used to deform *any* metric. To prove any result of some geometrical interest, we have to guess geometric assumptions which could warrant convergence to some "preferred" metric. The archetype of results in the field remains Hamilton's 3-dimensional breakthrough.

Theorem 4.1 [50]. *Any metric of positive Ricci curvature on a 3-manifold can be deformed along the Ricci flow to some metric of constant (positive) sectional curvature.*

What is specific to dimension 3 is the fact that there the Ricci curvature determines the whole Riemannian curvature. In higher dimensions, appears the part of the Riemann curvature which is conformally invariant, the *Weyl curvature*. It is not clear how the Ricci curvature influences it as a field.

Nevertheless, the next contribution was the proof (cf. [70], or, for somewhat weaker statements [56] and [78]) that, provided the curvature of the initial metric is sufficiently close to being a positive constant, in any dimension the Ricci flow converges to a metric with constant sectional curvature. "Closeness" is here measured pointwise by the algebraic deviation of the curvature tensor from having constant sectional curvature [69]:

$$\mathcal{D}_g = R_g - \frac{2}{n(n-1)} \operatorname{Scal}_g \operatorname{Id}_{\Lambda^2} = R_g - \frac{1}{n(n-1)} \operatorname{Scal}_g R_{(S^n, \operatorname{can})} .$$

From an algebraic point of view, the introduction of the deviation is natural enough since $\frac{2}{n(n-1)} \operatorname{Scal} \operatorname{Id}_{\Lambda^2}$ is the projection of the Riemann curvature onto the 1-dimensional O_n-irreducible factor corresponding to the scalar curvature, which is the trace of the Riemann curvature viewed as an operator. In particular $\mathcal{D}$ and $\operatorname{Id}_{\Lambda^2}$ are orthogonal for the natural scalar product on $S^2 \Lambda^2 T^* M$. If, on 3-manifolds, $\mathcal{D}$ can be recovered from the tracefree Ricci curvature, in higher dimensions, it also involves the Weyl curvature we introduced earlier.

As a straightforward corollary of this, one gets effective (classical) sphere theorems under pinching assumptions on the (only pointwise) ratios of the sectional curvature. More interesting, the "weak pinching" assumption in [69] is weak enough to allow some negative curvature. It can be noted though that this statement is not sharp, and the pinching constants in it have no geometrical meaning.

The idea of the proof of such a result relies on parabolic inequalities satisfied by the norm of the deviation along an integral curve of the Ricci flow such as

$$\frac{\partial}{\partial t}(|\mathcal{D}_g|^2 \operatorname{Scal}_g^{-\alpha}) + \Delta(|\mathcal{D}_g|^2 \operatorname{Scal}_g^{-\alpha})$$

$$- 2(\alpha - 1) \operatorname{Scal}_g^{-1}(d \operatorname{Scal}_g, d|\mathcal{D}_g|^2 \operatorname{Scal}_g^{-\alpha}) \leq P(\mathcal{D}_g, \operatorname{Scal}_g, z_g), \quad (4)$$

where $\alpha \in [1, 2]$ and P is a homogeneous polynomial of degree 3.

It is not too difficult to observe that $P(\mathcal{D}_g, \operatorname{Scal}_g, z_g)$ is non-positive provided $|\mathcal{D}_g| \operatorname{Scal}_g^{-1}$ is small enough. The parabolic maximum principle then implies that $\sup_M |\mathcal{D}|^2 \operatorname{Scal}^{-2}$ is a non-increasing function along any integral curve issued from a metric with sufficiently weakly-pinched curvature.

Although $\sup_M |d \operatorname{Scal}|^{-2}$ does not behave monotonically along the flow, this can be proved for a proper linear combination of $\operatorname{Scal}^{-1} |d \operatorname{Scal}|^2$, Scal_g^2 and $|\operatorname{Ric}_g - \frac{1}{n} \operatorname{Scal}_g g|^2$, provided the Ricci curvature is non-negative. This is a straightforward consequence of the assumed pinching which, as we already showed, is preserved along an integral curve. This leads to a control on the gradient of the scalar curvature, involving only the scalar curvature itself.

A straightforward application of the maximum principle to the (parabolic) equation satisfied by the scalar curvature shows that the maximum solution is defined on a compact interval. This provides an upper bound on the size of the interval as soon as the scalar curvature of the initial metric has some positive lower bound. But the whole curvature $|R|^2 = |\mathcal{D}|^2 + \frac{2}{n(n-1)}$ Scal^2 blows up at the finite upper bound, say T, of the maximal existence interval so that $\sup_M |\mathcal{D}|^2$ Scal^{-2} being non increasing, $\max_M$ Scal must blow up at T. This and the previous result on $|d$ Scal $|$ are good enough to prove, using Myers' theorem, that the ratio Scal$_{max}$ / Scal$_{min}$ tends (uniformly) to 1 as t approaches the upper bound T.

We can now apply the maximum principle to the equation (1) for some $\alpha < 2$, large enough for P to remain non positive along an integral curve of the Ricci flow, and conclude that $\lim_{t \to T} |\mathcal{D}_g|^2$ Scal$_g^{-2} = 0$. This ends, up to an appropriate rescaling, (a skeleton of) the proof of (a weak version of) the claimed result.

Local Use of the Ricci Flow: Regularization. In fact there are interesting properties of the Ricci flow which only rely on its local existence. In the same way as the ordinary "heat equation" is known as an efficient "regularizer" (e.g., in the Weierstrass approximation theorem), J. Bemelmans, M. Min-Oo and E. Ruh [6] proved that deforming a metric along the Ricci flow provides geometers with a good "smoother". In particular, the supremum norm of a covariant derivative of the curvature of any order of the regularized metric is uniformly bounded by the C^0-norm of the curvature of the initial metric, the factor only depending on the dimension, the order of the covariant derivative and how good the C^0-approximation on the metric is. This has been extended to the case of metrics of bounded sectional curvature on complete manifolds by W. X. Shi [85]. This result is of obvious interest in Riemannian geometry since it seems a common rule that only C^0 assumptions (or integral assumptions) on the curvature are considered reasonable.

Local Use of the Ricci Flow: L^p-Pinching. Another interesting use of the local Ricci flow is to derive L^p-versions of known C^0-pinching theorems. For example Min-Oo and Ruh (cf. [75], but see also [38], [99] for similar L^p statements) proved as a corollary of C^0-pinching theorems that existence of a sufficiently L^2-pinched metric on a manifold of non zero total scalar curvature implies the existence of a metric of constant sectional curvature. The proof is quite straightforward: from the evolution equation for $|\widetilde{\mathcal{D}}|^2 = |R - \frac{\text{scal}}{n(n-1)} R_{(S^n, \text{can})}|^2$ on any interval $[0, T]$, one gets the parabolic inequality

$$\left(\frac{\partial}{\partial t} + \Delta \right) |\widetilde{\mathcal{D}}|^2 \leq C |\widetilde{\mathcal{D}}| ,$$

which, through Moser iteration procedure, implies

$$\max_{[\frac{2}{3}T, T] \times M} |\widetilde{\mathcal{D}}|^2 \leq C \, T^{-1} \int_{\frac{1}{3}T}^{T} \text{Vol}^{-1} \int_M |\widetilde{\mathcal{D}}|^2 .$$

From the same evolution equation one also gets the ordinary differential inequality

$$\frac{\partial}{\partial t} \int_M |\widetilde{\mathcal{D}}|^2 \leq C(n) \int_M |\widetilde{\mathcal{D}}|^2 ,$$

i.e., the exponential L^2-bound $\int_M |\widetilde{\mathcal{D}}|^2 \le e^{Ct} \int_M |\widetilde{\mathcal{D}}|^2_{|t=0}$.

These two together reduce the proof of the L^2-pinching theorem we claim to the classical L^∞-statement.

Along the same lines, the proof of the L^2-almost flat theorem according to which any manifold M with an L^2-flat enough curvature (i.e., such that $\mathrm{Diam}_g^4 \, \mathrm{Vol}_g^{-1} \int_M |\mathrm{Riem}_g|^2 < \varepsilon$, with ε depending only on the dimension and an upper bound on $\mathrm{Diam}_g^2 \, |\mathrm{Sect}_g|$) is diffeomorphic to a compact quotient of a nilpotent Lie group by a discrete subgroup of isometrics reduces to Gromov's almost flat manifold theorem.

4.3 Sharp (Classification) Results

There is a special feature – sharpness – of Hamilton's 3-sphere theorem which we did not encounter again in any of the developments we reported so far, and which we want to discuss now.

Surfaces. The Ricci flow has been used on compact surfaces to deform any metric to constant curvature (cf. [51, 24]). This gives an approach to uniformization through deformation. The proof in the case of negative curvature reduces to a straightforward application of the maximum principle when, in the case of positive curvature, it involves a geometric version of Yau's Harnack inequality for solutions of the heat equation, and the control along an integral curve of the "entropy" ($\int_M \mathrm{Scal} \, | \log \mathrm{Scal} \, |$) of the probability measure associated to the curvature, $\mathrm{Scal} \, d\mathrm{Vol}$. For considerations around the Harnack inequality for the Ricci flow in higher dimensions, see [52].

4-Manifolds. As early as 1986, Hamilton and Margerin gave two different (sharp) characterizations of the smooth 4-sphere, the only oriented 4-manifold of constant positive sectional curvature by Synge's theorem. There is no hope positive Ricci curvature will do the job in this dimension (e.g., $(S^2 \times S^2, \mathrm{can})$ has positive Ricci curvature).

Theorem 4.2 [53]. *Any compact oriented 4-manifold with positive curvature operator is diffeomorphic to S^4.*

Positive curvature operator is a stronger assumption than positive sectional curvature. Indeed, in $\Lambda^2 TM$, there are skew 2-vectors which are not decomposed ones, the only ones which correspond to (geometric) 2-planes.

Here is an other sharp characterization of the smooth canonical 4-sphere

Theorem 4.3 [68] Chap. 4, [70]. *Any compact oriented manifold of dimension 4 which is more weakly pinched than $(\mathbb{C}P^2$, Fubini-Study) is diffeomorphic to S^4.*

(Here, "weak pinching" is measured by the norm $\| \mathrm{Scal}^{-1} \, \mathcal{D} \|_{C^0}$.)

In both cases, the statements are not only sharp, but rigidity is proved in the case of equality. Only quotients of S^4, $\mathbb{C}P^2$, $S^3 \times \mathbb{R}$ by a finite group of fixed-point free isometries of the standard metrics may achieve the weak pinching of ($\mathbb{C}P^2$, Fubini-Study), when $\mathbb{R}^4$, $\mathbb{R}^2 \times S^2$ and $S^2 \times S^2$ must be added to the list of manifolds admitting metrics with non-negative curvature operators.

If both proofs consist in pushing the initial metric along the Ricci flow and studying convergence at infinity, they are different in spirit. Systematizing his 3-dimensional approach, Hamilton studies the behavior under the O.D.E. associated with the parabolic equation satisfied by the curvature of some convex set in the bundle $S^2 \Lambda^2 T^* M$ (the so-called "pinching set"). Margerin uses as much as possible the algebraic properties of the curvature algebra in dimension 4 in particular existence of a further irreducible splitting under $SO(4)$, involving the Hodge $*$-operator, to prove, with the help of the (scalar) maximum principle, that the weak pinching behaves well along the integral curve of the Ricci flow, even if one starts (almost) as far as ($\mathbb{C}P^2$, F-S) or (S^1, α can) $\times$ (S^3, β can) for $(\alpha, \beta) \in \mathbb{R}_+^2$.

Although these two proofs require a much more refined approach than previously quoted results, this indicates the metric deformation procedure is a powerful tool to explicitly work out a (partial) classification of differential structures, in dimensions 3 and 4 at least,, where we exhibit a nice "absolute" minimum with (S^n, can) and a total description of the limiting cases.

Higher Dimensions: A Sharp Characterization of the Canonical Metric on the Sphere. Using once more the algebraic structure of the curvature algebra, Margerin more recently proved the following, a first step towards a classification for large enough dimensions.

Theorem 4.4 [68], Chap. 5]. *There exists $n_0 \in \mathbb{N}$ such that for all $n > n_0$ any compact n-manifold whose curvature weak-pinching is stronger than that of (S^1, can) $\times$ (S^{n-1}, can) is diffeomorphic to a finite quotient of the standard-sphere. In the equality case, one can also obtain quotients of $\mathbb{R} \times S^{n-1}$.*

The proof still relies on straight ideas and elementary tools like the (scalar) maximum principle for parabolic equations, but the algebraic study is pretty elaborated, mainly because one has to deal with all (or almost all) dimensions at a time. There is of course no reason why n_0 should not be 3. In the present state, it can be taken equal to 3676, a dimension which has no geometrical meaning... and is likely to be improved by a closer look at the arithmetics going into the proof. In any case, some more geometric ideas seem necessary to go down to dimension 4.

4.4 Complex Geometry

Let us conclude with some applications of the Ricci flow to Kähler geometry. Up to now, we have a bit arbitrarily skipped this aspect. In [22], H.D. Cao, taking advantage of previous deep work by S.T. Yau, proved that the Ricci flow can be used to deform any Kähler metric to some metric in the same class whose Ricci curvature will be any prescribed form of type (1,1) in the first Chern class of the manifold. In particular, when the first Chern class is zero, the Kähler metric can be deformed to a Ricci-flat metric, giving a somewhat new proof of the Calabi conjecture.

The Ricci flow also applies in the case where the first Chern class is negative, allowing to deform the initial Kähler metric to a Kähler-Einstein one. As expected, this case is even simpler.

Even a local use of the Ricci flow on Kähler manifolds may lead to interesting conclusions. One can easily see that non-negative curvature operators are preserved along the Ricci flow (see [69]), and, as noted by Hamilton [53] the image of the curvature, viewed as an endomorphism of $\Lambda^2 T^* M$, defines an invariant Lie subalgebra Im R – w.r.t. both space and deformation parameters – of $\Lambda^2 T^* M$ on some interval $(0, \varepsilon)$. By the (independent) work of M. Berger and Y. T. Siu, an irreducible compact Kähler manifold which is not Hermitian symmetric and with non-negative curvature operator has holonomy group U_m, a property which is open among Kähler metrics, and preserved by a (small) Ricci deformation.

By our previous remark, Im(R) is then isomorphic to $u(m)$ and R remains non-negative on some interval $[0, \varepsilon]$. This means that the restriction of the curvature to $\Lambda^{1,1}$ is positive. Note then that positivity of the bisectional curvature is, according to the proof of the Frankel conjecture due to Mori, Siu and Yau, a characterization of $\mathbb{C}P^m$. Using the splitting theorem of Howard, Smyth and Wu, one gets a complete classification of compact Kähler manifolds with non-negative curvature operator – a first step towards the classification of manifolds with non-negative bisectional curvature, (cf. [23]). This question has been settled earlier in dimension 3 by S. Bando [5], as a rather straightforward corollary of the maximum principle applied to the Ricci flow.

The complete solution, due to N. Mok [76], requires more work. Howard-Smyth and Wu's splitting theorem reduces the conjecture to considering compact Kähler manifolds of non-negative holomorphic bisectional curvature with Ricci curvature positive at some point. Non negativity of the holomorphic bisectional curvature is shown to be preserved along the Ricci flow. This also implies that the Ricci curvature and the holomorphic sectional curvature are positive everywhere. For such a metric, Mok uses Mori's results on rational curves and proves that the set of all tangent directions to rational curves of minimal degree is a proper holonomy-invariant subset of the projectivized tangent bundle. This is verified through the evolution equation for the bisectional curvature. Berger's holonomy group classification concludes the proof.

4.5 Conclusion

We hope we succeeded in giving the reader an idea of how powerful a tool metric deformations can be, an idea of the results they led to, but also of the different strategies which have been developed to prove these results. Many aspects have been skipped, basically by lack of space.

It would be unfair to close this, without mentioning a variant of the Ricci flow which can be introduced on Hermitian (holomorphic) vector bundles. On (Mumford-)stable bundles, any Hermitian structure can be deformed along the Kobayashi curvature (a section of the endomorphism bundle, namely, the trace w.r.t. the polarization of the base of the curvature of the bundle) to a Hermite-Einstein metric (i.e., one whose Kobayashi curvature is a constant multiple of the identity). This is a deep theorem due to S. K. Donaldson, K. K. Uhlenbeck and S. T. Yau (see [71]) which throws new light on the study of moduli spaces (of holomorphic bundles), and remains a cornerstone of a rapidly expanding field (e.g., [86]).

5. Moduli Spaces of Geometric Objects

In recent years, a good deal of work has been devoted by geometers to the study of deformations of objects subject to certain constraints, and to the possibility of putting some specific structures on these "moduli spaces" (see [Penner] for a general account). The most classical moduli space is that of complex structures on a Riemann surface, often called *the* moduli space. Moduli spaces correspond to equivalence classes of objects, hence are obtained after dividing out by all groups acting on the situation. These groups are, as we shall see, most of the time infinite dimensional.

Before getting to specific moduli spaces, let us emphasize how the interest of theoretical physicists for these questions revived the whole theory. They brought in new points of view (as their insistance of taking full advantage of the gauge invariance), new problems (since they needed specific pieces of information in very precise situations), and revived or invented concepts (such as twistors in Penrose's programme). The impact in Mathematics of these new developments is spectacular, in particular in Differential Topology.

5.1 A Gauge Approach to Teichmüller Space

Some Backgound. It was a major discovery when Riemann noticed that on orientable surfaces holomorphic atlases may be defined. The study of equivalence classes of such atlases on a compact surface Σ gave birth to the so-called *Riemann moduli space* $\mathcal{R}\Sigma$.

The *uniformization theorem* translates the problem back into a group theoretical one, namely, to describe all possible discrete cocompact subgroups of the isometry groups of the three models of 2-dimensional geometry: the spherical one; the Euclidean one; and the hyperbolic one. The situation can be summarized as follows: the holomorphic atlas on S^2 is rigid (up to equivalence by a diffeomorphism); on T^2, this amounts to describing all possible lattices in $\mathbb{R}^2$, giving rise to the so-called *modular domain*; for surfaces Σ of higher genus, $\mathcal{R}\Sigma$ is an analytic space of (complex) dimension $3\gamma(\Sigma) - 3$.

It was later discovered that the study of $\mathcal{R}\Sigma$ is made easier if one considers first the *Teichmüller space* $\mathcal{T}\Sigma$ of the surface Σ, obtained by dividing out only by diffeomorphisms *isotopic to the identity* (later, we denote by $\mathcal{D}_0\Sigma$ the group they form, whereas the full group is denoted by $\mathcal{D}\Sigma$). and then the action of the discrete group $\mathcal{D}_0\Sigma/\mathcal{D}\Sigma$ on $\mathcal{T}\Sigma$.

Teichmüller space has several different aspects, a fact that makes its study fascinating: there is obviously a *complex* one; there is also an *algebraic* one since it can be viewed as the space of representations of the fundamental group in $\mathrm{PSl}_2(\mathbb{R}R)$; there is finally a geometric one as moduli space of constant curvature metrics (notice that this is an example of a symplectic reduction since, in this context, the curvature can be interpreted as a moment map).

Recently, Anthony Tromba and Arthur Fischer (cf. [97] for a systematic account) reestablished most of the results obtained over the years on $\mathcal{T}\Sigma$ by taking a (real) gauge approach to the problem. This paralleled the way theoretical physicists, engaged in string theories and 2-dimensional conformal field theories, were also looking at these objects. In the next paragraphs, we present some of the results which have been obtained by this direct differential geometric approach.

The Space of Almost Complex Structures on a Surface. One possible way of getting at Teichmüller space is to view it as the quotient by $\mathcal{D}_0\Sigma$ of the space $\mathcal{J}\Sigma$ of almost complex structures on Σ. Thanks to Liouville's local uniformization theorem, any family of complex structures on $T_x\Sigma$ for $x \in U$, an open set in Σ, can be "integrated" into a germ of complex chart, the converse direction being evident. The group $\mathcal{D}_0\Sigma$ acts by usual pull-back on elements of $\mathcal{J}\Sigma$, a submanifold of the space $\Gamma(T^*\Sigma \otimes T\Sigma)$ of sections of the endomorphism bundle.

One is interested in all natural structures that one can define on $\mathcal{J}\Sigma$ and which are invariant under this gauge action. The first one which comes to mind is an almost complex structure **J**. Indeed, the space $T_J\mathcal{J}\Sigma$ identifies itself with $\{H \mid H \in \Gamma(T^*\Sigma \otimes T\Sigma), \; HJ + JH = 0\}$, and the map $H \mapsto HJ$ is indeed an almost complex structure on $\mathcal{J}\Sigma$, which is invariant under the action of $\mathcal{D}\Sigma$. It therefore induces an almost complex structure on $\mathcal{T}\Sigma$, which does coincide with the usual one.

Moreover, **J** leaves the orbits under $\mathcal{D}_0\Sigma$ invariant, hence induces an almost complex structure on the "horizontal" space. This structure therefore descends to Teichmüller space. Uwe Abresch exhibited local complex coordinate charts on $\mathcal{J}\Sigma$ proving that **J** is integrable. By lifting the almost complex structure on $\mathcal{T}\Sigma$, one can prove in this way that it is integrable, providing therefore a "real" approach to the complex structure of Teichmüller space.

The Space of Hyperbolic Metrics on a Surface. In 2 dimensions, there is also a metric approach to the notion of complex structure. Indeed, to any conformal class of Riemannian metrics on an oriented surface one can associate an almost complex structure, hence a complex structure by Liouville's theorem. This suggests another possible definition of Teichmüller space, namely to view it as $\mathcal{P}\Sigma \backslash \mathcal{M}\Sigma / \mathcal{D}_0\Sigma$

where $\mathcal{M}\Sigma$ denotes the space of Riemannian metrics on the surface Σ, and $\mathcal{P}\Sigma$ the multiplicative group of positive functions on Σ.

Although useful, this identification turns out ot be substantially improved if one completes it with the following construction which can be made on all surfaces of genus ≥ 2. Indeed, in each conformal class of such a surface, one finds a unique hyperbolic metric, i.e., a metric with curvature -1. Moreover, isometries of hyperbolic metrics are never isotopic to the identity. Therefore, the space $\mathcal{M}_{-1}\Sigma$ of hyperbolic metrics is in 1-1 correspondence with $\mathcal{J}\Sigma$, and since $\mathcal{D}_0\Sigma$ acts without fixed point on it, $\mathcal{T}\Sigma$ can be identified with $\mathcal{M}_{-1}\Sigma/\mathcal{D}_0\Sigma$.

Two new natural structures can then be derived from this description. First, the space $\mathcal{M}\Sigma$ has a natural one-parameter family of Riemannian metrics, namely, for $h, k \in T_g\mathcal{M}\Sigma = \Gamma(S^2T^*\Sigma)$ and $\alpha \in \mathbb{R}$, one defines $(h, k)_g^{(\alpha)} = \int_{\Sigma}(\text{Trace}(H_g \circ K_g) - \alpha\,\text{Trace}(H_g)\,\text{Trace}(K_g))\,v_g$, where H_g denotes the linear map associated to the bilinear tensor field h by the Riemannian metric g, and v_g the volume element associated to g. This metric is also gauge invariant. It induces a well-defined metric on $\mathcal{M}_{-1}\Sigma$ which, when pushed down to $\mathcal{T}\Sigma$, can be verified to coincide with the classical *Petersson-Weil metric*.

The other structure relies on the following theorem due to R. Schoen and S. T. Yau.

Theorem 5.1. *There is exactly one harmonic diffeomorphism in each homotopy class of degree one maps between two negatively curved surfaces.*

Indeed, in our setting, let us fix a point j_0 in $\mathcal{T}\Sigma$, and a hyperbolic metric g_0 inducing this complex structure. To any point $j \in \mathcal{T}\Sigma$, one can associate the diffeomorphism $\varphi \in \mathcal{D}_0\Sigma$ provided by Theorem 5.1 which is harmonic as a map between (Σ, g_0) and (Σ, g), where the hyperbolic metric g is chosen to induce the complex structure j. In this way, we define a canonical metric attached to j, namely, φ^*g. This construction is equivariant under $\mathcal{D}_0\Sigma$, so that the Dirichtlet energy $E_{g_0}(\varphi)$ can be viewed as a map ψ from $\mathcal{T}\Sigma$ to $\mathbb{R}$. This map turns out to be a Morse function with controlled growth whose only critical point is j_0. This provides a "real" proof of the following classical result.

Theorem 5.2. *The Teichmüller space of a surface with genus ≥ 2 is a cell.*

Further Structures on Teichmüller Spaces. We already introduced on $\mathcal{T}\Sigma$ a metric and a complex structure, so that it is natural to ask whether these two structures define a Kählerian one. It is a classical result that it is indeed the case. Proving that the 2-form associated to the complex structure by the metric is closed is not that easy. Once again, it turns out to be more efficient to treat the problem upstairs in the infinite dimensional setting. For that purpose, $\mathcal{J}\Sigma$ is the most appropriate since, on it, it is possible to define the connection $(X, Y) \mapsto D_XY = D_X^0Y + J(X \circ Y + Y \circ X)$ where D^0 denotes the vector space connection in $\Gamma(T^*\Sigma \otimes T\Sigma)$, and $\circ$ is the composition of linear maps. One directly verifies that $D\mathbf{J} = 0$, hence an "easy" proof of the fact that $\mathcal{T}\Sigma$ is a Kählerian manifold after one has checked that D induces the Levi-Civita connection for the Petersson-Weil metric on $T\Sigma$.

This simplification extends to the calculation of the curvature of the Petersson-Weil metric on $T\Sigma$, which was proved to be negative by Ahlfors after a brute force calculation using convoluted identities between Green's kernels.

5.2 The Moduli Space of Instantons on 4-Dimensional Manifolds

The introduction of moduli spaces of self-dual connections on 4-dimensional manifolds is exemplary of the new type of interactions that developed recently between mathematicians and theoretical physicists. The object itself was introduced by physicists in their efforts to derive quantum approximations of some classical gauge theories. For that purpose, they needed to evaluate a partition function whose first term involved integrating over this moduli space. What was first at stake was to determine its dimension, and a natural volume form on it.

To start with, the base space was taken to be S^4, a conformal compactification of $\mathbb{R}^4$. This led to the very successful programme of Penrose, translating self-dual connections on a bundle over S^4 into holomorphic connections on the pull-back bundle over $\mathbb{C}P^3$, which is the total space of a Hopf fibration over S^4 viewed as $\mathbb{H}P^1$. (For a general account of this, see [3], [4].) The proper general setting for this translation is to consider the *twistor space* JM of the base space M, which is a complex 3-manifold provided M is 4-dimensional and the Riemannian metric on M is half-conformally flat, a condition to which we turn in Sect. 5.3.

The case of general 4-dimensional base manifolds requires a good deal of non-linear analysis. It was developed since the early 80's by cumulative work by K. K. Uhlenbeck and C. H. Taubes. S. K. Donaldson turned the whole approach around, recognizing that all these constructions could be used to produce invariants of 4-dimensional differentiable manifolds (cf. [31], [32], or for a survey [54]). This was later generalized to lead to great progress in the understanding also of 3-dimensional topology (see for example [93]). For the sake of presentation, we will concentrate in this section on the four-dimensional case. (Excellent systematic presentations of the necessary background can be found in [65], [37], and [33].)

The Yang-Mills Setting. Let G be a compact Lie group. We consider a G-bundle E over a compact 4-dimensional manifold M. We denote the space of G-connections over E by $\mathcal{A}E$. It is an affine space modeled after the vector space $\Omega^1(M, \mathrm{aut}_G E)$, where $\mathcal{G}E$ denotes the infinitesimal automorphism bundle of E, a sub-bundle of $E^* \otimes E$. The *gauge group* $\mathcal{G}E$ is the group of sections of the automorphism bundle GE of E. We are interested in the solutions of the Yang-Mills equations, i.e., the critical points of the *Yang-Mills functional* which to a connection A associates $\mathcal{YM}(A) = \frac{1}{2}\int_M \|F^A\|^2 v_g$, where F^A denotes the *curvature* of the connection A, $\|\ \|$ the norm deduced from an invariant scalar product on the Lie algebra of G and a scalar product g on M taken to be Riemannian (although M is thought to be space-time in the physical context). The Yang-Mills equations form a system of partial differential equations of second order in the connection expressing that the curvature is harmonic. If the Lie group G is non-abelian, the Yang-Mills equations are non-linear what makes their study more subtle... hence more interesting.

The Yang-Mills functional is gauge invariant forcing an infinite dimensional group of invariance on the Yang-Mills equations. Moreover, it depends only on the conformal class of the metric g on M. This extra invariance is responsible for a lack of compactness in the space of solutions, since solutions may concentrate at points as if the metric is concentrated there in a conformal way.

When M is oriented, one can give a topological lower bound for $\mathcal{YM}$ in terms of the so-called *instanton number* of E. Absolute minima for $\mathcal{YM}$ are shown to be solutions of an algebraic equation in the curvature, expressing that one of the two-irreducible components of the curvature w.r.t the action of the Hodge $*$-operator vanishes. These solutions are called *self-dual* or *anti-self-dual* connections, or else *instantons*.

Complete solutions of the Yang-Mills equations for bundles over (S^4, can) using completely algebraic terms have been found by M. F. Atiyah, V. I. Drinfeld, N. Hitchin, and Yu. Manin. The main achievements of the 80's have been to give general criteria for the existence of solutions over more general 4-dimensional manifolds, and then to derive many deep topological conclusions.

Let us present some results in the case of *simple* 4-manifolds, i.e., simply connected manifolds with negative definite intersection form. Many other results have been obtained, most of the time by constructing the appropriate moduli space, and gathering information on it.

Theorem 5.3 [89]. *On any* SU_2*-bundle with non-negative instanton number over a compact oriented 4-manifold with negative definite intersection form, there always exist self-dual connections.*

The proof of Taubes takes full advantage of the conformal invariance of the problem. He conformally changes the metric g on M to make it look like that of a standard sphere on a very large portion of M. In this way, on any bundle with instanton bundle 1, he can "graft" an instanton from the standard sphere. The connection obtained in this way is almost self-dual. The assumption on the intersection form allows then to apply an iteration scheme under which the non-linear part of the curvature can be dominated precisely because the connection is almost self-dual. The process finally converges to a self-dual connection. The case of manifolds with non definite intersection form is more involved (see [90]).

Moduli Spaces of Instantons as Topological Spaces. The structure of the moduli space of instantons on a simple manifold is interesting in itself. For instanton number k (and group SU_2), it is of dimension $8k - 3$. The fact that it is orientable and that its end has a collar structure diffeomorphic to a product of M by an interval were crucial for S. K. Donaldson's first breakthrough. Indeed, S. K. Donaldson shows that, for a generic metric on M, the moduli space of instantons of the simple manifold M provides an oriented cobordism between M and the connected sum of as many copies of $\mathbb{C}P^2$ as one can find harmonic 2-forms on M with self-intersection number -1. This leads to the following theorem.

Theorem 5.4 [31]. *Any simple manifold has a diagonal intersection form over the integers.*

This result came as a great surprise since it showed that compact 4-dimensional differentiable manifolds are much more constrained than 4-dimensional topological ones. S. K. Donaldson developed this method further with great success (cf. [32]) using the moduli space as an invariant coding the differentiable structure of M inside the space $\mathcal{A}E/\mathcal{G}E$, which is only a homeomorphism invariant. By intersecting the moduli space of instantons with properly chosen homology classes in $\mathcal{A}E/\mathcal{G}E$, he was able to define new polynomials attached to the differentiable structure of M. This led to the discovery of many inequivalent differentiable structures on a number of 4-dimensional manifolds, in particular to infinitely many on $\mathbb{R}^4$, as was proved by S. K. Donaldson and M. H. Freedman,

The Yang-Mills functional can also be used to study the topology of $\mathcal{A}E/\mathcal{G}E$ if it can be proved that a Morse approach is legitimate. Taubes showed that it is indeed the case in a series of papers requiring very detailed control of the non-linear behaviour of Yang-Mills fields [91, 92].

Moduli Spaces of Instantons as Riemannian Manifolds. As we mentioned earlier, physicists are interested in more refined structures on moduli spaces of instantons. A systematic study of the natural metrics they inherit from the gauge-invariant L^2-metric on $\mathcal{A}E$ has been carried out by D. Groisser and T. Parker (cf. [44, 45]). They prove the following.

Theorem 5.5. *The moduli space of instantons over the* SU_2-*bundle with instanton number 1 on the standard sphere* (S^4, can) *is radially symmetric and conformally flat with positive curvature. It has finite diameter and finite volume. Its boundary is isometric to* $(S^4, 4\pi^2 \text{ can})$.

In many cases, the pointed moduli space obtained by fixing a frame at a point has a natural structure. For SU_2-bundles over a simple manifold, these moduli spaces are $8k$-dimensional. The natural metric they inherit are naturally hyperkählerian as shown in [55], using for example a quotient construction from a linear space.

5.3 Moduli Spaces of Special Riemannian Metrics

Special classes of Riemannian metrics are most of the time distinguished by curvature properties. In the case of surfaces, the only natural condition to impose is that the curvature, then a fonction, be constant, and we recover the Riemannian interpretation of the uniformization theorem. In higher dimensions, the Riemann curvature tensor R becomes a more involved object. It acquires its full generality in dimension 4. Indeed, it is the first dimension in which, in its decomposition into irreducible components w.r.t. the natural action of the orthogonal group, R has three components: the first one, corresponding to a trivial representation, is a multiple of the identity operator on exterior 2-forms, and contains the scalar curvature; the second one, corresponding to a representation isomorphic to that of traceless symmetric tensors, contains the traceless part of the Ricci curvature; the third one is what is left, and is the so-called *Weyl conformal curvature tensor* since it is invariant under conformal changes of the metric.

Natural classes of metrics correspond to the vanishing of some of these components: curvature tensors whose second components vanish are called *Einstein*; curvature tensors whose third components vanish are *conformally flat*. It is on those special Riemannian manifolds that we will concentrate our attention in this section.

Einstein Metrics. For a long time, examples of Einstein metrics have been rather scarce. The theory of isotropy irreducible homogeneous spaces provided a large family, of a rather algebraic nature. We now have at our disposal other families of Einstein metrics thanks to the special meaning that the Ricci curvature has on Kähler manifolds and to the solution of the Calabi conjecture, and also thanks to the examples explained in Sect. 4.4.

The general theory of moduli spaces $\mathcal{E}M$ of Einstein metrics (with normalized volume 1) on a compact manifold M has been initiated by N. Koiso [64]. (One can also consult [11] for a general account.) In recent years, substantial progress have been made in our understanding of the global structure of $\mathcal{E}M$. It seems to be confirmed that Einstein metrics having positive, vanishing, or negative Einstein constants behave quite differently. First of all, it is known that they belong to different connected components of $\mathcal{E}M$. This is further reflected in the structure of $\mathcal{E}M$. Here is one of the interesting statements obtained when M is 4-dimensional.

Theorem 5.6 [2]. *Let M be a compact oriented 4-manifold. The closure in the L^2-topology of $\mathcal{E}M$ is the union of $\mathcal{E}M$, $\mathcal{E}_oM$ (consisting of orbifold singular Einstein metrics on M) and $\mathcal{E}_cM$, consisting of Einstein metrics on the union of cusps obtained from M by collapsing part of it. It is compact and Hausdorff when the Einstein constant λ is positive, non-compact and Hausdorff when λ is zero, and in both cases without cusps. When $\lambda < 0$, the closure is non-compact and Hausdorff but cusps may occur, bringing lower dimensional spaces among the possible limits.*

Kähler-Einstein Metrics. On an m-dimensional Kähler manifold, the Ricci curvature can be directly interpreted as the curvature of the line bundle $\Lambda^m T_{\mathbb{C}}^* M$ with the metric it naturally inherits. The cohomology class it defines is therefore necessarily $2\pi c_1(M)$, the first Chern class of the bundle $T_{\mathbb{C}}M$. Solving the Einstein equation in this case amounts to solving a Monge-Ampère equation. An extension of the solution of the Calabi conjecture gives many examples of Einstein metrics on complex manifolds with negative definite or vanishing first Chern class. A case of special interest has been that of K3 surfaces (complex surfaces with vanishing irregularity and vanishing first Chern class), where a complete description of the moduli space of Einstein metrics is now available, including the description of the

L^2-metric on it. Notice that on a K3 surface, any Einstein metric is automatically hyperkählerian (as follows from the limiting case of the Thorpe-Hitchin inequality).

Theorem 5.7. *The moduli space of Einstein metrics on a K3 surface is isometric to a quotient of the symmetric space* $SO_{3,19}/SO_3 \times SO_{19}$.

In recent years, substantial results have been obtained concerning the existence of Kähler-Einstein metrics when the first Chern class is positive. There, the situation is much more intricate since obstructions connected to the non-triviality of the connected component of the identity of the group of holomorphic transformations are known to exist. The case of surfaces has been solved somewhat satisfactorily by the work of G. Tian and S. T. Yau [96].

(Half)-Conformally Flat Metrics. Conformal flatness of the metric is, as we said, measured by the vanishing of the Weyl curvature tensor W. It is the necessary and sufficient condition for the natural almost complex structure on the space JM be integrable. (For a general survey, see [39].) The 4-dimensional theory can be refined when M is oriented. Indeed, the almost complex structure on $\mathcal{J}^-M$, the space of almost complex structures inducing the opposite orientation as that of M, is integrable when W^+, one of the two irreducible components of W under the action of SO_4, vanishes. One then says that the metric is *half-conformally flat*, or sometimes *self-dual*.

Recently, moduli spaces of half-conformally flat metrics on 4-manifolds have been described quite completely (cf. [62, 59]). Their dimension can be explicitly computed. Moreover, may more examples of manifolds admitting half-conformally flat manifolds have been produced, whereas, for some time, the belief was that such metrics should be scarce. The best result in this direction is the following.

Theorem 5.8 [94]. *Let M be an arbitrary compact oriented* 4-*manifold. There exists* $k \in \mathbb{N}$ *such that* $M \sharp k \, \mathbb{C}P^2$ *has a half-conformally flat metric.*

This provides a huge flexibility in the construction of 3-dimensional manifolds as twistor space of these 4-dimensional half-conformally flat manifolds. One of the striking consequences of that fact is the following result.

Theorem 5.9 [94]. *Any finitely presented group is the fundamental group of a* 3-*dimensional complex manifold.*

This result has to be compared with the many restrictions that one now knows on the fundamental group of Kähler manifolds. This shows once more that, besides $\mathbb{C}P^3$ and the 3-dimensional flag manifold, twistor spaces form a family of complex manifolds which is very different from Kähler ones.

References

1. Abresch, U.: Constant mean curvature tori in terms of elliptic functions. J. Reine Ang. Math. **374** (1987) 169–192
2. Anderson, M.: The L^2-structure of moduli spaces of Einstein metrics on 4-manifolds. Geom. Functional Anal. **2** (1992) 29–89
3. Atiyah, M.F.: Geometry of Yang-Mills fields. Lezioni Fermiane. Acad. Naz. dei Lincei, Sc. Norm. Sup. Pisa, 1979
4. Atiyah, M.F., Hitchin, N.J., Singer, I.M.: Self-duality in four-dimensional Riemannian geometry. Proc. Roy. Soc. London **362** (1978) 425-461
5. Bando, S.: On 3-dimensional compact Kähler manifolds of non-negative bisectional curvature. J. Differential Geom. **19** (1984) 283–297
6. Bemelmans, J., Min, Oo, M., Ruh, E.: Smoothing Riemannian metrics. Math. Z. **188** (1984) 69–74
7. Benoist, Y., Foulon, P., Labourie, F.: Flots d'Anosov à distributions stable et instable différentiables. J. Amer. Math. Soc. **5** (1992) 33–74
8. Bérard, P.: Variétés riemanniennes isospectrales non isométriques. Astérisque **177–178** (1989) 127–154
9. Bérard, P.: Transplantation et isospectralité I. Math. Ann. **292** (1992) 547–559
10. Besse, A.L.: Manifolds all of whose geodesics are closed. Ergebnisse der Mathemathik, vol. 98. Springer, Berlin 1973
11. Besse, A.L.: Einstein manifolds. Ergebnisse der Mathematik, vol. 10. Springer, Berlin Heidelberg New York 1987
12. Besson, G.: On the multiplicity of the eigenvalues of the Laplacian. In: Geometry and analysis on manifolds, Katata/Kyoto. Lecture Notes in Mathematics, vol. 1339 (1987) 32–53
13. Besson, G., Courtois, G., Gallot, S.: Volume minimal des espaces localement symétriques. Inventiones Math. **103** (1991) 417–445
14. Besson, G., Courtois, G., Gallot, S.: Les variétés hyperboliques sont des minima locaux de l'entropie topologique. Inventiones Math. (to appear)
15. Bourguignon, J.P.: L'équation de la chaleur associée à la courbure de Ricci, d'après R.S. Hamilton. In: Séminaire Bourbaki 1985–1986, Exposé 653. Astérisque **145–146** (1987) 45–62
16. Bowen, R.: Periodic orbits for hyperbolic flows. Amer. J. Math. **94** (1972) 1–30
17. Brooks, R.: The spectral geometry of a tower of coverings. J. Differential Geom. **23** (1986) 97–107
18. Brooks, R.: Constructing isospectral manifolds. Amer. Math. Monthly **95** (1988) 823–839
19. Burger, M.: Spectre du laplacien, graphes et topologie de Fell. Comment. Math. Helv. **63** (1988) 226–252
20. Burns, K., Gerber, M.: Real analytic Bernoulli geodesic flows on S^2. J. Ergodic Th. Dynamical Systems **9** (1989) 27–45
21. Buser, P.: Geometry and Spectra of Compact Riemann Surfaces. Birkhäuser, Bâle 1992
22. Cao, H.D.: Deformation of Kähler metrics to Kähler-Einstein metrics on compact Kähler manifolds. Inventiones Math. **81** (1985) 359–372
23. Cao, H.D.: Compact Kähler manifolds with non-negative curvature operator. Inventiones Math. **83** (1986) 553–556
24. Chow, B.: The Ricci flow on the 2-sphere. J. Differential Geom. **33** (1991) 325–334
25. Colin de Verdière, Y.: Spectre du laplacien et longueurs des géodésiques périodiques I. Compositio Math. **27** (1973) 83–106; idem: II, ibidem, 159–184
26. Colin de Verdière, Y.: Sur la multiplicité de la première valeur propre non nulle du laplacien. Comment. Math. Helv. **61** (1986) 254–270

27. Colin de Verdière, Y.: Construction de laplaciens dont une partie finie du spectre est donnée. Ann. Sci. Ecole Norm. Sup. Paris **20** (1987) 599–615
28. DeTurck, D.: Deforming metrics in the direction of their Ricci tensors. J. Differential Geom. **18** (1983) 157–162
29. DeTurck, D., Gordon, C.: Isospectral deformations I: Riemannian structures on two-step nilspaces. Commun. Pure Appl. Math. **40** (1987) 367–387; idem: II, Trace formulas, metrics, and potentials, ibidem **42** (1989) 1067–1095
30. DeTurck, D., Gluck, H., Gordon, C., Webb, D.: The inaudible geometry of nilmanifolds. Inventiones Math. **111** (1992) 271–284
31. Donaldson, S.K.: An application of gauge theory to the topology of 4-manifolds. J. Differential Geom. **18** (1983) 279–315
32. Donaldson, S.K.: Connections, cohomology and the intersection form of 4-manifolds. J. Differential Geom. **24** (1986) 275–342
33. Donaldson, S.K., Kronheimer, P.B.: The geometry of four-manifolds. Oxford Univ. Press, Oxford, 1990
34. Duistermaat, J.J., Guillemin, V.: The spectrum of positive elliptic operators and periodic bicharacteristics. Inventiones Math. **29** (1975) 39–79
35. Eberlein, P.: When is a geodesic flow of Anosov type ? I. J. Differential Geom. **8** (1973) 437–463
36. Foulon, P., Labourie, F.: Sur les variétés compactes asymptotiquement harmoniques. Inventiones Math. **109** (1992) 97–112
37. Freed, D.S., Uhlenbeck, K.K.: Instantons and four-manifolds. Math. Sci. Res. Inst. Publications **1** (1984) Springer, Berlin Heidelberg New York
38. Gao, L.Z.: Convergence of Riemannian metrics: Ricci and $L^{n/2}$-pinching. J. Differential Geom. **32** (1990) 349-382
39. Gauduchon, P.: Variétés riemanniennes autoduales. Séminaire Bourbaki, Exposé **767** (1993) 1–27
40. Ghys, E.: Rigidité différentiable des groupes fuchsiens. Preprint, Ec. Norm. Sup. Lyon, 1992
41. Gordon, C., Wilson, E.: Isospectral deformations of compact solvmanifolds. J. Differential Geom. **19** (1984) 241–256
42. Gordon, C., Webb, D., Wolpert, S.: Isospectral plane domains and surfaces via Riemannian orbifolds. Inventiones Math. **110** (1992) 1–22
43. Green, L.: Surfaces without conjugate points. Trans. Amer. Math. Soc. **76** (1954) 529–546
44. Groisser, D., Parker, T.: The Riemannian geometry of the Yang-Mills moduli space. Commun. Math. Phys. **112** (1987) 663–689
45. Groisser, D., Parker, T.: The geometry of the Yang-Mills moduli space for definite manifolds. J. Differential Geom. **29** (1989) 499–544
46. Gromov, M.: Manifolds of negative curvature. J. Differential Geom. **13** (1978) 223–230
47. Gromov, M.: The foliated Plateau problem: I. Geom. Functional Anal. **1** (1992) 14–79, ; idem: II, ibidem, 253–320
48. Hadamard, J.: Les surfaces à courbures opposées et leurs lignes géodésiques. J. Math. Pures Appl. **4** (1898) 27–74
49. Hamenstädt, U.: A new description of the Bowen-Margulis measure. J. Ergodic Th. Dynamical Systems **9** (1989) 455–464
50. Hamilton, R.S.: 3-Manifolds with positive Ricci curvature. J. Differential Geom. **17** (1982) 255–306
51. Hamilton, R.S.: The Ricci flow for surfaces. Contemporary Math. **71** (1988) 237–262
52. Hamilton, R.S.: The Harnack estimate for the Ricci flow. J. Differential Geom. **37** (1993) 225–243
53. Hamilton, R.S.: Four-manifolds with positive curvature operator. J. Differential Geom. **24** (1986) 153–179

54. Hitchin, N.J.: The Yang-Mills equations and the topology of 4-manifolds. In: Séminaire Bourbaki 82-83, Exposé 606. Astérisque **105–106** (1983) 167–178
55. Hitchin, N.J., Karlhede, A., Lindström, U., Rocek, M.: Hyperkähler metrics and supersymmetry. Commun. Math. Phys. **108** (1987) 535-589
56. Huisken, G.: Ricci deformation of the metric on a Riemannian manifold. J. Differential Geom. **17** (1985) 47–62
57. Hsiang, W.Y., Palais, R.S., Terng, C.L.: The topology of isoparametric submanifolds. J. Diffferential Geom. **27** (1988) 423–460
58. Ikeda, A.: Riemannian manifolds p-isospectral but not $(p+1)$-isospectral. In: Geometry of manifolds. Acad. Press, New York 1989
59. Itoh, M.: Moduli of half conformally flat structures. Math. Ann. (to appear)
60. Katok, A.: Four applications of conformal equivalence to geometry and dynamics. J. Ergodic Th. Dynamical Systems **8** (1988) 215–240
61. Katok, A., Spatzier, R.: Differential rigidity of hyperbolic Abelian actions. Preprint, Math. Sci. Res. Inst., Berkeley 1992
62. King, A.D., Kotschick, D.: The deformation theory of anti-self-dual conformal structures. Math. Ann. **294** (1992) 591-609
63. Knieper, G.: Spherical means on compact Riemannian manifolds of negative curvature. Habilitationsschrift, Institut für Mathematik, Augsburg 1992
64. Koiso, N.: Rigidity and infinitesimal deformability of Einstein metrics. Osaka J. Math. **29** (1982) 643–668
65. Lawson, H.B.Jr.: The theory of gauge fields in four dimensions. C.B.M.S. Regional Conference Series **58** (1985)
66. Ledrappier, F.: Harmonic measures and Bowen-Margulis measures. Israel J. Math. **71** (1990) 275–288
67. Lyons, T., Sullivan, D.: Function theory, random paths and covering spaces. J. Differential Geom. **19** (1984) 299–323
68. Margerin, Ch.: Pour une théorie de la déformation en géométrie métrique: Chap. 1: Intégrabilité locale; Chap. 2: Intégrabilité globale; Chap. 4: Un résultat optimal en dimension 4; Chap. 5: Un résultat optimal asymptotique. Preprint, Ecole Polytechnique, 1992
69. Margerin, Ch.: Pointwise-pinched manifolds are space forms. Proc. Amer. Math. Soc. Symp. Pure Math. (Arcata 1984) **44** (1986) 307–328
70. Margerin, Ch.: A sharp theorem for weakly-pinched 4-manifolds. C.R. Acad. Sci. Paris **303** (1986) 877–880
71. Margerin, Ch.: Fibrés stables et métriques d'Hermite-Einstein. Séminaire Bourbaki, Exposé 683. Astérisque **152–153** (1987) 263–283
72. Margulis, G.A.: On some applications of ergodic theory to the study of manifolds of negative curvature. Funkt. Anal. i Prilozhen. **3** (1969) 89–90
73. McKean, H.: Selberg's trace formula as applied to a compact Riemann surface. Commun. Pure Appli. Math. **25** (1972) 225–246
74. Min Oo, M., Ruh , E.: Curvature deformations. In: Curvature and topology of Riemannian manifolds. Lectures Notes in Mathematics, vol. 1201. Springer (1986) 180–190
75. Min Oo, M., Ruh, E.: L^2-curvature pinching. Comment. Math. Helvetici **65** (1990) 36–51
76. Mok, N.: The uniformization theorem for compact Kähler manifolds of non-negative holomorphic bisectional curvature. J. Differential Geom. **27** (1988) 179–214
77. Nadirashvili, N.S.: Multiple eigenvalues of the Laplace operator. Math. USSR Sbornik **61** (1988) 225–238
78. Nishikawa, S.: Deformation of Riemannian manifolds. Proc. Amer. Math. Soc. Symp. Pure Math. (Arcata 1984) **44** (1986) 343–352
79. Osgood, B., Phillips, R., Sarnak, P.: Compact isospectral sets of surfaces. J. Functional Anal. **80** (1988) 212–234

80. Ouyang, H., Pesce, H.: Déformations isospectrales sur les nilvariétés de rang deux. C.R. Acad. Sci. Paris **314** (1992) 621–623
81. Pansu, P.: Le flot géodésique des variétés riemanniennes à courbure négative. Séminaire Bourbaki, Exp. 738, Astérisque **202** (1991) 269–298
82. Penner, R.C.: Calculus on moduli spaces. In: Geometry of group representations. Contemporary Math., Amer. Math. Soc., Providence, **74** (1988) 277–293
83. Prat, J.J.: Etude et convergence angulaire du mouvement brownien sur une variété à courbure négative. C.R. Acad. Sci. Paris **290** (1975) 1539–1542
84. Séminaire Palaiseau: Géométrie des surfaces K3: Modules et périodes. Astérisque **126**, 1983
85. Shi, W.X.: Deforming the metric on complete Riemannian manifolds. J. Differential Geom. **30** (1989) 223–301
86. Simpson, C.T.: Constructing variations of Hodge structures using Yang-Mills theory and applications to uniformization. J. Amer. Math. Soc. **1** (1988) 867–918
87. Sunada, T.: Riemannian coverings and isospectral manifolds. Ann. Math. **121** (1985) 169–186
88. Sunada, T.: Fundamental groups and laplacians. In: Geometry and analysis on manifolds Katata/Kyoto. Lecture Notes in Mathematics, vol. 1339. Springer (1987) 248–277
89. Taubes, C.H.: Self-dual Yang-Mills connections on non selfdual 4-manifolds. J. Differential Geom. **17** (1982) 139–170
90. Taubes, C.H.: Self-dual connections on manifolds with indefinite intersection matrix. J. Differential Geom. **19** (1984) 345–391
91. Taubes, C.H.: Long range forces and topology of instanton moduli spaces. In: Colloque en l'honneur de Laurent Schwartz. Astérisque **132** (1985) 243–255
92. Taubes, C.H.: The stable topology of moduli spaces. J. Differential Geom. **29** (1989) 163–230
93. Taubes, C.H.: Casson's invariant and gauge theory. J. Differential Geom. **31** (1990) 547–599
94. Taubes, C.H.: The existence of anti-self-dual conformal structures. J. Differential Geom. **36** (1992) 163-253
95. Thorbergsson, G.: Isoparametric foliations and their buildings. Ann. Math. **133** (1991) 429–446
96. Tian, G., Yau, S.T.: Kähler-Einstein metrics on complex surfaces with $c_1 > 0$. Commun. Math. Phys. **42** (1987) 175–203
97. Tromba, A.J.: Teichmüller theory in Riemannian geometry. Lect. in Math. ETH Zürich. Birkhäuser, Basel 1992
98. Wolpert, S.: The eigenvalue spectrum as moduli for compact Riemann surfaces. Bull. Amer. Math. Soc. **83** (1977) 1306–1308
99. Yang, D.: L^p-pinching and compactness theorems for compact Riemannian manifolds on curvature: I. Ann. Sci. Ecole Norm. Sup. Paris **25** (1992) 77–105; idem: II, ibidem, 179–199

Mathematical Aspects
of Computer Aided Geometric Design

Wolfgang Boehm, Josef Hoschek, and Hans-Peter Seidel

1. Introduction

Computer Aided Geometric Design (CAGD) is concerned with the design, computation, and representation of curved objects on a computer. Therefore, not surprisingly, CAGD has traditionally had strong ties to some classical mathematical disciplines such as approximation theory (approximation by polynomial and piecewise polynomial functions), differential geometry (parametric surfaces), algebraic geometry (algebraic surfaces), functional analysis and differential equations (surface design by minimizing functionals), and numerical analysis. In addition, work in CAGD also requires a solid background in computer science.

Since 1982 we have had the opportunity to organize conferences on CAGD at the Oberwolfach Institute. The discussions and the "thinking together" on topics such as knot insertion, blossoming, continuity conditions, and on applications in medicine (CT reconstruction) and in technique (ship and carbody design), have greatly influenced further work in these areas and have set the course for several important new developments.

When representing shape on a computer there are various questions that have to be addressed first [52]:

Continuous or Discrete? Stimulated by the success of raster graphics, where an image is simply considered as the union of a large number of squares (pixels), and supported by cheap memory, advances in graphics hardware, and progress in image processing, the discrete representation of 3D objects as the union of a large number of cubes (voxels) is now common in areas such as medical imaging, geographical modeling, and visualization. However, whenever the exact representation of curved surfaces is required, a voxel-based approach is insufficient. Therefore, in almost all technical applications, a continuous representation of shape is required.

One Piece or Many? We might choose to model the entire shape S via a single system of equations. That works fine for very simple shapes. But designers of more complicated shapes want to be able to modify one region of their design without affecting the rest of it. The usual way to make this possible is to break up the shape S into pieces. We model each piece by its own system of equations, and we guarantee that the pieces join smoothly by enforcing certain constraints on the systems of equations that determine the joining pieces. The word *spline*, in its most general sense, means a piecewise model of a shape in which the pieces have been constrained to join with some level of smoothness.

Parametric or Implicit? Classical differential geometry offers two standard approaches to construct smooth curves and surfaces. A *parametric surface S* is the image of an open set $U \subset \mathbb{R}^2$ under a differentiable regular map while an *implicit surface* $S = F^{-1}(c)$ is the counter image of a regular value $c \in \mathbb{R}$ under a differentiable map $F : \mathbb{R}^3 \to \mathbb{R}$. Note that both the parametric and implicit model of shape encode some extra information, over and above the shape S itself. For parametric surfaces it is easy, e.g., to evaluate a point on the surface, while for objects defined by a collection of implicit surfaces it is easy to decide if a given point lies inside, outside, or on the boundary of the object. Both types of extra information sometimes come in handy.

What Class of Functions? Another decision that faces us is what class of functions to select from when choosing a function to model our shape S, either parametrically or implicitly. The simplest functions are *polynomial functions*. Both the parametric model $G(u) = (u, u^2)$ and the implicit model $H(x, y) = y - x^2$ of the parabola are polynomials. Note that polynomials can be built up using only addition, subtraction, and multiplication.

If we add division to the set of legal operations, we get *rational functions*. Polynomial and rational functions are currently the main building blocks for the representation of shape within CAGD. They are easy to handle and store while at the same time providing sufficient flexibility for the representation of complex shapes. While algebraic, analytic, or C^∞-functions are also possible in principle, they haven't been used extensively so far.

2. Polynomials and Polar Forms

Perhaps the simplest class of curves and surfaces is the class of parametric polynomials. In the context of CAGD these curves and surfaces are best studied with the help of a classical mathematical tool, the polar form [30, 94, 104]: Given a polynomial $F : \mathbb{R}^s \to \mathbb{R}^t$ of degree n the corresponding (multiaffine) polar form is defined as the unique symmetric multiaffine map $f : (\mathbb{R}^s)^n \to \mathbb{R}^t$ satisfying $f(u, ..., u) = F(u)$. For a monomial

$$F(u) = \sum_{i=0}^{n} \mathbf{a}_i u^i \, ,$$

e.g., the corresponding polar form is given as

$$f(u_1, \dots, u_n) = \sum_{i=0}^{n} \mathbf{a}_i \binom{n}{i}^{-1} \sum_{\substack{S \subseteq \{1,\dots,n\} \\ |S|=i}} \prod_{j \in S} u_j \, .$$

A coordinate-free formula for the polar form is [94]

$$f(u_1, \ldots, u_n) = \frac{1}{n!} \sum_{\substack{S \subseteq \{1, \ldots, n\} \\ |S| = i}} (-1)^{n-i} i^n F\left(\frac{1}{i} \sum_{j \in S} u_j\right).$$

A more geometric interpretation of $f(u_1, \ldots, u_n)$ can be given as the intersection of the $(n - \mu)$-dimensional osculating flats to F at $u_1, \ldots, u_n$, where μ_i denotes the multiplicity of u_i. Among other things, polar forms make it easy to phrase smoothness conditions between adjacent polynomial segments or patches: Two parametric polynomials F and G are C^q-continuous at a parameter u iff their polar forms agree on all argument bags that contain at least $(n - q)$ many u's, i.e., $f(u^{n-q} u_1, \ldots, u_q) = g(u^{n-q} u_1, \ldots, u_q)$ for $u_1, \ldots, u_q \in \mathbb{R}^s$.

Polar forms allow for a particularly simple treatment of Bézier and B-spline curves and surfaces [30, 94, 102], arguably the most important curves and surfaces in CAGD, and have been helpful in developing new curve and surface schemes [23, 89 103].

3. Bézier Curves

Polynomial representations have been studied extensively throughout the mathematical literature. Numerical analysis, e.g., has traditionally used orthogonal polynomials as well as the monomial, Lagrange, or Hermite basis. By far the most important representation for CAGD applications is the Bernstein Bézier representation

$$F(u) = \sum_{i=0}^{n} B_i^{\Delta,n}(u) \mathbf{b}_i, \quad \mathbf{b}_i \in \mathbb{R}^t,$$

where

$$B_i^{\Delta,n}(u) = \binom{n}{i} \left(\frac{u - s}{t - s}\right)^i \left(\frac{t - u}{t - s}\right)^{n-i}, \quad i = 0, \ldots, n$$

are the *Bernstein polynomials* w.r.t. the interval $\Delta = [s, t]$. We remark that the Bernstein polynomials are positive on $[s, t]$, form a partition of unity, and satisfy the recursion $B_i^{\Delta,n}(u) = \frac{u-s}{t-s} B_{i-1}^{\Delta,n-1}(u) + \frac{t-u}{t-s} B_i^{\Delta,n-1}(u)$. The points $\mathbf{b}_i \in \mathbb{R}^t$ are the control or Bézier points. They form the control or Bézier polygon. In a CAGD context these curves have first been used by Paul de Faget de Casteljau at Citroen and by Pierre Bézier at Renault about some 30 years ago.

The importance of Bézier curves within CAGD stems from the following two facts: First, the shape of the curve closely mimics the shape of the control polygon, i.e., the coefficients have geometric significance. Secondly, the Bernstein basis is extremely stable: As recently shown by Farouki and Rajan [40], the condition number of the simple roots of a polynomial is smaller in the Bernstein basis than in almost any other basis.

Bézier curves satisfy the following shape properties [9, 13, 39, 41, 64] (see Fig. 1):

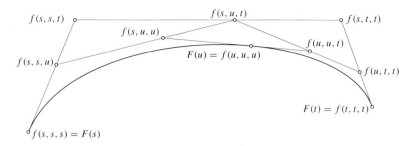

Fig. 1. The de Casteljau Algorithm in the case $n = 3$

Convex Hull Property: A Bézier curve is contained in the closed convex hull of its Bézier polygon.

Endpoint Interpolation: A Bézier curve interpolates the end points of its control polygon and is tangent to the control polygon there. A similar statement holds for higher order derivatives.

Variation Diminishing Property: The number of intersection points of a Bézier curve with an affine hyperplane H is bounded by the number of intersection points of H with the control polygon. Intuitively this means that a Bézier curve doesn't wiggle more than its control polygon.

Affine Invariance: The relationship between a Bézier curve and its control polygon is invariant under affine transformations.

An algorithm for the stable evaluation of Bézier curves is the recursive *de Casteljau Algorithm*

$$\mathbf{b}_i^0(u) = \mathbf{b}_i, \quad i = 0, \dots, n$$

$$\mathbf{b}_i^l(u) = \frac{t-u}{t-s}\,\mathbf{b}_i^{l-1}(u) + \frac{u-s}{t-s}\,\mathbf{b}_{i+1}^{l-1}(u), \quad l = 1, \dots, n, \ i = 0, \dots, n-l.$$

which produces a triangular array of points $\mathbf{b}_i^r(u) \in \mathbb{R}^t$ and in particular computes the curve value

$$\mathbf{b}_0^n(u) = F(u).$$

Note that the de Casteljau Algorithm uses convex combinations throughout and hence is very stable. Using polar forms, a closed form solution to the de Casteljau recursion is given by [94]

$$\mathbf{b}_i^l = f(\underbrace{s, \dots, s}_{n-l-i}, \underbrace{u, \dots, u}_{l}, \underbrace{t, \dots, t}_{i}) = f(s^{n-l-i}\, u^l\, t^i).$$

This follows from the fact that

$$F(u) = f(u^n) = \frac{t-u}{t-s} f(u^{n-1}\ s) + \frac{u-s}{t-s} f(u^{n-1}\ t) = \dots$$

$$= \sum_{i=0}^{n} B_j^{\Delta,n}(u)\ f(s^{n-i}\ t^i),$$

i.e.

$$\mathbf{b}_i = f(s^{n-i}\ t^i),$$

and from

$$f(s^{n-l-i}\ u^l\ t^i) = \frac{t-u}{t-s} f(s^{n-l-i+1}\ u^{l-1}\ t^i) + \frac{u-s}{t-s} f(s^{n-l-i}\ u^{l-1}\ t^{i+1}).$$

In particular, $\mathbf{b}_0^n(u) = f(u^n) = F(u)$.

A closer look at the above equations demonstrates that the de Casteljau Algorithm produces much more than just evaluation. In particular, it allows to *subdivide* a given Bézier curve at an arbitrary parameter u. According to the formula above, the Bézier points of the left curve segment (w.r.t. $[s, u]$) are then given by

$$\mathbf{b}_i^{left} = f(s^{n-i}\ u^i) = \mathbf{b}_0^i(u),$$

while the Bézier points of the right curve segment (w.r.t. $[u, t]$) are given by

$$\mathbf{b}_i^{right} = f(u^{n-i}\ t^i) = \mathbf{b}_i^{n-i}(u).$$

Note that these points appear along the sides of the de Casteljau triangle. Bézier subdivision exhibits quadratic convergence [21] and thus can be used to draw the curve [68]. Another side effect of the de Casteljau Algorithm is the computation of the tangent $T_u F$ which is given by the points $f(s\ u^{n-1})$ and $f(u^{n-1}\ t)$ at the $(n-1)$-st level of the algorithm.

Perhaps the most surprising property of the de Casteljau Algorithm, however, is its symmetry: If we modify the above de Casteljau recursion to

$$\mathbf{b}_i^0(.) = \mathbf{b}_i,$$

$$\mathbf{b}_i^r(u_1 \cdots u_r) = \frac{t-u_r}{t-s} \mathbf{b}_i^{r-1}(u_1 \cdots u_{r-1}) + \frac{u_r - s}{t-s} \mathbf{b}_{i+1}^{r-1}(u_1 \cdots u_{r-1}),$$

the result $\mathbf{b}_0^n(u_1 \cdots u_n)$ is symmetric in $u_1, \dots, u_n$. Therefore $\mathbf{b}_0^n(u_1 \cdots u_n) = f(u_1 \cdots u_n)$ is the polar form of F [30, 94].

We conclude this section with a brief discussion of how to join adjacent Bézier curve segments to produce an overall smooth curve. Consider two Bézier curves $F(u) = \sum_i B_i^{\Lambda,n}(u)\mathbf{b}_i$ over $\Lambda = [r, s]$ and $G(u) = \sum_i B_i^{\Delta,n}(u)\mathbf{c}_i$ over $\Delta = [s, t]$. The C^q-conditions from Section 2 then translate into

$$\mathbf{c}_i = f(s^{n-i}\ t^i) = \mathbf{b}_{n-i}^i(t), \quad i = 0, \dots, q,$$

with $\mathbf{b}_{n-i}^i(t)$ computed by running the de Casteljau Algorithm for F at $u = t$. For C^2-continuous cubics this translates, e.g., into the requirement that the Bézier points of F and G form a so-called A-frame with the additional requirement that all the ratios along this A-frame are equal to the ratio $(t-s) : (s-r)$ of the corresponding parameter intervals (see Fig. 4).

4. B-Splines

Using the results of the preceding section, it is possible (and has actually been done in practice), to build a CAD package based on Bézier curves. However, this approach requires to explicitly keeping track of all continuity constraints. A better solution is to eliminate these continuity constraints once and for all, and to choose a basis where the necessary constraints are already built in. For the class of piecewise polynomial curves, *B-splines* provide such a basis [26, 98].

A B-spline curve of degree n over a non-decreasing knot sequence $T = (t_i)_{i \in \mathbb{Z}}$ with $t_i < t_{i+n+1}$ is defined as

$$F(u) = \sum_i N_i^n(u) \, \mathbf{d}_i \,, \quad \mathbf{d}_i \in \mathbb{R}^t \,,$$

where N_i^n are the *normalized B-splines* over T, defined recursively as

$$N_i^0(u) = \begin{cases} 1 & u \in [0, 1] \\ 0 & \text{otherwise} \end{cases}$$

and

$$N_i^l(u) = \frac{u - t_i}{t_{i+l} - t_i} N_i^{l-1}(u) + \frac{t_{i+l} - u}{t_{i+l+1} - t_{i+1}} N_{i+1}^{l-1}(u).$$

The points $\mathbf{d}_i \in \mathbb{R}^t$ are the control or de Boor points.

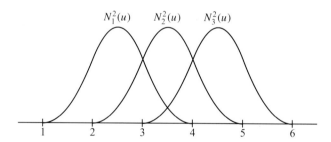

Fig. 2. Quadratic B-splines over a uniform knot vector

The normalized B-splines are positive and form a global partition of unity. In addition, they have local support ($N_i^n(u) = 0$ for $u \notin [t_i, t_{i+n+1})$), and it can be shown that they are $C^{n-\mu}$-continuous at a knot of multiplicity μ. Hence, in the case of single knots, B-splines are C^{n-1}-continuous everywhere.

B-spline curves have similar shape properties as Bézier curves [9, 13, 39, 64].

Convex Hull Property: A B-spline curve is contained in the convex hull of its de Boor points. Moreover, if $u \in [t_j, t_{j+1})$, then $F(u) \subset [\mathbf{d}_{j-n}, \dots, \mathbf{d}_j]$ (local convex hull property).

Multiple Control Points: If n control points $\mathbf{d}_{j-n+1} = \dots = \mathbf{d}_j = \mathbf{d}$ coincide, then $F(t_{j+n}) = \mathbf{d}$, i.e., the curve interpolates to this point and is tangent to the control polyton there.

Collinear Control Points: If $n + 1$ control points $\mathbf{d}_{j-n}, \ldots, \mathbf{d}_j$ lie on a line L, then $F([t_j, t_{j+1}]) \subseteq L$, i.e., the curve contains a line segment.

Multiple Knots: If n knots $t_{j+1} = \ldots = t_{j+n} =: t$ coincide, then $F(t) = \mathbf{d}_j$, i.e., the curve interpolates to this point and is tangent to the control polygon there.

Variation Diminishing Property: The number of intersection points of a B-spline curve with an affine hyperplane H is bounded by the number of intersection points of H with the control polygon. Intuitively this means that a B-spline curve doesn't wiggle more than its control polygon.

Affine Invariance: The relationship between a B-spline curve and its control polygon is invariant under affine transformations.

We also remark that Bézier curves are a special case of B-splines: For the special knot vector

$$T = (\underbrace{s, \ldots, s}_{n} < \underbrace{t, \ldots, t}_{n})$$

the normalized B-splines N_i^n over T coincide with the Bernstein polynomials $B_i^{\Delta,n}$ w.r.t. $\Delta = [s, t]$. Therefore, piecewise Bézier curves may be considered as B-spline curves with n-fold knots.

B-splines can be evaluated recursively by the de Boor Algorithm [26]: For $u \in [t_j, t_{j+1})$ we recursively compute

$$\mathbf{d}_i^0(u) = \mathbf{d}_i, \quad i = j - n, \ldots, j$$

$$\mathbf{d}_i^l(u) = \frac{t_{i+n+1} - u}{t_{i+n+1} - t_{i+l}} \, \mathbf{d}_i^{l-1}(u) + \frac{u - t_{i+l}}{t_{i+n+1} - t_{i+l}} \, \mathbf{d}_{i+1}^{l-1}(u),$$

$$r = 1, \ldots, n, \quad i = j - n, \ldots, j - l,$$

thus again obtaining a triangular array of points $\mathbf{d}_i^r(u) \in \mathbb{R}^l$, and in particular computing the curve value

$$\mathbf{d}_{j-n}^n(u) = F(u).$$

This follows from the fact that, similar to the Bézier case, a closed form solution to the de Boor recursion is given by [94]

$$\mathbf{d}_i^l(u) = f_j(t_{i+l+1} \cdots t_{i+n} \, u^l)$$

where f_l is the polar form of the restriction F_j of F to the segment $[t_j, t_{j+1})$. In particular, $\mathbf{d}_{j-n}^n(u) = f_j(u^n) = F(u)$, and

$$\mathbf{d}_i = f_j(t_{i+1} \cdots t_{i+n}).$$

Note, that this equation yields a dual functional for B-splines. Among other things this leads to a simple proof of the Curry-Schoenberg Theorem [26, 98] that every piecewise polynomial can be represented as linear combination of B-splines over a suitably chosen knot vector.

Even more important than evaluation is knot insertion. The problem is as follows: Given a B-spline curve $F(u) = \sum_i N_i^n(u)\mathbf{d}_i$ over a knot vector $T = (t_i)_{i \in \mathbb{Z}}$

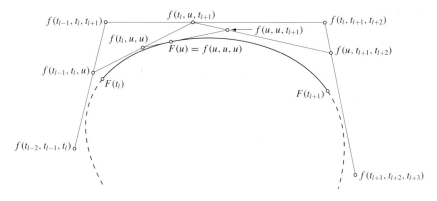

Fig. 3. Knot insertion and the de Boor Algorithm for the case $n = 3$

plus a new knot t, say $t_j \leq t < t_{j+1}$, compute the coefficients $\mathbf{d}_i^*$ of the B-spline representation of F over the refined knot sequence $\hat{T}$ that is obtained from T by inserting the new knot t. Writing t as an affine combination of t_i and t_{i+n},

$$t = \frac{t_{i+n} - t}{t_{i+n} - t_i} t_i + \frac{t - t_i}{t_{i+n} - t_i} t_{i+n} \,,$$

the above formula for the B-spline coefficients yields [102]

$$\begin{aligned}
\mathbf{d}_i^* &= f_j(t_{i+1} \cdots t_j \ t \ t_{j+1} \cdots t_{i+n-1}) \\
&= (1 - \frac{t - t_i}{t_{i+n} - t_i}) \ f_j(t_i \cdots t_j \ t_{j+1} \cdots t_{i+n-1}) \\
&\quad + \frac{t - t_i}{t_{i+n} - t_i} \ f_j(t_{i+1} \cdots t_j \ t_{j+1} \cdots t_{i+n}) \\
&= \frac{t_{i+n} - t}{t_{i+n} - t_i} \mathbf{d}_{i-1} + \frac{t - t_i}{t_{i+n} - t_i} \mathbf{d}_i \,.
\end{aligned}$$

This is exactly the Boehm Algorithm [10]. Comparison of the Boehm Algorithm with the de Boor Algorithm also shows that evaluation can be thought of as multiple knot insertion. Other knot insertion algorithms are discussed in [6, 19].

The opposite to knot insertion is knot deletion. Knot deletion is used for the approximation of a given B-spline curve by a new curve with fewer segments. Knot deletion is studied in detail by Lyche and Morken in [71]. Hoschek's approach to approximation uses the parametrization as a design parameter: The shape of an approximation curve will be changed if the parameter values of the approximation points are changed during the approximation process. Best approximations are considered in [31].

5. Geometric Continuity for Curves

So far we have focused almost exlusively on parametric continuity, i.e., continuity of a given parametrization. However, instead of looking at the continuity of a specific parametrization we have to look at the continuity of the shape of the curve which is independent under possible reparametrization. Thus we say that two curves join with *geometric continuity*, if the curves are parametrically continuous under reparametrization. In other words, two curves F and G are G^k-continuous at a parameter s if there exists a reparametrization φ (which preserves the given orientation of the curve) such that $F(\varphi(u))$ and $G(u)$ are C^k-continuous at s. Using the chain rule of differentiation it is easily seen that G^k-continuity is equivalent to the well-known β-constraints [7]

$$G'(s) = \beta_1 F'(s)$$
$$G''(s) = \beta_1^2 F''(s) + \beta_2 F'(s)$$
$$G'''(s) = \beta_1^3 F'''(s) + 3\beta_1 \beta_2 F''(s) + \beta_3 F'(s)$$

$$\vdots$$

$$G^{(k)}(s) = \sum_{j=0}^{k} \sum_{\substack{i_1 + i_2 + .. + i_k = j \\ i_1 + 2i_2 + .. + ki_k = k}} \frac{k!}{i_1!(1!)^{i_1} \cdots i_k!(k!)^{i_k}} \beta_1^{i_1} \cdots \beta_k^{i_k} F^{(j)}(s).$$

where $\beta_k = \varphi^{(k)}(s)$, and $\beta_1 > 0$ since φ preserves the given orientation. The parameters β_i are also called *shape parameters*. We pause to remark that G^1-continuity is equivalent to a continuous unit tangent, while G^2-continuity is equivalent to a continuous curvature vector.

How can we exploit the conditions above to join two Bézier curves with a given order of geometric continuity? Consider two Bézier curves $F(u) = \sum_{i=0}^{n} B_i^{A,n}(u)\mathbf{b}_i$ over $A = [r, s]$ and $G(u) = \sum_{i=0}^{n} B_i^{\Delta,n}(u)\mathbf{c}_i$ over $\Delta = [s, t]$. In order to join F and G with G^1-continuity, it is necessary and sufficient that the three Bézier points $\mathbf{b}_{n-1}$, $\mathbf{b}_n = \mathbf{c}_0$, $\mathbf{c}_1$ are collinear and satisfy the ratios

$$\overline{\mathbf{b}_{n-1}\mathbf{b}_n} : \overline{\mathbf{c}_0\mathbf{c}_1} = (s - r) : \beta_1(t - s).$$

Things become slightly more complicated for G^2-continuity (see Fig. 4). It can be shown [11] that F and G are G^2-continuous at s precisely if the two lines $\mathbf{b}_{n-2}\mathbf{b}_{n-1}$ and $\mathbf{c}_1\mathbf{c}_2$ intersect in a common point $\mathbf{d}$ (i.e. are coplanar and form a so-called A-frame) and satisfy the ratios

$$\overline{\mathbf{b}_{n-2}\mathbf{b}_{n-1}} : \overline{\mathbf{b}_{n-1}\mathbf{d}} = (s - r) : \gamma_1(t - s) \quad \text{and} \quad \overline{\mathbf{d}\mathbf{c}_1} : \overline{\mathbf{c}_1\mathbf{c}_2} = (s - r) : \gamma_2(t - s)$$

for parameters γ_1, γ_2 with $\gamma_1\gamma_2 = \beta_1^2$. Thus instead of directly using the shape parameters β_1 and β_2, it is sometimes advantageous to use the shape parameters β_1 and γ_1 instead [11, 8]. Note that for $\beta_1 = \gamma_1 = \gamma_2 = 1$ we obtain the special case of parametric continuity as discussed in Section 3. Generalizations to higher

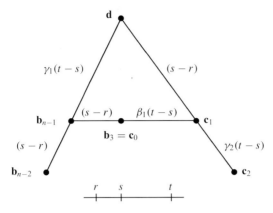

Fig. 4. A-frame construction for a G^2-joint between two cubic Bézier curves

degree and higher order of continuity are straightforward, although the resulting conditions become quite messy.

If the β-constraints at the different knots are fixed it is again possible to construct locally supported basis functions that have the required continuity constraints built in [7, 11, 49, 77]. The resulting β-splines, sometimes also called γ-splines, if the A-frame construction above is used, are a direct generalization of B-splines and many properties carry over from B-splines to β-splines almost word by word.

6. Rational Curves

Rational curves are important for two reasons: First, they allow the exact representation of conic sections, and second, they come up naturally when projecting a 3D polynomial curve onto the screen. A rational Bézier curve is given as [39, 42, 86]

$$F(u) = \frac{\sum_{i=0}^{n} B_i^n(u) w_i \mathbf{b}_i}{\sum_{i=0}^{n} B_i^n(u) w_i}, \quad w_i \in \mathbb{R}, \quad \mathbf{b}_i \in \mathbb{R}^l.$$

The points $\mathbf{b}_i \in \mathbb{R}^l$ are again called Bézier points and the scalars w_i are called *weights*. If all weights are equal to one we obtain non-rational Bézier curves as discussed in Section 3.

A geometric interpretation is to consider F as the projection of a polynomial curve in $\mathbb{R}^{l+1}$ onto the affine hyperplane $w = 1$. This interpretation makes it obvious that many properties and algorithms (evaluation, subdivision, computation of tangents, etc.) carry over from polynomial to rational Bézier curves with positive weights almost word by word.

It is easy to see that the shape of a rational Bézier curve is invariant under a change of weights

$$\hat{w}_i = \alpha^i w_i \quad \text{with } \alpha = \sqrt[n]{\frac{w_0}{w_n}}.$$

This change of weights results in $\hat{w}_0 = \hat{w}_n = w_0$, and after dividing all weights by w_0 we obtain a curve in *standard form* $w_0 = w_n = 1$.

116   Wolfgang Boehm, Josef Hoschek, and Hans-Peter Seidel

If rational Bézier curves are to be used in a design context it may be best to get rid of the weights altogether, since the effect of manipulating the weights on the shape of the curve is generally hard to predict. A preferable alternative to the weights are the weight points (Farin points)

$$\mathbf{q}_i = \frac{w_i \mathbf{b}_i + w_{i+1} \mathbf{b}_{i+1}}{w_i + w_{i+1}}.$$

The weight points lie on the edge of the given control polygon and changing a weight point has a fairly predictable effect on the curve. In combination with the cross ratio, weight points also simplify the construction of G^2-curves [88].

Of particular interest are rational quadratics

$$F(u) = \frac{B_0^2(u)\mathbf{b}_0 + B_1^2(u)w_1\mathbf{b}_1 + B_2^2(u)\mathbf{b}_2}{B_0^2(u) + w_1 B_1^2(u) + B_2^2(u)}.$$

Using barycentric coordinates λ_i of $F(u)$ w.r.t. $\mathbf{b}_i$, it is easy to see that F satisfies the implicit equation

$$\lambda_1^2 - 4w_1^2 \lambda_0 \lambda_2 = 0$$

which defines an algebraic curve of degree 2, i.e., a conic. Furthermore, this equation demonstrates that the two rational curves F and F^c (which is obtained from F by replacing the weight w_1 by $-w_1$) actually define two complementary segments of the same conic. Moreover, it is easily verified that the points $F(u)$, $F^c(u)$, and $\mathbf{b}_1$ are always collinear.

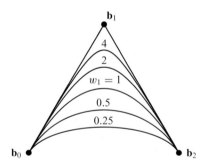

Fig. 5. Effect of the weight w_1 on the shape of a rational quadratic

The implicit equation of F also demonstrates that the affine type of the conic (ellipse, parabola, hyperbola) only depends on the weight w_1. Since the affine type of the conic is determined by the number of zeroes in the denominator of a rational quadratic, we get an ellipse for $w_1 < 1$, a parabola for $w_1 = 1$, and a hyperbola for $w_1 > 1$ (see Fig. 5).

Let us now turn to the opposite question of parametrizing part of a given conic by a rational Bézier curve. Let $\mathbf{b}_0$, $\mathbf{b}_2$ be two given points on the conic, and let $\mathbf{b}_1$ be the intersection of the corresponding tangents. We then obtain a rational parameterization by simply taking these points as Bézier points and by adjusting the weight w_1 such as to satisfy the above implicit equation. If the intersection point $\mathbf{b}_1$ is not allowed to lie at infinity, and if the weights w_1 are required to be positive, this procedure requires at least three segments to parametrize the whole conic. Potential benefits of using control points at infinity are discussed by Piegl.

A simple way to produce smooth piecewise rational curves is to extend the defining function from Bézier curves to B-splines, i.e., to consider projections of B-spline curves in $\mathbb{R}^{t+1}$ onto the affine hyperplane $w = 1$ [86]. This yields non uniform rational B-splines (NURBS). While NURBS provide a unified format for both conics and splines, they do not exploit the potential flexibility of piecewise rational curves to its full extent. In order to do so, it is necessary to join rational curves with geometric continuity.

7. Tensor Product Surfaces

By far the most popular surfaces in computer aided geometric design are tensor product surfaces: Given a curve scheme $F(u) = \sum_{i=0}^{n} B_i(u)\,\mathbf{b}_i$, $\mathbf{b}_i \in \mathbb{R}^t$, the corresponding tensor product scheme is defined as [26]

$$F(u, v) = \sum_{i=0}^{n} \sum_{j=0}^{m} B_i(u)\,B_j(v)\,\mathbf{b}_{ij}\,, \quad \mathbf{b}_{ij} \in \mathbb{R}^t\,,$$

which can also be written as

$$F(u, v) = \sum_{i=0}^{n} B_i(u)\,\mathbf{b}_{iv} \quad \text{with} \quad \mathbf{b}_{iv} = \mathbf{b}_i(v) = \sum_{j=0}^{m} B_j(v)\,\mathbf{b}_{ij}\,.$$

The last equation demonstrates that tensor-product surfaces may be considered as curves of curves, and thus explains that we first have to understand curves in order to understand tensor-product surfaces.

Most relevant for CAGD applications are TP Bézier and TP B-spline surfaces and many properties (with the notable exception of the variation diminishing property) and algorithms carry over from the underlying curve scheme to TP surfaces almost word by word [9, 39, 64]. Intersection algorithms are discussed in [4] and [81].

One major problem with TP surfaces is that certain operations are no longer local. Perhaps the most important of these operations is knot insertion: Insertion of a new knot in one coordinate direction really introduces a new *knot line* and refines the surface everywhere along the whole line. Thus, if further refinement is desired only in isolated regions (e.g. in surface fitting, animation), simple knot insertion will not be very satisfactory.

A solution to this problem is provided by the construction of hierarchical surfaces [45]. A hierarchical surface consists of a layer of difference surfaces, each of these defined over a successively finer grid. The difference surfaces are zero almost everywhere and only contribute in areas where local refinement is being sought. An

118   Wolfgang Boehm, Josef Hoschek, and Hans-Peter Seidel

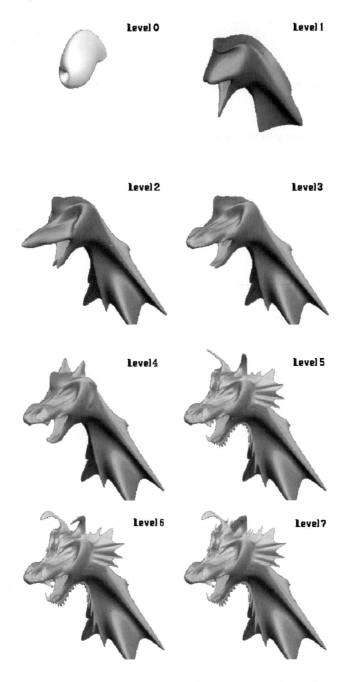

Fig. 6. Hierarchical B-splines: This sequence of images shows the various levels in the construction of a hierarchical B-spline surface. The final dragon head consists of 3088 bicubic patches and is fully specified by only 505 control points. (Courtesy D. Forsey, UBC)

example of a hierarchical B-spline surface is shown in Figure 6. A generalization of this technique (overlays) allows the addition of unaligned spline-based features upon a TP-surface that retain their surface characteristics even when the shape of the underlying surface is modified.

The concept of hierarchical splines is closely related to the concept of multi-resolution analysis and wavelets [18, 25, 73]. Let V_j, $j \in \mathbb{Z}$, be a sequence of closed subspaces of $L^2\mathbb{R}$, and let $\varphi \in V_0$. Then $((V_j)_j, \varphi)$ is a multiresolution analysis with scaling function φ if the following conditions are satisfied:

$$\ldots \subset V_{-2} \subset V_{-1} \subset V_0 \ldots \subset V_1 \subset V_2 \ldots$$

$$\overline{\bigcup_{j \in \mathbb{Z}} V_j} = L^2\mathbb{R} \quad \text{and} \quad \bigcap_{j \in \mathbb{Z}} V_j = \{0\}$$

$$F \in V_j \iff F(2^{-j} \cdot) \in V_0$$

$$F \in V_0 \iff F(\cdot - k) \in V_0$$

$\{\varphi(\cdot - k) | k \in \mathbb{Z}\}$ is an orthonormal basis for V_0.

In this situation the functions $\varphi_{j,k} = 2^{j/2}\varphi(2^j \cdot -k)$, $k \in \mathbb{Z}$ form an orthonormal basis for V_j.

Every multiresolution analysis $((V_j)_j, \varphi)$ allows the construction of wavelets. More precisely, there exists a function ψ, such that the wavelets $\psi_{j,k}$ defined by $\psi_{j,k}(x) = 2^{j/2}\psi(2^j x - k)$, $j, k \in \mathbb{Z}$, yield orthogonal decompositions

$$V_{j+1} = V_j \oplus W_j \quad \text{where} \quad W_j = clos_{L^2\mathbb{R}}\langle \psi_{j,k} \mid k \in \mathbb{Z}\rangle .$$

In particular we obtain the orthogonal decomposition $L^2\mathbb{R} = \bigoplus_j W_j$. In order to construct interesting examples one often starts with a scaling function φ that satisfies the refinement equation

$$\varphi(u) = \sum_{k \in \mathbb{Z}} c_k \varphi(2x - k) .$$

A very natural choice for φ are uniform B-splines, where the coefficients c_k of the refinement equation are given by knot insertion. The scaled functions $\varphi_{j,k}$ are uniform B-splines over a refined partition of size $h_j = 2^{-j}$. It is easily seen that B-splines satisfy the first four conditions for a multiresolution analysis, and although they are no longer orthonormal, B-splines still form a Riesz basis (equivalence of Norms in $l_2\mathbb{R}$ and $L^2\mathbb{R}$), so the existence of wavelets remains guaranteed.

8. Triangular Bézier Patches

Triangular Bézier patches have originally been introduced by de Casteljau using polar forms. They have since been studied widely and several excellent accounts on the subject are available.

Consider a triangle $\triangle = [\mathbf{t}_0, \mathbf{t}_1, \mathbf{t}_2]$ in $\mathbb{R}^2$. Given a point $\mathbf{u} \in \mathbb{R}^2$ there exist unique scalars $\lambda_0(\mathbf{u})$, $\lambda_1(\mathbf{u})$, and $\lambda_2(\mathbf{u})$ such that

$$\mathbf{u} = \sum_{i=0}^{2} \lambda_i(\mathbf{u})\mathbf{t}_i \quad \text{and} \quad \sum_{i=0}^{2} \lambda_i(\mathbf{u}) = 1 \,.$$

The coefficients $\lambda = (\lambda_0, \lambda_1, \lambda_2)$ are called *barycentric coordinates*. Given a triangle $\triangle$ the bivariate *Bernstein polynomials* $B_\beta^{\triangle,n}(\mathbf{u})$ w.r.t. $\triangle$ are defined as

$$B_\beta^{\triangle,n}(\mathbf{u}) = \binom{n}{\beta} \lambda_0^{\beta_0}(\mathbf{u})\lambda_1^{\beta_1}(\mathbf{u})\lambda_2^{\beta_2}(\mathbf{u}) \,, \quad |\beta| = \beta_0 + \beta_1 + \beta_2 = n \,.$$

Note that the Bernstein polynomials are positive on $\triangle$, form a partition of unity, and satisfy the recursion

$$B_\beta^{\triangle,n}(\mathbf{u}) = \sum_{i=0}^{2} \lambda_i(\mathbf{u}) B_{\beta-e^i}^{\triangle,n-1}(\mathbf{u}) \,.$$

In analogy to the curve case, the representation

$$F(\mathbf{u}) = \sum_{|\beta|=n} B_\beta^{\triangle,n}(\mathbf{u})\mathbf{b}_\beta \,, \quad \mathbf{b}_\beta \in \mathbb{R}^t$$

is then called the *Bernstein Bézier representation* of F. The coefficients $\mathbf{b}_i \in \mathbb{R}^t$ are the Bézier points and form the Bézier net. Similar to the TP case, many shape properties (again with the notable exception of the variation diminishing property) and algorithms carry over from the curve case almost word by word [38].

The de Casteljau Algorithm, e.g., now reads

$$\mathbf{b}_\beta^0(\mathbf{u}) = \mathbf{b}_\beta \,, \quad |\beta| = n$$

$$\mathbf{b}_\beta^l(\mathbf{u}) = \sum_{i=0}^{n} \lambda_i(\mathbf{u})\mathbf{b}_{\beta+e^i}^{l-1} \,, \quad l = 1, \dots, n \,, \quad |\beta| = n - l \,,$$

and exactly the same argument as in the curve case shows that

$$\mathbf{b}_\beta^l(\mathbf{u}) = f(t_0^{\beta_0} \ t_1^{\beta_1} \ t_2^{\beta_2} \ \mathbf{u}^l)$$

is a closed form solution [94]. In particular,

$$\mathbf{b}_\beta = f(t_0^{\beta_0} \ t_1^{\beta_1} \ t_2^{\beta_2})$$

are the Bézier points. Again, the de Casteljau Algorithm can be used for evaluation, subdivision, computing derivatives, and for computing the polar form (symmetry) [30, 48, 94].

Finally, the C^q-conditions between two adjacent Bézier patches $F = \sum_i B_\beta^{\Lambda,n}(\mathbf{u})\mathbf{b}_\beta$ over $\Lambda = [\mathbf{r}, \mathbf{s}, \mathbf{t}]$ and $G = \sum_i B_\beta^{\triangle,n}(\mathbf{u})\mathbf{c}_\beta$ over $\triangle = [\mathbf{r}, \mathbf{s}, \mathbf{t}]$ translate into [38]

$$\mathbf{c}_\beta = f(\mathbf{r}^{\beta_0}\ \mathbf{s}^{\beta_1}\ \mathbf{t}^{\beta_2}) = \mathbf{b}^{\beta_2}_{(\beta_0\beta_1 0)}(\mathbf{t})\,, \qquad \beta_2 = 0, \ldots, q$$

with $\mathbf{b}^{\beta_2}_{(\beta_0\beta_1\ 0)}(\mathbf{t})$ computed by running the de Casteljau Algorithm for F at $\mathbf{u} = \mathbf{t}$. In the special case $q = 1$ this means that the four Bézier points $f(\mathbf{r}^{\beta_0}\ \mathbf{s}^{\beta_1}\ \mathbf{q})$, $f(\mathbf{r}^{\beta_0+1}\ \mathbf{s}^{\beta_1}) = g(\mathbf{r}^{\beta_0+1}\ \mathbf{s}^{\beta_1})$, $f(\mathbf{r}^{\beta_0}\ \mathbf{s}^{\beta_1+1}) = g(\mathbf{r}^{\beta_0}\ \mathbf{s}^{\beta_1+1})$, $g(\mathbf{r}^{\beta_0}\ \mathbf{s}^{\beta_1}\ \mathbf{t})$ are coplanar and are an affine image of the domain configuration $\square(\mathbf{q}, \mathbf{r}, \mathbf{s}, \mathbf{t})$. A similar statement holds for the C^1-continuity of several triangular Bézier patches around a common vertex.

Triangular Bézier patches are frequently used for interpolation. We briefly discuss two split-schemes that are obtained by splitting a triangular Bézier patch into several 'mini-patches'.

The Clough-Tocher interpolant [105] is a C^1-continuous piecewise cubic that interpolates to C^1-data at the vertices of the macro-triangle and to a given cross-boundary derivative along each edge. The split is obtained by splitting the macro-triangle into three pieces along the centroid. The bounary Bézier points are determined by the interpolation conditions while the inner Bézier points are set to satisfy the desired C^1-conditions (even C^2-conditions) at the centroid.

The Powell-Sabin interpolant [90] is a C^1-continuous piecewise quadratic that interpolates to C^1-data at the vertices of the macro-triangle. In the generic case each macro triangle is split into six subtriangles by connecting every vertex and midpoint of an edge to the centroid. The Bézier points are set to interpolate the given data and to make the interpolant C^1-continuous throughout.

9. Multivariate B-Splines

In this section we are interested in the construction of smooth piecewise polynomial functions over arbitrary triangulations of the parameter plane. Given a triangulation T of $\mathbb{R}^2$ we are thus interested in the space $\Pi^q_{n,T}$ of all C^q-continuous piecewise polynomials of degree n over T [28]. Unfortunately, the structure of this space is quite complicated, and although there are several deep results for various special cases, even the dimension of $\Pi^q_{n,T}$ is unknown in general. For a recent survey on the dimension problem see [2].

If even the dimension of $\Pi^q_{n,T}$ is unknown, it will be hard to construct nice basis functions. What can be constructed, however, are functions spanning a subspace of $\Pi^q_{n,T}$ that is big enough to be useful in practical applications. A geometric way to construct smooth piecewise polynomial functions $M : \mathbb{R}^2 \to \mathbb{R}$ is by projecting a polyhedron $P \subset \mathbb{R}^t$ onto $\mathbb{R}^2$ and by defining $M(\mathbf{u})$ as the $(t-2)$-dimensional volume of the fibre $\pi^{-1}(\mathbf{u})$. This definition generalizes the geometric definition of univariate B-splines and hence the resulting functions are called *multivariate B-splines*. Depending on whether P is just any polyhedron, a box, or a simplex, the resulting multivariate B-splines are also called polyhedral, box, or simplex splines.

If the given triangulation T happens to be regular, then box splines are the natural choice. Box splines are the natural generalization of uniform B-splines and have a very rich structure. In particular, they have a stable recurrence and can be generated by subdivision [29]. In the CAGD context, box splines have first been considered by Sabin [97]. Computational aspects and algorithms for converting to piecewise Bézier representation through the use of masks have been given in [14]. Surface fitting with box splines is discussed in [20]. The first book completely devoted to box splines is [29].

If an irregular triangulation is given, then simplex splines have to be used. Simplex splines can be defined recursively as follows: Given the *knots* $\mathbf{t}_0, \ldots, \mathbf{t}_{n+2} \in \mathbb{R}^2$ one can show [74] that the recursion

$$M(.|\mathbf{t}_0, \mathbf{t}_1, \mathbf{t}_2) = \frac{\chi_{[\mathbf{t}_0, \mathbf{t}_1, \mathbf{t}_1)}}{\mathbf{d}(\mathbf{t}_0, \mathbf{t}_1, \mathbf{t}_1)}$$

$$M(\mathbf{u}|\mathbf{t}_0, \ldots, \mathbf{t}_{l+2}) = \sum_{i=0}^{l+2} \lambda_i M(\mathbf{u}|\mathbf{t}_0, \ldots, \hat{\mathbf{t}}_i, \ldots, \mathbf{t}_{l+2})$$

with $\mathbf{u} = \sum_i \lambda_i(\mathbf{u})\mathbf{t}_i$ and $\sum_i \lambda_i(\mathbf{u}) = 1$ is well-defined and yields a simplex spline M of degree n that is C^{n-1}-continuous if the knots are in general position.

The next problem is to construct a spline space. While this problem becomes rather trivial for box splines (taking translates), this problem is difficult for simplex splines: Given an arbitrary triangulation T, exactly what simplex splines should we consider?

A first solution to this problem has been given by Höllig [60] and by Dahmen/Micchelli [22], and a second solution has been obtained recently by Dahmen, Micchelli and Seidel [23]. Both constructions start with a triangulation T of $\mathbb{R}^2$ (or of a bounded domain). With every vertex $\mathbf{t}_i$ of T a *cloud of knots* $\mathbf{t}_{i,0}, \ldots, \mathbf{t}_{i,n}$ is assigned. Then a rule for selecting $\binom{n+2}{2}$ subsets of $n+3$ knots from the three clouds associated with a triangle is given. Each such subset then yields a simplex spline of degree n which is generically C^{n-1}. The linear span $\mathcal{E}_n$ of all these degree n simplex splines is then the spline space of interest. Note that both schemes produce splines over a refined partition T' of T.

The two schemes differ in the knot selection rule, in the ease of use, and in the class of surfaces that they are able to represent. The first scheme by Höllig and Dahmen/Micchelli allows to represent all polynomial surfaces, i.e., $\Pi_n \subset \mathcal{E}_n$, but the representation of arbitrary piecewise polynomials remained unsolved.

This defect is overcome by the new scheme [23]: The knot selection rule of the new scheme is based on the use of polar forms [103], and for a given triangle $\Delta = [\mathbf{t}_0, \mathbf{t}_1, \mathbf{t}_2)$ selects the simplex splines

$$M_\beta^{\Delta,n}(\mathbf{u}) = M(\mathbf{u}|V_\beta), \quad |\beta| = n$$

with

$$V_\beta = \{\mathbf{t}_{0,0}, \ldots, \mathbf{t}_{0,\beta_0}, \mathbf{t}_{1,0}, \ldots, \mathbf{t}_{1,\beta_1}, \mathbf{t}_{2,0}, \ldots, \mathbf{t}_{2,\beta_2}\}.$$

Furthermore, the new scheme not only allows the representation of polynomials, i.e., $\Pi_n \subset \mathcal{E}_n$, but also the representation of the much larger class of all C^{n-1}-continuous piecewise polynomials over T, i.e. $\Pi_{n,T}^{n-1} \subset \mathcal{E}_n$. Moreover, up to normalization, the coefficients in the resulting representation

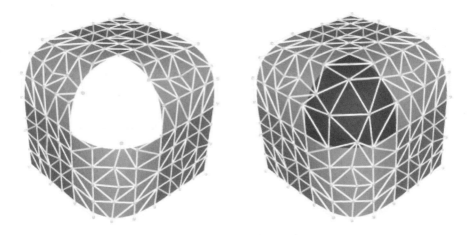

Fig. 7. Construction of a suitcase corner using multivariate B-splines: First, the given piece-wise polynomial surface around the hole is represented as linear combination of B-splines (left). This B-spline surface can then be extended to produce an overall C^{n-1}-continuous fill of the hole (right)

$$F(\mathbf{u}) = \sum_{\Delta, \beta} M_\beta^{\Delta, n}(\mathbf{u}) \mathbf{d}_\beta^\Delta$$

of a piecewise polynomial F as linear combination of simplex splines are given as

$$\mathbf{d}_\beta^\Delta = f_\Delta(\mathbf{t}_{0,0}, \dots, \mathbf{t}_{0,\beta_0-1}, \mathbf{t}_{1,0}, \dots, \mathbf{t}_{1,\beta_1-1}, \mathbf{t}_{2,0}, \dots, \mathbf{t}_{2,\beta_2-1})$$

by evaluating the polar form f_Δ of the restriction F_Δ of F to the triangle $\Delta = [\mathbf{t}_0, \mathbf{t}_1, \mathbf{t}_2]$ on a suitable sequence of knots. Note that this formula captures completely the analog formula for the de Boor points in the B-spline expansion of a univariate spline (Section 4), and also the formula for the Bézier points in the expansion of a polynomial surface as a triangular Bézier patch (Section 8).

10. Geometric Continuity for Surfaces

While the concept of geometric continuity is neat for curves, it is essential for surfaces. After all, even a simple sphere does not allow a global C^1-parametrization. Moreover, there are many practical applications where it is impossible to add a new patch to an existing patch complex with parametric C^q-continuity [51].

The concept of geometric continuity is again based on reparametrization: A patch complex is *geometrically G^q-continuous*, if it is parametrically C^q-continuous under local reparametrization, or, in other words, if the patch complex forms a C^q-manifold in the sense of differential geometry. Note that G^1-continuity is equivalent to a continuous tangent plane, and G^2-continuity is equivalent to continuity of the second fundamental form.

When discussing the concept of geometric continuity in a CAGD context, the following two fundamental problems have to be addressed: First, we need conditions on two adjacent patches to be G^q-continuous along their common boundary (patch/patch continuity problem), and second we need conditions on a set of patches to be G^q-continuous around a vertex (vertex enclosure problem).

Let us briefly elaborate on these two problems. It is obvious that the G^1-conditions between two adjacent patches $F(u, v)$ and $G(u, w)$ across a common boundary parametrized by u are equivalent to the existence of scalar-valued functions λ, μ, and ν such that, at each point on the boundary

$$\lambda \, F_u = \mu \, F_v + \nu \, G_w \quad \text{common tangent plane}$$

and

$$\mu \, \nu > 0 \quad \text{proper orientation}.$$

Additional constraints on λ, μ, ν are obtained if F and G happen to be polynomials: As first observed fully by Peters [84], Cramer's rule then requires λ, ν, μ to be low degree polynomials as well. If the above condition on a common tangent plane is rewritten as

$$G_w = \alpha F_u + \beta F_v$$

this implies that α, β will in general no longer be polynomial but rational.

In order to be useful in practice, the above conditions have to be translated into conditions on the control points. Sufficient conditions on the control points have been derived independently by various researchers, and several excellent accounts are available (see, e.g., [39, 51, 64]). Necessary and sufficient G^1-conditions are derived, e.g., in [33]. Additional G^2-conditions along a patch boundary are also quite managable. [82] shows that two patches are G^2 along a C^1-curve if they are G^1 and additionally share the same normal curvatures for the directions given by a (not necessarily continuous) tangential vector field along the curve.

Let us now turn to the vertex enclosure problem and let us assume that m boundary curves meet at a common vertex $\mathbf{p}$. We are interested in the conditions that this given curve mesh can be interpolated by a G^1-continuous patch complex with one patch per facet. An obvious necessary condition is of course that all tangents to the given curves at $\mathbf{p}$ are coplanar. However, this condition is by no means sufficient. Instead it turns out, rather surprisingly, that the existence of a solution depends on the distinction if m is odd or even. This is due to the fact that the existence of a solution hings on the invertability of a certain matrix that arises in determining the *twist constraints*, the major step in the interpolation process. This matrix is always invertible at odd points but may be rank deficient at even points [51]. One way to overcome these difficulties in the even case is to impose additional constraints on the input curve mesh. [94] shows that if all mesh curves conform to the same second fundamental form at $\mathbf{p}$, then a solution always exists. Another approach to overcome the vertex enclosure problem is to simply split every patch at its centroid (see Clough-Tocher and Powell-Sabin split in Section 8). This split results in enough degrees of freedom to solve the problem.

Combined solutions to the patch/patch continuity problem and to the vertex enclosure problem can then be applied to obtain a solution to the *polygonal hole problem* which requires to fill an *m*-sided hole within a given (typically rectangular) patch complex [50, 51, 55]. This situation typically arises when two or more rectangularly meshed surfaces are to be blended together. The next step is the construction of a geometrically continuous patch complex that interpolates to a given curve mesh. This problem comes up naturally in the context of interpolating polyhedral data and has been addressed by a large number of researchers. A survey and classification of various methods is given in [83] (see also [64]) and in [70]. We come back to this problem in the following section.

11. Surfaces from Polyhedra

The results of the preceeding section are essential for a solution to the following *surface-from-polyhedron problem*: Given a triangulated polyhedron P of arbitrary topological type, construct a smooth surface S that is at least G^1-continuous and mimics the shape of the input mesh.

The various approaches to solve this problem can roughly be classified into three different categories: The first approach solves the *parametric surface interpolation problem* and constructs a parametric surface S that interpolates to the vertices of the given polyhedron [70]. The second approach is based on *recursive subdivision* [16, 36]: Here the control mesh is refined recursively in each step, and finally converges to the limit surface S. Subdivision schemes can be both interpolatory and approximating, but do generally not yield an explicit parametrization of S. The third approach tries to construct *parametric approximating schemes* with the vertices of P as control points. In particular, the surface should then lie inside the convex hull of P. We elaborate on each of the three approaches briefly. A fourth approach using piecewise algebraic surfaces is discussed in Section 12.

Methods for *parametric surface interpolation* generally proceed through the following three steps [70]: First, a smooth mesh of curves that joins adjacent input vertices is constructed. Second, derivative information is built up along the curve mesh. Third, a surface that interpolates to the constructed curve mesh and derivative data is constructed. The third step is generally based on one of the following two methods: *Split-domain schemes* try to solve the vertex enclosure problem (Section 10) by splitting each triangular facet in three micro-facets (Clough-Tocher split). After splitting, each micro-patch is made to interpolate the data along each boundary independently of the data on the other two boundaries. The remaining degrees of freedom are used to satisfy the G^1-constraints. *Convex-combination schemes* create a (rational) surface for each facet as follows: First, a patch is created for each curve/opposite-vertex pair by passing curves from points along one of the boundaries to the opposite vertex (side-vertex method). All curves are constructed using two points and associated normals. Second, the three curve/opposite-vertex patches for each triangular facet are blended together (convex combination). Note that convex-combination schemes bypass the vertex enclosure problem altogether by allowing a singularity at the vertices. A comparison and survey of various schemes is given in [70]. It is worth mentioning that surface interpolating schemes,

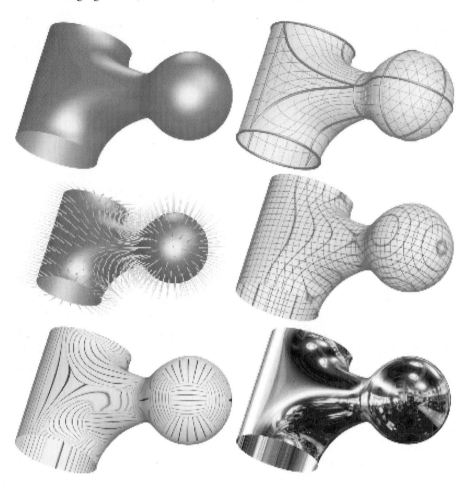

Fig. 8. Surface-from-Polyhedron Problem: The input to this method consists of point and normal information. First, a G^2-continuous minimum variation curve network has been generated. This curve network has then been interpolated to create a smooth surface. Free parameters have been adjusted as to minimize the overall curvature variation (nonlinear optimization). The quality of the resulting surface is demonstrated by the following pictures: Shaded image of the interpolating surface (top left), the interpolating G^2 mininum variation curve network (top right), normal vector field to the interpolating surface (middle left), a set of parallel section curves (middle right), reflection lines (for pairwise perpendicular illumination lines on the floor, walls, and ceiling) (down left), hight-quality rendering of the interpolating surface (down right). (Courtesy H. Moreton, Silicon Graphics Inc.)

even after satisfying all interpolation and smoothness constraints, generally have a considerable number of free parameters left. As pointed out in [70] altering these parameters can drastically influence the overall quality of the surface shape. Recent approaches have therefore used numerical optimization to ensure good settings of free parameters. Another approach is to relax the continuity constraints which allows for lower degree solutions and hence diminishes the problem of leftover free variables (see Color Plate 2, p. 133).

The main idea behind *recursive subdivision* is perhaps best explained for curves: Assume that the points $(\mathbf{d}_i^k)_{i \in \mathbb{Z}}$ are given, and define

$$\mathbf{d}_{2i}^{k+1} = \frac{3}{4}\,\mathbf{d}_i^k + \frac{1}{4}\,\mathbf{d}_{i+1}^k \quad \text{and} \quad \mathbf{d}_{2i+1}^{k+1} = \frac{1}{4}\,\mathbf{d}_i^k + \frac{3}{4}\,\mathbf{d}_{i+1}^k$$

by subdividing the polygon edges of level k in the ratio $1 : 2 : 1$. If this process is iterated, the resulting sequence of polygons converges to a differentiable curve F which is in fact a quadratic B-spline with control points $\mathbf{d}_i^0$ [95]. Similar subdivision algorithms are available for higher degree B-splines [37]. The above algorithm uses corner-cutting, i.e., in each iteration, it cuts off the corners of the control polygon. Corner-cutting schemes always satisfy the convex hull and the variation diminishing properties. [27] shows that corner cutting converges under very general assumptions. However, recursive subdivision is by no means restricted to corner cutting, but can produce interpolatory schemes as well [37, 75].

Subdivision has first been applied to the construction of smooth surfaces in [15, 34]. If all vertices are of order 4, and if all facets are 4-sided, these algorithms produce bi-quadratic and bi-cubic TP B-splines. Convergence and smoothness of the limit surface is discussed in [21, 76]. This discussion involves the study of extraordinary points, i.e., points of order $m \neq 4$. The Doo-Sabin construction, e.g., eliminates all extraordinary points already in the first step. Interpolating subdivision schemes for surfaces are constructed in [37]. A rather complete survey on stationary subdivision (including the relationship between subdivision and multiresolution analysis) is given in [16].

Generalized B-spline like *approximating schemes* that accept an arbitrary polyhedron P as input mesh have been constructed by Sabin [97], Höllig/Mögerle [61], and Loop/DeRose [69]. G-splines [61] fix the geometric continuity constraints and thus turn the whole problem into a (large sparse) linear system. The schemes by Sabin and Loop/DeRose are based on multi-sided patches. Loop/DeRose use S-patches, a generalization of triangular Bézier patches that is obtained by first embedding a regular n-gon into an n-simplex $\Delta \subset \mathbb{R}^{n-1}$, and then composing this embedding with a $(n - 1)$-variate Bézier simplex (Section 8).

A somewhat different approach has been taken recently by Peters [85] and Loop. Their algorithms proceed by first refining the input mesh through a small number of subdivision steps. This refined mesh, with extraordinary points well separated, is then used as control points to construct a low degree piecewise polynomial surface representation. The resulting surface schemes generalize quadratic box-splines and biquadratic TP B-splines.

All three classes of algorithms mentioned so far require as input a triangulated polyhedron (i.e. vertices plus adjacency information). In many applications this adjacency information is not explicitly available, but must be deduced from the given data. Two recent solutions to this problem are presented in [62] and [106] (see Fig. 9).

Fig. 9. Surface reconstruction from unorganized points: Starting from a set of unorganized points (left), a γ-neighbourhood graph is used to construct a polyhedral surface (middle). This polyhedron can then be used as input to produce a G^1-interpolating surface. (Courtesy R. Veltkamp, CWI)

We conclude this section with a brief discussion of *scattered data interpolation*, although conceptually the surface-from-polyhedron problem and the scattered data interpolation problem are quite different: The first problem requires the construction of a smooth manifold, while the second problem requires the construction of a function between manifolds. However, at least in the case of bivariate functional data this distinction has sometimes become blurred.

The basic *scattered data interpolation problem* is as follows: Given two manifolds S_1 and S_2, and a sequence of points $(\mathbf{u}_i, \mathbf{p}_i) \in S_1 \times S_2$, $i = 1, \ldots, m$, construct a map $F : S_1 \rightarrow S_2$ with $F(\mathbf{u}_i) = \mathbf{p}_i$. Typical examples are $(S_1, S_2) = (\mathbb{R}^2, \mathbb{R})$ (bivariate functional data), $(S_1, S_2) = (\mathbb{R}^3, \mathbb{R})$ (scalar field, or volumetric data), and $(S_1, S_2) = (S, \mathbb{R})$ with $S \subset \mathbb{R}^3$ a surface (surface-on-surface problem), but other combinations, e.g., $(S_1, S_2) = (\mathbb{R}^3, S^2)$ (unit vector field) are of practical interest as well.

Scattered data interpolation of bivariate functional data has been studied by numerous researchers, and we refer to [47] for a recent survey. A set of test functions and test data can be found in [46]. Several 2D methods can be extended to volumetric data. A set of 3D test data and test functions, together with a comparison of various methods is given in [78]. The surface-on-surface problem is considerably simplified if S_1 is a sphere [44]. A solution for arbitrary surfaces is presented in [5].

12. Algebraic Surfaces

So far we have dealt almost exclusively with parametric surfaces. As mentioned in the introduction, an alternative way to model smooth shapes is by means of implicit surfaces. If polynomials are used, this leads to algebraic curves and surfaces.

In a sense, of course, we have been dealing with algebraic surfaces all along: Any rational surface of degree n can also be represented in implicit form as an algebraic surface of degree n^2, and any rational TP surface of degree (n, m) can be represented as an algebraic surface of degree $2nm$ [1, 57, 99]. In fact, the warm reception of Sederberg's thesis [99], reflected the hope that, using implicitization, the strengths of both the parametric and implicit representation could be fully exploited – thus avoiding the inherent weaknesses of either representation.

Several approaches to implicitization have been taken by various researchers [58, 65]: The oldest and best-known method for eliminating variables is the resultant method, whose development goes back to the last century. The resultant of a system of $n + 1$ algebraic equations in $n + m$ variables $x_1, \ldots, x_n, y_1, \ldots, y_m$ is an expression in m variables $y_1, \ldots, y_m$ that vanishes for a set of values of the y_i if and only if the original system has a solution for the same values of the y_i. Thus the resultant gives the desired implicit equation.

One problem with implicitization is the potential introduction of extreneous factors. In other words, the resulting polynomial might be $F(x, y, z) = F_1(x, y, z) \cdot F_2(x, y, z)$ where $F_1(x, y, z)$ is the implicit equation, and $F_2(x, y, z)$ is unnecessary. Manocha and Canny [72] prove that the algebraic set defined by the polynomial obtained with the resultant is irreducible, provided the parametrization has no base points – that is, if the denominator and numberator functions do not vanish simultaneously for certain parameter values. Other methods for implicitization include Gröbner basis [57] and a recent method due to Wu and Ritt [65].

The opposite to implicitization is parametrization, and so it is natural to ask what algebraic surfaces allow a rational parametrization. This queston has received a great deal of attention in the algebraic geometry literature and is well-understood [1].

The simplest algebraic surfaces admitting a rational parametrization are of course quadratics. In order to parametrize a quadratic surface we can simply take a point on the surface and consider the pencil of lines through this point. Since every line in this pencil will intersect the surface in exactly one more point, a rational parametrization of the line pencil then also yields a rational parametrization of the surface of degree two. Note that the same method also works for so-called monoids, i.e., algebraic surfaces of degree n with an $(n - 1)$-fold singular point [1].

The related question, when exactly does a rational quadratic triangular Bézier patch or a rational bi-quadratic TP patch lie on a quadratic surface is a little bit more subtle. Let us consider, e.g., a rational quadratic Bézier patch. Recall from Section 8 that a rational quadratic Bézier patch is completely defined by its three boundary curves which are segments of a conic. In order for the patch to lie on a quadratic surface, all three conics must intersect in a common point and have a common tangent plane there. In particular, it is impossible to parametrize a full octant of a sphere with a single triangular Bézier patch of degree $n < 4$.

A larger class of surfaces that also allows for a dual parametric/implicit representation are the Dupin cyclides [35]. Cyclides are algebraic surfaces of degree 4, that can be defined by the equation

$$(x^2 + y^2 + z^2 - \mu^2 + b^2)^2 = 4(ax - c\mu)^2 + 4b^2 y^2 ,$$

and also admit a parametrization as rational bi-quadratics [91, 92]. Cyclides are particularly suited for blending applications [12, 17, 92]. The use of cyclides for the construction of free form surfaces is discussed in [80]. Generalized cyclides are considered in [32].

Another avenue of research in surface design tries to bypass conversion between parametric and implicit representations altogether and simply considers a surface a system of $n - 2$ nonlinear algebraic equations in n variables. In the CAGD context this approach is usually referred to as the *dimensionality paradigm* [57]. A significant body of algorithmic infrastructure is meanwhile available for surfaces defined using the dimensionality paradigm. Examples include algorithms for computing intersections, local parametrizations, local curvatures, and global approximations (see Color Plate 1, p. 133).

Other researchers abandon the parametric representation altogether and start out by designing with algebraic surfaces from scratch. Almost all these approaches use the idea of an *algebraic surface patch*, due to Sederberg [99, 100, 101]: Instead of representing an algebraic surface $F(x, y, z) = 0$ in monomial form, the underlying polynomial $F : \mathbb{R}^3 \to \mathbb{R}$ is represented as a trivariate Bézier polynomial w.r.t. a tedrahedron $\Delta \subset \mathbb{R}^3$,

$$F(\mathbf{u}) = \sum_{|\beta|=n} B_\beta^{\Delta, n}(\mathbf{u}) w_\beta ,$$

with real-valued coefficients $w_\beta \in \mathbb{R}$. An algebraic surface patch is then defined as the intersection of the zero set $F(\mathbf{u}) = 0$ with the tedrahedron Δ. The shape of an algebraic surface patch is completely controled by the coefficients $w_i \in \mathbb{R}$, so that these coefficients are also referred to as *control points* of the algebraic patch. It is possible to derive conditions on the weights which ensure that an algebraic surface patch is single sheeted, and that two adjacent patches join smoothly (see Fig. 10).

In order to be useful in practice, a design scheme that uses algebraic patches has to ensure that the necessary smoothness conditions between adjacent patches are satisfied automatically. Furthermore, the overall shape of the resulting piecewise algebraic surface should be described by few intuitive parameters.

What is needed then is a solution to the following surface-from-polyhedron problem for algebraic surfaces: Given a set of data points $\mathbf{p}_1, \ldots, \mathbf{p}_m \in \mathbb{R}^3$ and a triangulation T of these points, construct a mesh of low degree algebraic surface patches such that the resulting composite surface is at least G^1-continuous and mimics the shape of the input mesh.

Several solutions to this problem have become available recently: Dahmen constructs a surrounding simplicial hull (built-up from tedrahedra) around the given mesh and uses a Powell-Sabin split to split each algebraic patch into six subpatches. The approach produces a G^1-continuous piecewise implicit quadratic, but is restricted to the case where the original triangulation of the data set allows a transversal system of planes. In order to overcome these limitations Guo [53] extends the approach to cubics and uses a Clough-Tocher split for each macro-patch. A different approach is taken by Bajaj/Ihm [3]: They first construct a smooth mesh

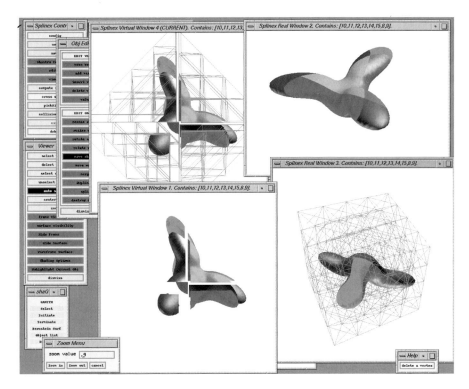

Fig. 10. A quartic algebraic surface in Bézier form used for a 3-way smooth join. (Courtesy C. Bajaj, Purdue University)

of space curves that interpolates to the given data and then interpolate to the mesh with one patch per facet. However, even if the underlying polyhedron is convex, this approach requires implicit patches of degree at least 5. Again using Dahmen's idea of a surrounding simplicial hull, recent work by Dahmen/Thamm-Schaar [24] and finally overcomes both splitting, and the use of patches with high degree, and produces a cubic G^1-interpolant with one cubic algebraic patch per triangle. In addition, this construction is completely local.

We conclude this section with a brief discussion of the use of algebraic surfaces for constructing blend surfaces. The basic idea is probably illustrated best with a simple 2D example: Consider two lines $L_1 = 0$ and $L_2 = 0$ together with two points $\mathbf{p}_1$ and $\mathbf{p}_2$ on each of them. Let $L_3 = 0$ be the line joining $\mathbf{p}_1$ and $\mathbf{p}_2$. It is easy to see that the equation

$$C = (1 - \lambda)L_1 L_2 + \lambda L_3^2 = 0, \quad \lambda \in [0, 1],$$

defines a family of conics, each of which produces a G^1-blend between $L_1 = 0$ and $L_2 = 0$ at $\mathbf{p}_1$ and $\mathbf{p}_2$. Various researchers have generalized this approach to surfaces, to higher order primitives, and to higher order smoothness [56, 59, 96, 107, 108].

13. Surface Interrogation

In order to avoid problems during the construction (milling) process it is strongly desirable to detect any unwanted surface irregularities as early as possible during the design stage. However, even with the aid of a graphics workstation the detection of surface irregularities is far from trivial. The display of a simple shaded picture, e.g., is absolutely insufficient, since most shading algorithms tend to smooth irregularities out, rather than detect them.

A much better tool is the computation and display of curvature. In particular, this helps to detect unwanted flat spots and wiggles. Common methods include the display of color-coded Gaussian curvature (as surface texture) and the display of the lines of curvature (principal directions). Another method to display Gaussian curvature is through the use of focal surfaces [54]. Consider the normal congruence $C(u, v) = F(u, v) + d\,\mathbf{n}(u, v)$, $d \in \mathbb{R}$, where $\mathbf{n}$ is the normal vector of the surface F. This normal congruence can be considered as the set of lines that touch two given surfaces, the focal surfaces S of the normal congruence. An explicit parameterization of the focal surfaces is then given by

$$S(u, v) = F(u, v) + \kappa_i^{-1}\,\mathbf{n}(u, v,), \quad i = 1, 2,$$

where κ_1, κ_2 are the principal curvatures. We remark that the sphere is the only surface for which the two sheets of the focal surface degenerate into a point, and the Dupin cyclides (see Section 12) are the only surfaces whose focal surfaces degenerate into curves.

More elaborate methods for surface interrogation simulate the reflection of light and lead to the computation of orthotomics, isophotes, and reflection lines: Isophotes (lines of constant intensity under diffuse reflection) [87] are given by

$$\mathbf{n} \cdot \mathbf{l} = \text{const},$$

where $\mathbf{n}$ is the normal vector and $\mathbf{l}$ is the unit vector in the direction of the light source. Since the isophotes of a C^q-surface are themselves C^{q-1}, isophotes can be used to detect any unwanted jumps in the derivatives between two adjacent patches.

Orthotomics [63] describe the reflection of a point light source at a curved surface: Given a surface $F(u, v)$ and a point $\mathbf{a}$, not on F nor on any of its tangent planes, the corresponding 2-orthotomic is given by

$$O(u, v) = \mathbf{a} + 2[(F(u, v) - \mathbf{a}) \cdot \mathbf{n}(u, v)] \cdot \mathbf{n}(u, v).$$

In order to obtain better pictures it is often useful to replace the factor 2 by some $k \in \mathbb{N}$, thus obtaining a so-called k-orthotomic. The usefulness of orthotomics stems from the fact that a k-orthotomic of F has a singularity at precisely those parameter values, where the Gaussian curvature of F vanishes. Thus k-orthotomics can be used to decide if a surface is convex.

Reflection lines [66] simulate the reflection of parallel light lines on a curved surface. Given a surface F, a point $\mathbf{a}$, and a light line L, the corresponding reflection line is the reflection of L on the surface that can be seen from **a**. By displaying reflection lines for several parallel light lines even small irregularities on a surface can then be detected.

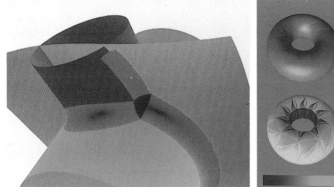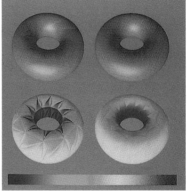

Color Plate 1. Variable-radius blends are difficult to describe exactly and delicate to trim. The picture shows three variable-radius blending surfaces meeting smoothly at a vertex where two cylinders and a bicubic surface intersect. The surfaces have been defined using the dimensionality paradigm. (Courtesy C. Hoffmann, Purdue University).

Color Plate 2. Comparison of two different interpolants to a sampling of the torus: On the left is a quartic surface created by a G^1 split scheme. On the right is an approximately G^1-continuous cubic surface. The maximum discontinuity in surface normal is about 6.1 degrees. Below each shaded image appears a false coloring of the surface to indicate the Gaussian curvature. Blue is used in regions of negative curvature, green for zero curvature, and red for positive curvature. (Courtesy S. Mann, University of Waterloo).

Color Plate 3. Telephone, pen, and glasses modeled with the Evans & Sutherland CDRS system. The image illustrates the fidelity with which common objects can be created and represented using today's technology. (Courtesy T. Jensen, Evans & Sutherland).

Color Plate 4. Car with reflection lines. The car has been completely designed and rendered with the Mercedes-Benz SYRKO system. Reflection lines demonstrate the high quality of the surface. (Courtesy R. Klass, Mercedes-Benz AG).

References

[1] S. Abhyankar: Algebraic Geometry for Scientists and Engineers. Mathematical Surveys and Monographs, vol. 35. American Mathematical Society, 1990

[2] P. Alfeld, L.L. Schumaker, and M. Sirvent: On dimension and existence of local bases for multivariate spline spaces. J. Approx. Th. **70**, 243–264 (1992)

[3] C.L. Bajaj and I. Ihm: Smoothing polyhedra using implicit algebraic splines. In: Proc. SIGGRAPH'92, pp. 79–88, 1992

[4] R.E. Barnhill, G. Farin, M. Jordan, and B. Piper: Surface/surface intersection. Computer-Aided Geom. Design **4**, 3–16 (1987)

[5] R.E. Barnhill, K. Opitz, and H. Pottmann: Fat surfaces: a trivariate approach to triangle-based interpolation on surfaces. Computer-Aided Geom. Design **9**, 365–378 (1992)

[6] P.J. Barry and R.N. Goldman: Algorithms for progressive curves: extending B-spline and blossoming techniques to the monomial, power, and newton dual bases. In: Knot Insertion and Deletion Algorithms for B-Spline Modeling, R.N. Goldman and T. Lyche (eds.). SIAM, 1992

[7] B.A. Barsky: The Beta-spline: a local representation based on shape parameters and fundamental geometric measures. PhD thesis, Univ. of Utah, Salt Lake City 1981

[8] B.A. Barsky and T.D. DeRose: Geometric continuity of parametric curves: Constructions of geometrically continuous splines. IEEE Comp. Graph. Appl. **10**, 60–68 (1990)

[9] R.H. Bartels, J.C. Beatty, and B.A. Barsky: An Introduction to Splines for Use in Computer Graphics and Geometric Modeling. Morgan Kaufmann Publishers, 1987

[10] W. Boehm: Inserting new knots into a B-spline curve. Comput. Aided Design **12**, 199–201 (1980)

[11] W. Boehm: Curvature continuous curves and surfaces. Computer-Aided Geom. Design **2**, 313–323 (1985)

[12] W. Boehm: On cyclides in geometric modelling. Computer-Aided Geom. Design **7**, 243–255 (1990)

[13] W. Boehm, G. Farin, and J. Kahmann: A survey of curve and surface methods in CAGD. Computer-Aided Geom. Design **1**, 1–60 (1984)

[14] W. Boehm, H. Prautzsch, and P. Arner: On triangular splines. Constr. Approx. **3**, 157–167 (1987)

[15] E. Catmull and J. Clark: Recursively generated B-spline surfaces on arbitrary topological meshes. Comput. Aided Design **10**, 350–355 (1978)

[16] A.S.W. Cavaretta, W. Dahmen, and C.A. Micchelli: Stationary Subdivision. Memoirs of the AMS, American Mathematical Society, Providence, 1991

[17] V. Chandru, D. Dutta, and C.M. Hoffmann: Variable radius blending using dupin cyclides. In: Geometric Modeling for Product Engineering, M. Wozny, J. Turner, and K. Preiss (eds.). North-Holland, Amsterdam 1990, pp. 39–57

[18] C. Chui: An Introduction to Wavelets. Academic Press, Boston 1992

[19] E. Cohen, T. Lyche, and R.F. Riesenfeld: Discrete B-splines and subdivision techniques in computer aided geometric design and computer graphics. Computer Graphics and Image Processing **14**, 87–111 (1980)

[20] M. Daehlen and T. Lyche: Box splines and applications. In: Geometric Modeling. Springer, 1991, pp. 35–93

[21] W. Dahmen: Subdivision algorithms converge quadratically. J. Comp. Appl. Math. **16**, 145–158 (1986)

[22] W. Dahmen and C.A. Micchelli: On the linear independence of multivariate B-splines I. Triangulations of simploids. SIAM J. Numer. Anal. **19**, 993–1012 (1982)

[23] W. Dahmen, C.A. Micchelli, and H.-P. Seidel: Blossoming begets B-splines built better by B-patches. Math. Comp. **59**, 97–115 (1992)

[24] W. Dahmen and T.-M. Thamm-Shaar: Cubicoids: modeling and visualization. Computer-Aided Geom. Design **10**, 89–108 (1993)

[25] I. Daubechies: Ten Lectures on Wavelets. SIAM, Philadelphia 1992

[26] C. de Boor: A Practical Guide to Splines. Springer, New York 1978

[27] C. de Boor. Corner cutting always works. Computer-Aided Geom. Design **4**, 125–131 (1987)

[28] C. de Boor: Multivariate piecewise polynomials. Acta Numerica, pp. 65–109, 1993

[29] C. de Boor, K. Höllig, and S. Riemenschneider: Box Splines. Springer, New York 1993

[30] P. de Casteljau: Formes à Pôles. Hermes, Paris 1985

[31] W. Degen: Best approximation of parametric curves by splines. In: Mathematical Methods in CAGD and Image Processing, T. Lyche and L.L. Schumaker (eds.). Academic Press, Boston, MA., 1992, pp. 171–184

[32] W. Degen: Generalised cyclides for use in CAGD. In: Mathematics of Surfaces IV, A. Bowyer (ed.). Oxford University Press, 1993

[33] T. DeRose: Necessary and sufficient conditions for tangent plane continuity of Bézier surfaces. Computer-Aided Geom. Design **7**, 165–180 (1990)

[34] D. Doo and M.A. Sabin: Behaviour of recursive division surfaces near extraordinary points. Comput. Aided Design **10**, 356–360 (1978)

[35] D. Dutta, R.R. Martin, and M.J. Pratt: Cyclides in surface and solid modeling. IEEE Comp. Graph. Appl. **13**, 53–59 (1993)

[36] N. Dyn: Subdivision schemes in computer aided geometric design. In: Advances in Numerical Analysis II, Wavelets, Subdivision Algorithms and Radial Functions, W.A. Light (ed.). Oxford University Press, 1991, pp. 36–104

[37] N. Dyn, D. Levin, and J. Gregory: A butterfly subdivision scheme for surface interpolation with tension control. ACM Trans. Graph. **9**, 160–169 (1990)

[38] G. Farin: Triangular Bernstein-Bézier patches. Computer-Aided Geom. Design **3**, 83–127 (1986)

[39] G. Farin: Curves and Surfaces for Computer Aided Geometric Design. Academic Press, 1993

[40] R. Farouki and V. Rajan: On the numerical condition of polynomials in Bernstein form. Computer-Aided Geom. Design **4**, 191–216 (1987)

[41] I. Faux and M. Pratt: Computational Geometry for Design and Manufacture. Ellis Horwood, 1979

[42] J. Fiorot and P. Jeannin: Courbes et Surfaces Rationelles. Masson, Paris 1989

[43] T.A. Foley: A shape preserving interpolant with tension controls. Computer-Aided Geom. Design **5**, 105–118 (1988)

[44] T.A. Foley et al.: Interpolation of scattered data on closed surfaces. Computer-Aided Geom. Design **7**, 303–312 (1990)

[45] D.R. Forsey and R.H. Bartels: Hierarchical B-spline refinement. Computer Graphics (Proc. SIGGRAPH '88) **22**, 205–212 (1988)

[46] R. Franke: Scattered data interpolation: Tests of some methods. Math. Comp. **38**, 181–200 (1982)

[47] R. Franke and G. Nielson: Scattered data interpolation: A tutorial and survey. In: Geometric Modeling: Methods and Applications, H. Hagen and D. Roller (eds.). Springer, 1991, pp. 131–160

[48] R.N. Goldman: Subdivision algorithms for Bézier triangles. Comput. Aided Design **15**, 159–166 (1983)

[49] T.N.T. Goodman: Properties of beta-splines. J. Approx. Th. **44**, 132–153 (1985)

[50] J. Gregory: N-sided surface patches. In: The Mathematics of Surfaces, J. Gregory (ed.). Clarendon Press, 1986, pp. 217–232

[51] J. Gregory: Smooth parametric surfaces and *n*-sided patches. In: Computation of Curves and Surfaces, W. Dahmen, M. Gasca, and C.A. Micchelli (eds.). Kluwer Academic Publishers, 1990, pp. 457–498

[52] L. Guibas and L. Ramshaw: Computer Graphics – Mathematical Foundations, Class Lecture Notes. Technical report, Stanford University, 1991
[53] B. Guo: Modeling arbitrary smooth objects with algebraic surfaces. PhD thesis, Computer Science, Cornell University, 1991
[54] H. Hagen: Surface interrogation algorithms. IEEE Comp. Graph. Appl., pp. 53–60, 1992
[55] J. Hahn: Filling polygonal holes with rectangular patches. In: Theory and Practice of Geometric Modeling, W. Straßer and H.-P. Seidel (eds.). Springer, 1990, pp. 81–92
[56] E. Hartmann: Blending of implicit surfaces with functional splines. Comput. Aided Design **22**, 500–506 (1990)
[57] C.M. Hoffmann: Geometric & Solid Modeling. Morgan Kaufmann Publishers, San Mateo, California, 1989
[58] C.M. Hoffmann: Implicit curves and surfaces in CAGD. IEEE Comp. Graph. Appl. **13**, 79–88 (1993)
[59] C.M. Hoffmann and J. Hopcroft: Quadratic blending surfaces. Comput. Aided Design **18**, 301–306 (1986)
[60] K. Höllig: Multivariate splines. SIAM J. Numer. Anal. **19**, 1013–1031 (1982)
[61] K. Höllig and H. Mögerle: G-splines. Computer-Aided Geom. Design **7**, 197–207 (1989)
[62] H. Hoppe, T. DeRose, T. Duchamp, J. McDonald, and W. Stuetzle: Surface reconstruction from unorganized points. In: Proc. SIGGRAPH'92, pp. 71–78, 1992
[63] J. Hoschek: Smoothing of curves and surfaces. Computer-Aided Geom. Design **2**, 97–105 (1985)
[64] J. Hoschek and D. Lasser: Grundlagen der geometrischen Datenverarbeitung. Teubner, 1992
[65] D. Kapur and Y.N. Lakshman: Elimination methods: An introduction. In: Symbolic and Numerical Computation: An Integration, D. Kapur and J. Mundy (eds.). Academic Press, New York 1992
[66] E. Kaufmann and R. Klass: Smoothing surfaces using reflection lines for families of splines. Comput. Aided Design **20**, 312–316 (1988)
[67] R. Klass and B. Kuhn: Fillet and surface intersections defined by rolling balls. Computer-Aided Geom. Design **9**, 1992
[68] J. Lane and R.F. Riesenfeld: A theoretical development for the computer generation of piecewise polynomial surfaces. IEEE Trans. Pattern Anal. and Mach. Intell. **2**, 35–45 (1980)
[69] C.T. Loop and T.D. DeRose: Generalized B-spline surfaces of arbitrary topology. In: Proc. SIGGRAPH'90, pp. 347–356, 1990
[70] M. Lounsbery, S. Mann, and T.D. DeRose: An overview of parametric scattered data fitting. IEEE Comp. Graph. Appl. **12**, 45–52 (1992)
[71] T. Lyche and K. Morken: A data reduction strategy for splines with applications to the approximation of functions and data. Inst. Math. Applics. Numer. Anal. **8**, 185–208 (1988)
[72] D. Manocha and J. Canny: Implicit representations of rational parametric surfaces. J. Symbolic Computation **13**, 485–510 (1992)
[73] Y. Meyer: Ondelettes et Operateurs I. Hermann, Paris 1990
[74] C.A. Micchelli: A constructive approach to Kergin interpolation in $\mathbb{R}^k$, multivariate B-splines and Lagrange interpolation. Rocky Mt. J. Math. **10**, 485–497 (1980)
[75] C.A. Micchelli and H. Prautzsch: Computing curves invariant under halving. Computer-Aided Geom. Design **4**, 113–140 (1987)
[76] A. Nasri: Polyhedral subdivision methods for free-form surfaces. ACM Trans. Graph. **6**, 29–73 (1987)
[77] G. Nielson: Some piecewise polynomial alternatives to splines under tension. In: Computer Aided Geometric Design, R.E. Barnhill and R.F. Riesenfeld (eds.). Academic Press, 1974, pp. 209–235

[78] G. Nielson and J. Tvedt: Comparing methods of interpolation for scattered volumetric data. In: State of the Art in Computer Graphics, R.A. Earnshaw and D.F. Rogers (eds.). Springer, 1993

[79] H. Nowacki, P.D. Kaklis, and J. Weber: Curve mesh fairing and GC^2 surface interpolation. Mathematical Modelling and Numerical Analysis **26**, 113–135 (1992)

[80] A.W. Nutbourne and R.R. Martin: Differential Geometry Applied to Curve and Surface Design. Ellis Horwood, Chichester 1988

[81] N.M. Patrikalakis: Surface-to-surface intersections. IEEE Comp. Graph. Appl. **13**, 89–95 (1993)

[82] J. Pegna and F.E. Wolter: Geometric criteria to guarantee curvature continuity of blend surfaces. Trans. ASME, J. of Mech. Design **114**, 201–210 (1992)

[83] J. Peters: Local smooth surface interpolation: a classification. Computer-Aided Geom. Design **7**, 191–195 (1990)

[84] J. Peters: Smooth interpolation of a mesh of curves. Constr. Approx. **7**, 221–247 (1991)

[85] J. Peters: Smooth free-form surfaces over irregular meshes generalizing quadratic splines. Computer-Aided Geom. Design **10**, 347–362 (1993)

[86] L. Piegl and W. Tiller: Curve and surface constructions using rational B-splines. Comput. Aided Design **19**, 485–498 (1987)

[87] Th. Pöschl: Detecting surface irregularities using isophotes. Computer-Aided Geom. Design **1**, 163–168 (1984)

[88] H. Pottmann: Locally controllable conic splines with curvature continuity. ACM Trans. Graph. **10**, 366–377 (1991)

[89] H. Pottmann: The geometry of tchebycheffian splines. Computer-Aided Geom. Design **10**, 181–210 (1993)

[90] M.J.D. Powell and M. Sabin: Piecewise quadratic approximations on triangles. ACM Trans. Math. Software **3**, 316–325 (1977)

[91] M.J. Pratt: Applications of cyclide surfaces in geometric modeling. In: Mathematics of Surfaces III, D.C. Handscomb (ed.). 1989

[92] M.J. Pratt: Cyclides in computer aided geometric design. Computer-Aided Geom. Design **7**, 221–242 (1990)

[93] H. Prautzsch: A round trip to B-splines via de Casteljau. ACM Trans. Graph. **8**, 243–254 (1989)

[94] L. Ramshaw: Blossoms are polar forms. Computer-Aided Geom. Design **6**, 323–358 (1989)

[95] R.F. Riesenfeld: On Chaikin's algorithm. Computer Graphics and Image Processing **4**, 304–310 (1975)

[96] A. Rockwood and J. Owen: Blending surfaces in geometric modeling. In: Geometric Modeling, Algorithms and New Trends, G. Farin (ed.). SIAM, 1987, pp. 367–384

[97] M.A. Sabin: The Use of Piecewise Forms for the Numerical Representation of Shape. PhD thesis, Hungarian Academy of Sciences, Budapest, Hungary, 1976

[98] L.L. Schumaker: Spline Functions: Basic Theory. Wiley, New York 1981

[99] Th. W. Sederberg: Implicit and Parametric Curves and Surfaces for Computer-Aided Geometric Design. PhD thesis, Purdue University, West Lafayette, IN, 1983

[100] Th.W. Sederberg: Techniques for cubic algebraic surfaces I. IEEE Comp. Graph. Appl. **10**, 14–25 (1990)

[101] Th.W. Sederberg: Techniques for cubic algebraic surfaces II. IEEE Comp. Graph. Appl. **11**, 12–21 (1991)

[102] H.-P. Seidel: A new multiaffine approach to B-splines. Computer-Aided Geom. Design **6**, 23–32 (1989)

[103] H.-P. Seidel: Symmetric recursive algorithms for surfaces: B-patches and the de Boor algorithm for polynomials over triangles. Constr. Approx. **7**, 257–279 (1991)

[104] H.-P. Seidel: An introduction to polar forms. IEEE Comp. Graph. Appl. **13**, 38–46 (1993)

[105] G. Strang and G. Fix: An Analysis of the Finite Element Method. Prentice Hall, 1973
[106] R.C. Veltkamp: The γ-neighborhood graph. Computational Geometry, Theory and Applications **1**, 227–246 (1992)
[107] J. Warren: Blending algebraic surfaces. ACM Trans. Graph. **8**, 263–278 (1989)
[108] J.R. Woodwark: Blends in geometric modelling. In: The Mathematics of Surfaces II, R.R. Martin (ed.). University Press, Oxford 1987

On the Place of Mathematics in Culture[*]

Armand Borel

The place of mathematics in our society strikes me as a rather peculiar one. It is certainly somewhere in the catalogue of our activities, even generally, if sometimes ruefully, perceived as an item of growing importance, it being obvious that the evermore encroaching technology we have to deal with has a mathematical underpinning. But where should it be classified? Is it viewed as part of culture? Hardly so: a "cultured person" is expected to have a knowledge and appreciation of literature, history, art, music, past civilizations, etc, but not necessarily of science and not at all of mathematics. To qualify for such a label, it is mandatory to know something about such towering figures as Plato, Michel-Angelo, Goethe, V. Hugo, Peter the Great, Napoléon, etc, but it is no liability to be blissfully ignorant about the greatest mathematicians, such as L. Euler, C. F. Gauss, D. Hilbert or H. Poincaré, if not a source of some pride. In his book: "*The Two Cultures*", C. P. Snow argued indeed that the concept of culture should comprise a scientific culture besides the traditional humanistic one, on equal footing with it. But even there, mathematics is cursorily treated and I, for one, could not see whether it was in his mind part of either.

A great variety of opinions has been (and is being) held about mathematics, from high praise to downright contempt: "Queen of Science", reflecting the often held view that mathematics is the language of science and that in fact no discipline is properly speaking a science until it can be expressed mathematically, or "Servant of Science", implying that its main, if not sole, function and justification is the solution of problems raised in natural sciences, technology, etc, to help us deal with the real world. It has, however, also been maintained that it has an artistic component. On the occasion of a Thesis defense (in 1845), the candidate asserted that mathematics was "art and science", to which an opponent retorted it was "only art, not science". A long line of thinkers has also viewed the intensive practice of mathematics with some suspicion, finding it had a numbing effect on the mind,

[*] Faculty lecture given at the Institute for Advanced Study on October 16, 1991. It overlaps with an earlier lecture published first in German: *Mathematik: Kunst und Wissenschaft*, Themenreihe XXXIII, C. F. v. Siemens Stiftung, München 1982 and then, in a slightly modified form, in an English translation: *Mathematics: Art and Science*, Mathematical Intelligencer 5, n° 4 (1983), 9–17, Springer-Verlag. I shall allow myself to borrow from it without further references, a course of action which would be frowned upon in the case of a mathematical paper, and for which I apologize to anyone who has read it and still remembers it.

making it ill prepared for nobler pursuits such as theology or philosophy (a criticism I can live with). A few have not hesitated to view mathematics as a mere collection of tautologies: after all, the argument goes, you start from self-evident truths, called axioms, and manipulate them with elementary logical operations. So what can you expect to get?[1]

Needless to say, I shall try to present a more positive view of mathematics than that last one. As a starter, I would like to compare it with an iceberg. For people with no lost love for mathematics, this may seem to evoke remoteness, coldness, a forbidding and threatening object to stay away from. I have in mind a different point of comparison, however, namely that it consists of a small visible part and a much bigger invisible one. By visible part I mean the one which interacts with other activities of obvious importance, such as natural sciences, technology, computer science, etc, and is clearly useful, worth financial support, even by tax payers' money. There is no doubt that this visible part is at the origin of mathematics, which was born in antiquity from the need to solve very practical problems, such as measuring of land or quantities, book accounting, keeping track of the movements of celestial bodies, then engineering, building canals, bridges, etc. But as mathematics grew, it started to acquire a life of its own. Mathematicians began to think about problems regardless of whether they had any applications. In short, they engaged in "pure mathematics" or "mathematics for mathematics' sake". This is what I refer to as the invisible part of mathematics, i.e. invisible to the layman, sorry, to the layperson. When apprised of that distinction, the outsider is often surprised and wonders as to how worthwhile such an activity can be: are there really still problems to be solved? How do you find them? Are they just mind teasers, such as you see in some Sunday papers, or is there a hierarchy? Are some problems more important than others? If so, what are your criteria? Our first answer is that we indeed have criteria, and they are mostly aesthetical in nature. We weave patterns of certain ideas, as a painter weaves patterns of forms and colors, a composer of sounds, a poet of words, and we are acutely sensitive to elegance, harmony in proofs, in statements, and the handsome development of a theory. Mathematicians have often waxed eloquent on this, so, for instance, Poincaré in 1897:

In addition to this it provides its disciples with pleasures similar to painting and music. They admire the delicate harmony of the numbers and the forms; they marvel when a new discovery opens up to them an unexpected vista; and does the joy that they feel not have an aesthetic character even if the senses are not involved at all? ... For this reason I do not hesitate to say that mathematics deserves to be cultivated for its own sake, and I mean the theories which cannot be applied to physics just as much as the others.[2]

and also earlier:

If we work, it is less to obtain those positive results the common people think are our only interest, than to feel that aesthetic emotion and communicate it to those able to experience it.[3]

This point of view does not present the whole picture, however, as we shall see; but let us accept it for the moment and view pure mathematicians as artists. Then we are artists with a singular privilege: we write only for our peers, without attempting to reach a broader public. Hardly what is expected from a painter, sculptor or composer. I know this difference did not escape at least one well-known Princeton composer, Milton Babbitt, who, over thirty years ago, in a since then famous article: *"Who cares if you listen?"*[4], requested precisely that freedom for himself and fellow composers. He felt contemporary music had reached such a level of complexity that the layperson cannot have access to it without special preparation, as is the case with modern science. So why not follow the example of science and free the composer of contemporary music from the obligation to try to reach a wider public, who is anyhow reluctant and does in any case have "its own music... to read by, eat by, dance by and to be impressed by". Still, this point of view does not seem to have prevailed, even among composers. But our writing is indeed mostly addressed to fellow workers, which has of course contributed mightily to our invisibility.

However, we do not live solely on theorems and thin air. We also need some money and other amenities for more mundane needs. Now the people or institutions "with the power of the purse" might wish to separate the visible from the invisible. They might tell us, we do not really mind your weaving patterns of ideas accessible only to people of our ilk, but why should we subsidize you for doing so? Why can't we just pay for services rendered to the visible part, whose usefulness we can gauge, at least up to a point, and leave it up to you to devote some or all or none of your free time to "pure mathematics" as a hobby?

The answer is a fundamental fact, whose importance can hardly be overemphasized: *it is not possible to separate the two.* At a given moment, it is of course easy to point out that a particular part of mathematics has no outside use whatsoever, but to predict it never will is very hazardous, and many who did had later to eat their own words, even if only posthumously. John v. Neumann, who certainly cannot be accused of having ignored the applications of mathematics, once said in an address to Princeton alumni[5]:

But still a large part of mathematics which became useful developed with absolutely no desire to be useful, and in a situation where nobody could possibly know in what area it would become useful: and there were no general indications that it even would be so.... This is true of all science. Successes were largely due to forgetting completely about what one ultimately wanted, or whether one wanted anything ultimately; in refusing to investigate things which profit, and in relying solely on guidance by criteria of intellectual elegance....
    And I think it extremely instructive to watch the role of science in everyday life, and to note how in this area the principle of laissez faire has led to strange and wonderful results.

This surprising relevance of the apparently irrelevant has often been commented upon, for instance by our first Director, A. Flexner in an article: *Usefulness of useless knowledge*"[6], a theme recently revived by our present Director, or the physicist E. Wigner in an often quoted paper: *The unreasonable effectiveness of mathematics in the natural sciences*[7]. This theme could easily be the topic of a full, even entertaining, lecture but I shall have to be satisfied with a few examples. In sketching them, I shall use some mathematical concepts only vaguely defined, if at all, but a full understanding is not necessary. All I want to convey is a sense of the unexpectedness, or unreasonableness if you will, of the application to physics of a mathematical theory created for purely internal reasons. For the first two, I shall follow the steps of E. Witten in his own Faculty Lecture, since they are always the first ones to come to mind.

The most venerable one is the use of the conics by J. Kepler. The conics are the plane curves obtained by intersecting a cone over a circle with a plane. Apart from some degenerate cases, they are the familiar ellipses, parabolae and hyperbolae, the theory of which was developed by the Greeks in about the 4th century B.C., as a chapter of geometry.

At the beginning of the 17th century, Kepler had at his disposal a lot of observational data on the movements around the sun of the planets known at the time (five of them), and he was trying to find a general law describing their main features. According to a view generally accepted for over a thousand years, the universe, having been created by God, had to obey perfect laws, which should then be expressible in terms of the geometric figures hailed as the most perfect ones, namely circles, spheres and the five Platonic solids. Kepler did try to construct very complicated models using those, but they did not fit the data. Finally, he cast down his eyes on the lowly ellipses, which had no claim to such divine perfection, but it worked. He postulated that a planet moved in a plane on an ellipse, having the sun as one of its focal points, a hypothesis which turned out to agree with the observations and became the first Kepler law. Two other properties of ellipses allowed him to describe other features of the orbits of the planets, leading to the second and third Kepler laws[8]. Later, I. Newton derived these laws from his law of gravitation, providing a striking confirmation of the latter.

In that example, the time lag between the mathematical theory and the application is about two thousand years. In our next one, Riemannian geometry and general relativity, it amounts to about sixty years. In fact the story starts a bit earlier, with a fundamental contribution by C. F. Gauss, published in 1827, to "differential geometry", i.e. the study of curves and surfaces in three-space by the tools of analysis. Imagine two-dimensional mathematicians living on a surface, who have no feeling for a third dimension, though they may conceive it intellectually, in other words, who are in the same relationship with a third dimension as we are with a fourth spatial dimension. Assume this surface to be made of flexible material. It may therefore take various positions in space and be more or less curved. For instance, if it is a piece of my handkerchief, it can be flat, or very curved, say if I wrap it around my fist. Our mathematicians cannot see the difference just by measuring lengths and angles on the surface, since we precisely assume they are not altered by these deformations. However, Gauss showed the existence of a quantity they can compute at every point and which bears some relation to curvature, call it the Gaussian curvature. It would still not help them to distinguish between the various shapes I can give to my handkerchief. However, if I would ask: "Do you live on my handkerchief or on a portion of a sphere?" they could answer because the Gauss curvature is zero in the former case, but not in the latter. This discovery, which Gauss found striking enough to call "Theorema egregium", led to the division of the differential geometry of surfaces in space into two parts: "Inner geometry", dealing with the properties of surfaces depending only on measures of lengths and angles on the surface, and "Outer geometry", describing how surfaces sit in space. Shortly thereafter came the discovery of non-euclidean geometry, in which Euclid's parallel axiom is not satisfied, and in 1854, starting from Gauss' discovery, B. Riemann presented an extraordinarily bold generalization of the known geometries. It deals not only with surfaces, or three-space, but with continua of any dimension. In such a continuum a "metric" is given, which allows one to measure lengths and angles. Inspired by the Theorema egregium he defines a notion of curvature: at a given point, he considers the Gaussian curvature there of all surfaces going through it. This "Riemannian curvature" is now a function of the point, as before, and moreover of the direction of the surface. This opened up the possibility of vast generalizations, though it remained to be seen how substantial they would be. Once the paper of Riemann was published (1868) mathematicians pursued these ideas and saw quickly that they led indeed to a far reaching broadening of geometry.

By the way, these two episodes give a glimpse into two of the ways a mathematical mind can work. Gauss' theorem is not that hard to prove, once noticed, but it took remarkable insight to look for it and to realize its importance. The second one displays the power of abstraction and generalization: Starting from Gauss' observation, Riemann takes flight to unsuspected heights and opens up completely new vistas.

I now come to the application. From about 1908 on, A. Einstein was trying to develop a general theory of gravitation, not restricted to uniform motions, as special relativity was; but he was bogged down by the lack of mathematical tools. In 1912, he became professor at the E.T.H. in Zürich, where he found among his new colleagues a good friend and former classmate, the mathematician Marcel Grossmann. He appealed to him, telling him: *"Grossmann, you have to help me, otherwise I'll go crazy"*[9]. Einstein explained his difficulties, and Grossmann led him to Riemannian geometry as well as to subsequent developments such as tensor calculus and covariant derivative. This help was indeed what Einstein needed, and the first paper on general relativity was a joint one, Grossmann writing the mathematical part and Einstein the physical. It still took Einstein three more years of extraordinary effort to complete his theory, but it remained (and still is) firmly embedded in four-dimensional Riemannian geometry, without which it cannot even be formulated, forcing the physicists to learn a theory they could safely ignore until then.

These two examples illustrate the more or less usual relationship between physics and mathematics: The physicist has a problem, needs a theorem or a framework and the mathematician either makes it to order or, as in these cases, points out, maybe with some cockiness, that what is needed is available and has been in stock for a long time. But for me, as a pure mathematician, it is even more exciting when mathematics leads to completely new and fundamental physical insights. As an example, let me mention the discovery of the positron. In 1928 P. A. M. Dirac had set up equations for the motion of the free electron which were compatible with the requirements of relativity and quantum mechanics. However, these equations also admitted solutions with the same mass as the electron but with the opposite electric charge. As no such particle had been observed, and as it was assumed that all elementary particles were known, these "physically meaningless" solutions were viewed as a flaw of the theory, and to modify it so as to eliminate them became a matter of great concern. All attempts failed; so, after about two years, Dirac made up his mind that maybe the mathematics was right and then postulated the existence of such a particle. It was discovered later, christened the positron, and provided the first example of an antiparticle or of antimatter. Since then, as far as I know, it is still a dogma in nuclear physics that with each particle there should be an antiparticle.

I have confined myself to theoretical physics, but these unexpected uses of mathematics have been ubiquitous in science or technology, some of the most recent ones in areas such as cryptography or error-correcting codes, which until then had felt rather safe from the latest developments in pure mathematics. As tempting as it is to pursue this theme further, I shall leave the visible part of my iceberg for most of the rest of this lecture and try to make the invisible one a bit more concrete, more visible.

As the quotations from Poincaré make clear, the feeling that mathematics is an art is strong. The mathematician G. H. Hardy once wrote that mathematics, if socially justifiable, could only be so as an art[10]. This analogy is strengthened when we think about the way we work. You should not believe that mathematicians are just thinking machines who always proceed in steps clearly planned with implacable logic. This impression is often given by papers. Those are organized for maximal efficiency of the exposition; omitting all the false leads, they often proceed in an order inverse to that which led to the discovery. In the case of a very clear-headed and far-seeing mathematician, such efficiency may indeed reflect the usual way of his thinking. But frequently the going is rougher: at the start, we often do not know whether a given proposition is likely to be true or false, whether we should try to prove or disprove it, do not see a clear way to go about it, make various attempts without knowing whether the goal is reachable with the means at our disposal, as if trying to find our way in deep fog in unfamiliar surroundings. In such situations, considerations of elegance are secondary and pragmatism reigns. We want to find the solution or the proof by whatever available means. Then, if we are lucky, all of a sudden a breakthrough occurs, sometimes so unexpected that the word which comes to mind is "inspiration", not unlike that of a composer or poet. We have well-known statements by Gauss and Poincaré on these unexpected flashes of insight. After having found the solution of a problem he had been working off and on for ten years, Gauss wrote[11]:

Finally, just a few days ago, success – but not as a result of my laborious search, but only by the grace of God I would say. Just as it is when lightning strikes, the puzzle was solved;

I myself would not be able to show the threads which connect that which I knew before, that with which I had made my last attempt, and that by which it succeeded.

More recently, the Japanese mathematician K. Oka, reflecting on the genesis of his great discoveries (in the function theory of several complex variables), said that for a long time he was at a loss as to how to proceed, did not know where to go, and added that one should be patient: if one waits long enough, the solution which lies in our subconscious is likely eventually to emerge[12]. This unforeseeable character is also present elsewhere in science. In his Faculty Lecture here last year, J. Bahcall, speaking about one of the great puzzles facing astrophysicists, said we do not really know what we are looking for, but we will know it when we have found it. As he was saying this, I was reminded of an almost identical statement made I think by the French sculptor A. Rodin "What I am looking for I know after I have found it".

This dependence on ideas, the flow or occurrence of which we cannot control, makes us share with creative artists, composers, poets, painters, a gnawing worry of drying up. For instance, in 1913, when Einstein was about to leave the E.T.H. for Berlin, to head a Max Planck Institute, he confided to a friend:

The gentlemen from Berlin speculate on me as on a prized egg-laying hen. I do not know whether I can still lay eggs.[13]

For us mathematicians, this worry is worsened by the overwhelming evidence of an unfortunate correlation between aging and drying up. When I think of my colleagues in Historical Studies, I am green with envy, and greener by the year. There you seem to get started around forty, hit your stride at 50 or 60, and from then on you enjoy a majestic *crescendo*. One of my colleagues in that School, upon retiring at seventy, showed me his files and told me he had there material for twenty volumes. Well, for us, it is not quite the same. True, life does not stop after 35, as I used to think, many, many years ago. Enough mathematicians have done extremely deep work later, say around 45–50, to give some comfort. First rate achievements may even come later but less and less frequently, so much so that an eminent emeritus colleague once wrote, also, many, many years ago, that such achievements [of elder mathematicians] "fill us each time with astonishment and admiration". Altogether the correct overall musical marking to describe life for mathematicians after sixty, say, would be *diminuendo* rather than *crescendo* though it may be *sostenuto* for some privileged ones.

So far pure mathematics has been depicted solely as an art, but this analogy goes only so far and does not account for some of its other features. Though it is the work of individuals, it is also a collective effort in which the contributions of these individuals strongly depend on one another. Art and music have developed in various cultures in many widely different ways, but there is essentially only one mathematics. Often, people discover a theorem or establish a new theory quite in-dependently at about the same time. A given problem has usually just one solution. Even when artists or composers face a common problem, say what should take the place of the object in painting or of traditional harmony in music, they come up with very different solutions. Also, in mathematics, many papers or even books are written jointly. One rarely hears about joint works in painting or music. Haydn and Mozart knew one another socially, performed music together, admired one another, composed in related styles, but never collaborated on a piece of music. There are of course the Haydn quartets by Mozart, but no Haydn-Mozart quartet or symphony. Similarly, it would be hard for me to imagine two mathematicians as close to one another as Braque and Picasso in the six years preceding World War I, seeing one another almost daily, looking at and discussing their work, borrowing from one another, painting at that time in styles so similar, that it is sometimes difficult at first sight to decide who had painted what, and not eventually producing a joint work... unless they became embroiled in a priority fight.

There is also a notion of progress. Mathematics is strictly cumulative. Once a theorem has been proved and the proof is accepted, it is there forever. Future developments may change views about its interest, supersede it, absorb it in a more general one, it still remains. The work of a generation will add to and improve upon that of the previous ones, even if it is not a comparable intellectual achievement. To come back to my second example, Gauss and Riemann are both giants and I would not want to attempt a comparison: still I would maintain that Riemannian geometry is of a greater scope and impact on mathematics than Gauss' theory of surfaces; and I could go on describing further developments, such as H. Weyl's first example of a gauge theory or E. Cartan's theory of connections, which have further increased the power of the theory and the range of its applications. The old saying, quoted by I. Newton, which goes back to Bernard de Chartres early in the twelfth century, namely, that by standing on the shoulders of giants, one may see more and further than they did, is definitely true in mathematics. It is not even always necessary to stand on the shoulders of giants: by standing on those of two worthy predecessors, it is often possible to see further than they did simply by relating their works in a previously unsuspected way.

I know that an idea of progress has also existed in art and music. The notion that a given generation, having the benefit of the experiences of earlier ones, would be in a position to do better, was rather commonly accepted, notably around the end of last century. In fact, the Bernard de Chartres aphorism has been quoted under various guises and attributions in many contexts. For instance, Renaissance painters, equipped with the laws of perspective, were (mistakenly) rated higher than their "primitive" predecessors. Or a history of music would show a steady progress from modal music, to polyphony and to the crowning achievement, harmony, relegating modal music as complex as Indian or Balinese music to the rank of the expression of a primitive or semi-civilized culture. But the stormy developments in the twentieth century, a better knowledge and appreciation of the past and of many world cultures have pretty much done away with such views. In fact, the opposite opinion has been held, too. How often new developments, departing from accepted norms, have been viewed with alarm, as a sign of decadence. We need only remember how various schools of paintings were greeted at first or look at Slonimsky's "*Lexicon of Musical Invective*" to see that many of the now widely respected composers did not fare well at all at first. Saying this makes me realize however that my depiction of the progress in mathematics has been a bit too rosy and one-sided. New is not always better or not always perceived as such. Also, we are not exempt from controversies (some rather bitter, I shall mention one later) about the value of a new theory or point of view. Some were also denounced as dangerous trends, leading to the degeneration of our beloved mathematics, but these controversies are usually rather short lived, and make way to a consensus, mainly for a good reason I shall come to in a moment.

It would be difficult to account for these features of mathematics by viewing it only as an art, with some additional guidelines stemming from the needs of other disciplines. Rather, they are among the attributes of a science and also point out to the existence of a world of mathematical concepts, problems, theorems, to which mathematicians constantly add, collectively or individually. Saying this however leads one right into a question which has been debated for ages and will presumably be so as long as there are mathematicians: namely, what is the locus of that mathematical reality? Do we create mathematics step by step, or does it preexist us and we merely discover it little by little, as if exploring an unknown country? Both views have had and still have their advocates. Those of the latter may appeal to a religious belief or to a philosophical tradition going back to Plato. The Platonic view has been held by many, for instance, by G. H. Hardy[14]:

I believe that mathematical reality lies outside us, that our function is to discover or observe it, and that the theorems which we prove, and which we describe grandiloquently as our "creations", are simply our notes of our observations. This view has been held, in one form or another by many philosophers of high reputation, from Plato onwards....

K. Gödel was also of that opinion and more recently R. Penrose expressed similar feelings[15]:

Is mathematics invention or discovery? When mathematicians come upon their results are they just producing elaborate mental constructions which have no actual reality, but whose power and elegance is sufficient simply to fool even their inventors into believing that these mere mental constructions are 'real'? Or are mathematicians really uncovering truths which are, in fact, already 'there' – truths whose existence is quite independent of the mathematicians' activities? I think that, by now, it must be quite clear to the reader that I am an adherent of the second, rather than the first, view, at least with regard to such structures as complex numbers and Mandelbrot set.

For a religious mathematician, the locus of that mathematical reality will usually be God's mind. Such were the points of view, e.g., of C. Hermite[16]:

There exists, if I am not mistaken, an entire world which is the totality of mathematical truths, to which we have access only with our mind, just as a world of physical reality exists, the one like the other independent of ourselves, both of divine creation.

Or of G. Cantor, who even pushed this belief much further, viewing himself as a messenger of God: the theory he had published had been revealed to him by God and it was his mission to be a good messenger and spread it out. Maybe he was moved to such an extreme position not only by his deep faith but also in some measure by the predicament he was in: he had developed a very daring, even revolutionary, theory, that of transfinite numbers, which dealt with actual infinities (as opposed to the usual "potential infinities"), even introducing different orders of actual infinities. His theory met with considerable opposition from many mathematicians, one of them attacking him savagely, and was also viewed with suspicion by some philosophers and theologians. The objections of the official Church mattered a great deal to Cantor; so he took great pains to discuss them, and he was greatly relieved when it was pronounced that his theory of infinity did not contradict accepted religious doctrine[17].

In this connection, it is difficult not to mention C. S. Ramanujan, who attributed to the family goddess his mathematical gifts and some of his formulae, communicated to him in dreams[18]. More recently, a colleague of mine, in an introduction to a series of lectures on his own work, pointed out he had been preoccupied by a certain problem for many years and then startled his audience by stating: *"Why has God created the exceptional series?"*

But even a mathematician who believes mathematics is purely a human creation, as I do, has the obscure feeling that it exists somewhere out there. And I catch myself time and again talking as if it really does. To come to terms with that impression, I shall simply take the view, which is quite common, that this is a cultural phenomenon: if a concept is such that we are convinced it exists in the minds of others in the same way as it does in ours, so that we can discuss it, argue about it, then this very fact translates to a feeling of an objective existence, outside, and independent of, a particular individual. This attitude is not peculiar to mathematics, but a common experience in many aspects of our lives. So there are many religions in the world and a person will as a rule believe in at most one; in fact, the stronger the belief, usually the more exclusive it will be. For that person, the preferred religion will of course be an example of a collection of concepts, thoughts, stories reflecting an objective reality of non-human origin, while all the other religions just illustrate what I referred to, namely, a mental construct by a group of people, which they believe (erroneously) to have an objective existence. We also encounter this often in literature. Aren't we inclined to think that Sherlock Holmes has actually lived? After all, there is a Sherlock Holmes Museum, a Sherlock Holmes Society in England, among whose activities is the organization of trips, or should I say pilgrimages, to the site of the fight between S. Holmes and his arch-enemy, Dr. Moriarty, at Reichenbach Falls, in Switzerland. A few years ago, at a meeting of the learned American Philosophical Society in Philadelphia, held, as are all its meetings "to promote useful knowledge", there was a lecture on the tastes and achievements of S. Holmes in music. This phenomenon of a mental creation acquiring an objective reality is apparently also often experienced by writers, who create a character and then view themselves more or less as observers, having little or no control on the thoughts and acts of that character. For instance, Erskine Caldwell once said in an interview[19]:

I have no influence over them. I'm only an observer, recording. The story is always being told by the characters themselves. In fact, I'm often critical, or maybe ashamed, of what some of them say and do – their profanity or their immorality. But I have no control over it.

"But you do at least understand their motivations?"

I'm not an oracle by any means. I'm often at a loss to explain the desires and the motivations of my people. You'll have to find your explanation in them. They're their own creations.

Now it is very easy in mathematics, with its very precise language, to create a new object or concept, which will make sense in the same way to its creator as to whomever may be interested in it. Some of the properties of this mathematical object derive easily from the definitions, but others not at all. It may require tremendous efforts, over a long time, to pry them out, and then how can we escape the feeling that this object was there before and we just stumbled upon it? If moreover, the interest in those problems and the efforts to solve them are shared by others, the feeling of an objective existence becomes practically irresistible. It can still be argued as to whether there is underlying it a higher degree of reality, whether we create, or observe or reminisce. It seems to me that such a discussion can go on ad *infinitum*, without any prospect of a final convincing answer, and I leave it at that.[20]

Whatever its origin or locus, we have then at our disposal this enormous amount of concepts, theorems, open and solved problems, theories, a "mathematical reality", which has been amassed over more than two thousand years. It has for us as much objective reality as the natural world for a physicist or natural scientist. This analogy helps me to complete the first answer given to a question raised earlier: if one leaves out the applications, what are the internal guidelines of mathematics, in the invisible part of the iceberg? As already said, those of aesthetic nature are of paramount importance, but are not quite the only ones I think. The structure of this mathematical world brings a hierarchy between the open questions and makes a problem or a theory more important at a given time, and helps us to single out some of greater interest. After all, we accept easily the view that the investigation of nature imposes such criteria upon the natural scientist. That quest for knowledge is an unending one. Many years ago I read of a comparison of the amount of knowledge in science at a given time with a ball immersed in the sea of the unknown: When its radius increases, then so does its surface of contact with the unknown. It was, I believe, attributed to Poincaré, but I have not been able to trace it back. At any rate, this comparison applies perfectly well to our mathematical world, even though it is an intellectual one, in which we operate essentially with intellectual tools.

The feeling of a structure in this mathematical world is strengthened by what I would like to call our "belief in a myth", namely, the fundamental unity of mathematics. It is a myth because mathematics is much too big to be comprehended by one person and, at a given time, consists of many seemingly unrelated parts. But time and again totally unsuspected connections appear, two different topics, the respective experts of which had so far little in common, all of a sudden become part of a bigger one, under the impact of a new, usually more abstract, point of view.

This counterbalancing of unbridled expansion by contraction through unification makes many of us strong believers in the ultimate coherence of the whole of mathematics.

The analogy with natural sciences can even be pushed further. Pure mathematics also has an experimental side besides its theoretical one. The latter is the one usually associated with it: we strive to set up general theories, in which certain key theorems have many consequences. But we need first to build up the material to organize, or to explain. As already pointed out, when facing a new object or proposition, we often do not know in which direction to go, what should be proved or disproved and, when we do, what method might work. The only way, for most of us, to gain such intuition is to look at special cases and this is the experimental side. Although the exposition of mathematics usually goes from the general to the particular, the way to the discovery is often the opposite one. These experiments were traditionally performed with pencil and paper, but computers are used more and more, increasing considerably their scope and their impact on mathematics, also making the analogy with laboratory work closer.

There are even findings about our objects which surprise us enormously and make an impression not unlike that made on physicists by the discovery of an elementary particle. A few years ago, it was shown that the euclidean space in four dimensions carries several differentiable structures. I shall not try to define those terms. Let me just say that euclidean space is one of the most basic structures in mathematics. There is one in each dimension and it was known to have a unique differentiable structure except possibly in dimension 4. It was quite a shock when it was shown that this last case was indeed exceptional. The feeling of many mathematicians may have been akin to that of the physicist I. Rabi who, when apprised of the discovery of a new particle, the muon, somewhat unwelcome since it was unexpected and did not fit into any existing theory, exclaimed: "*Who ordered that?*". So our world takes a very concrete form for us. It does not need much of a provocation for me even to maintain that a definite mathematical object, say the n-dimensional sphere, "exists" at least as much as an elementary particle which lives all of 10^{-22} seconds, can be detected only by extremely complicated experiments and by a sophisticated interpretation of the experimental data.

By the way, I cannot resist pointing out that the discovery of these new differentiable structures provides a striking boost for the belief in the ultimate coherence of mathematics. It came about by comparing, and drawing consequences from, two theorems proved in completely different frameworks, mathematical physics and geometric topology, by totally unrelated techniques. In fact, nobody at the time was familiar with both.

This double or triple aspect of mathematics makes it also easier to approach the question of "aesthetics" in mathematics. It is often remarked there is quicker consensus in mathematics about the relative importance of theorems or of some of its parts than in other disciplines, in particular in art, so that it would be somewhat surprising if it were based purely on aesthetic feelings, since there usually are so many conflicting opinions in art about recent work. When we speak of beauty, elegance, we indeed express first of all an aesthetic judgment on the ideas, how they are put together and how original they are. But, combined with them, often implicitly, are judgments closer to a "bottom line" mentality. In considering a theorem, we are not only interested in its proof, but also in its consequences and applications, in its power. This also contributes to a feeling of beauty. It may even have a retroactive effect. For instance, to come back to Gauss' "Theorema Egregium", it is indeed beautiful in its own right, but the feeling of beauty is heightened by the knowledge that it was the starting point of Riemannian geometry. Strictly thinking, one should distinguish between beauty and importance, but we often lump them together as "aesthetic" judgments. So "success" is often implicitly part of beauty and, since it is difficult to argue with success, a consensus on "beauty" in this broad sense is more easily reached: if a new theorem solves an old outstanding problem, gives more power to a theory, brings new fruitful viewpoints, it will have to be accepted and will even eventually command admiration.

In stressing usefulness within mathematics I should not go too far and run the danger of being accused of adopting for mathematics a "profit oriented" attitude I had decried earlier. Such relative judgments are unavoidable, if only to guide one's own activity, but should not be absolute. It would also be easy to describe cases in which a very special topic, of interest at first only to a handful of specialists, sometimes even scorned by other mathematicians, turned out to have unexpected applications in a broader context. (Cantor's theory of transfinite numbers, alluded to earlier, is a case in point.) The eventual relevance of the apparently irrelevant also takes place within mathematics. The freedom for mathematicians with regard to science advocated by J. v. Neumann in my earlier quote has to be granted to them within pure mathematics, too.

In talking about pure mathematics, I have taken for granted that it is a legitimate object of study, regardless of applications. This view is now widely accepted, though not universally or unconditionally, but it took some time to emerge. The focus of mathematics in the late seventeenth and in the eighteenth centuries was analysis and its applications to mechanics. Number theory, originally the study of properties of the natural integers, the paragon of pure mathematics, was practiced by only a few (though outstanding) mathematicians and was rather commonly viewed as a minor topic, even though it had a distinguished pedigree, going back to the Greeks. The mathematician L. Euler, who worked on all aspects of pure mathematics and its applications, even writing a book on ship building, did publish several papers on number theory, but he felt it necessary to state in his introductory remarks that this was as justified as research on more applied topics, because it added to our knowledge, and in fact might eventually be useful even from a more practical point of view. In the nineteenth century, C. F. Gauss did not express himself publicly on such matters but only in a few letters, from which one sees that he valued mathematics way over some of the applied work his functions required him to do:

... all the measurements in the world do not compare with one Theorem, through which the science of the eternal truths is brought further.[21]

And he valued number theory above all:

The higher I put and have always put this part of mathematics above all the others – the more painful it is to me that – directly or indirectly, through external circumstances – I am so far from my favorite occupation.[22]

Curiously, the first statement in print claiming autonomy for pure mathematics may be due to someone who knew some mathematics, but was not at all a professional mathematician, namely, Johann Wolfgang von Goethe, who wrote:

Mathematics must declare itself, however, independently of everything external, must go its own great intellectual way and develop more purely than can happen when, as hitherto, it is linked up with the empirical and is aiming at gaining something from it or adjusting itself to it.

in an aphorism published in 1829 I shall come back to in a moment. Later, some mathematicians also came out of the closet to make similar claims. For instance, W. Hankel, in a book on complex numbers and quaternions published in 1867, wrote:

Needed to establish a general arithmetic is therefore a purely intellectual mathematics, free from intuition, a pure science of forms, in which not quantities or their representatives, the numbers, are related, but intellectual objects, objects of thoughts, which may correspond to actual objects or relations, but do not have to.[23]

That claim was however still confined to one part of mathematics, arithmetic or algebra, but a bit later, G. Cantor made more sweeping ones, to the effect that any mathematical concept is a legitimate object of study provided its definition is logically consistent. He viewed that freedom as the essence of mathematics and, to stress it, remarked he would prefer to speak of "free mathematics" rather than "pure mathematics", if he had the choice[24].

This view became more and more widespread, but not always without some reservation. Even H. Poincaré, in the 1897 lecture quoted earlier, issued a warning

If I may be allowed to continue my comparison with the fine arts, then the pure mathematician who would forget the existence of the outside world could be likened to the painter who knew how to combine colors and forms harmoniously, but who lacked models. His creative power would soon be exhausted.[25]

In order to make his point, Poincaré takes for granted, as an unquestionable truth, that a main goal of painting is representation of the outside world or, more broadly, that the latter is the main source of a painter's inspiration. But, in fact, this very tenet had been more and more questioned by some painters. In order to broaden the discussion, I would like to extend the distinction between invisible and visible to painting and music by letting the former refer to a practice based on the use of purely internal criteria and motivation and the latter to one in which the practitioners also obey imperatives from the outside, whose fulfillment may be more easily detected and appreciated by the outsider. In painting, those would be the duty to represent nature, people, idealized beauty, religious subjects, to exalt the mind, etc. It is that duty, viewed traditionally as the aim of painting, which had been more and more under attack. About twenty five years earlier, for instance, James

Whistler had called the famous portrait of his mother an "arrangement in grey and black", and proposed that it should stand or fall on its merits as an "arrangement", since the identity of the portrait, though of interest to him, could hardly matter to the public. A bit later, he stated

As music is the poetry of sound, so is painting the poetry of sight, and the subject matter has nothing to do with harmony of sound or color.[26]

In 1890, Maurice Denis began his first theoretical writing on painting with the since then famous sentence:

It is well to remember that a painting, before being a warhorse, a naked woman or some anecdote is essentially a plane surface covered with colors assembled in a certain order.[27]

In 1897, the very year of Poincaré's lecture, a painter in Munich, August Endell, was prophesying an art with forms that mean nothing, represent nothing, will be able to excite our souls as only music has been able to do with sounds[28].

By hindsight, we see that these artists were groping for one form or another of nonfigurative painting, developed a bit later, from about 1910 on, by Kandinsky, Kupka, Malevich and others[29]. For some, it became the only way to paint; others had a more balanced view, or kept with the traditional ideals, and at present we have the whole spectrum, representing all these outlooks.

In two of the above quotations, the artists refer to music as the model to emulate, the paragon of a "pure art" in which the primary guidelines are internal ones, i.e., in the language I have adopted here, consisting mainly of the invisible part. But here too this had not always been a prevalent view. In the Europe of the seventeenth and eighteenth centuries, composers had patrons and had to produce, on order, lots of functional music; if some of the great ones let themselves be carried away by their genius, there were often grumblings, not excluding even Bach or Mozart, about their music being too much of a display of technical virtuosity, ignoring feelings or melody, etc. Mozart himself lamented during his last illness that he had to leave when he saw prospects of improvement in his financial situation which would have allowed him to be no longer a slave to fashion, no longer chained to speculators and to follow his spirit freely and independently and write as his heart dictated[30]. His successors, Beethoven, Schubert and others did indeed take that right into their hands of course, but it seems that this view of music was presented systematically and explicitly first by the music critic Eduard Hanslick in 1854 in his treatise: "Vom Musikalisch-Schönen"[31]; there he stresses that music is an arrangement of sounds which may evoke feelings but is not obliged to express one, and he quotes many earlier statements of others insisting on the contrary that the primary duty of music is to express human feelings.

Such analogies have their limitations and I do not want to push them too far. Nevertheless it seems interesting that the growing feeling of the specificity of mathematics, not only in its techniques, but also in its goals, was paralleled by similar developments elsewhere. I have singled out painting and music, but it reflects an even broader cultural phenomenon of the time.

At this point, "pure mathematics", the invisible part of my iceberg, emerges as a discipline which is at the same time an art and a science of the mind, an intellectual science, both experimental and theoretical. It is now time to take again the visible one into consideration. I will do this by confessing a sin and trying to atone for it. It may have been a surprise that I could enlist Goethe as an ally in promoting pure mathematics. I did this by using a device not unknown to people with an axe to grind, namely, selective or out-of-context quotation. I read to you only the second half of an aphorism, the first half of which states:

Physics must show itself as separate from mathematics. The former must exist in a decided independence and must endeavour with all loving, respecting and pious strength to penetrate into nature and nature's holy life, wholly without concern for what mathematics achieves and performs for its part.[32]

and gives it a quite different meaning. On the one hand, it would seem I should welcome it as one more instance of the claims for specificity I was just quoting, but unfortunately it is emphatically not one I can go along with. I have indeed pleaded for the right to do pure mathematics, without worrying about applications, but have never intended to assert one should do so. Of course, if challenged, I might give a quite impressive list of achievements in pure mathematics in these last fifty years or so motivated by purely internal considerations without any visible output from outside. But it would be a very futile enterprise. A moment ago, I took some pride in relating an occasion in which pure mathematics led to new insights in physics, but physicists could point to many examples of the converse. In fact the interaction between physics and mathematics is at present at an all time high, as a two-way street, and I would be particularly ill-advised to deny it here, since this place is a focal point for it. Using for a last time the comparison with the iceberg, I should point out that the separation exists only when looking at it from the outside but not within the iceberg. Furthermore, this analogy, like all analogies, has its limitations. A severe one is its failure to give any account of the constant shifting and exchange between its two main parts. To complete my description of mathematics in the context of science I would rather say that it is both a queen and a servant of science or, more democratically speaking, an equal partner, freely exchanging ideas, theories, and problems with it for mutual benefit.

Mathematics appears now as a very complex, many faceted, structure. In fact, most of the opinions I mentioned at the beginning appear to carry some part of the truth: Queen of Science, servant of science, art, as well as an experimental and theoretical mental science in its own right, a gigantic, awesome product of the collective human mind. As a professional mathematician, I must confess I feel rather uncomfortable in just philosophizing about it, beating around the bush, rather than getting to some of its actual contents and saying more specifically why we are in awe of it. But, as two members of the first Institute Faculty, Marston Morse and John von Neumann, said over fifty years ago in a report on the School of Mathematics to the Board of Institute Trustees, mathematics is written in a unique language which cannot be translated into any other, so that one has to learn it to be in a position to understand mathematics. The situation seems to be improving though. Efforts to reach out, to make some cracks in the language barrier, are increasing. Some recent essays do manage, I think, to give to an interested reader a good idea of the substance and goals of mathematics, or at least of parts of it[33]. However, my purpose here was somewhat different. Rather, I tried to show that, in spite of its esoteric character, a number of features of its evolution and development, and of questions about its essence, trends, aesthetics, goals have their counterparts, sometimes with similar answers, in science or other human endeavors more commonly viewed as part of our culture. I hope to have in this way made more plausible my belief that mathematics is an integral and important part of culture, whether scientific, humanistic or artistic.

Notes

[1] For a discussion of, and references to, such dim views of mathematics, see A. Pringsheim, *Ueber den Wert und angeblichen Unwert der Mathematik*, Jahresbericht d. Deutschen Mathematiker-Vereinigung **13** (1904), 357–382.

[2] "Et surtout leurs adeptes y trouvent des jouissances analogues à celles que donnent la peinture et la musique. Ils admirent la délicate harmonie des nombres et des formes; ils s'émerveillent quand une découverte nouvelle leur ouvre une perspective inattendue; et la joie qu'ils éprouvent ainsi n'a-t-elle pas le caractère esthétique, bien que les sens n'y prennent aucune part?..."

"C'est pourquoi je n'hésite pas à dire que les mathématiques méritent d'être cultivées pour elles-mêmes et que les théories qui ne peuvent être appliquées à la physique doivent l'être comme les autres." [Address to the First International Congress of Mathematicians in Zürich: *Sur les rapports de l'analyse pure et de la physique*, (which he could not deliver in person), see the Proceedings of that Congress 81–90, Verlag von B. G. Teubner, Leipzig 1898; also reproduced under the title: *L'analyse et la physique*, in *La Valeur de la Science*, pp. 137–151, E. Flammarion, Paris 1905.]

[3] "Si nous travaillons, c'est moins pour obtenir ces résultats auxquels le vulgaire nous croit uniquement attachés, que pour ressentir cette émotion esthétique et la communiquer à ceux qui sont capables de l'éprouver." [*Notice sur Halphen*, Jour. Ecole Polytechnique, 60ème cahier, 1890; Oeuvres de G. H. Halphen, Vol. **1**, XVII-XLIII, Gauthier-Villars, Paris 1916.]

[4] High Fidelity **8**, N° 2, 1958, pp. 30–40, 126–127; also reproduced in: *Contemporary composers on contemporary music*, pp. 243–250, E. Schwartz and B. Childs, ed., Da Capo Press, New York 1967.

[5] J. v. Neumann, "*The role of mathematics in the science and in society*", address to Princeton Graduate Alumni, June 1954. Cf. *Collected Works*, 6 Vol., Pergamon, New York, 1961, Vol. VI, pp. 477–490.

[6] Harper's **179**, October 1939, pp. 544–552.

[7] Communications on Pure and Applied Mathematics **13** (1960), pp. 1–14.

[8] See e.g. M. Kline, *Mathematical Thought from Ancient to Modern Times*, Oxford University Press, N.Y. 1972, Chapter 12, 5, pp. 243-245.

[9] *"Grossmann, Du musst mir helfen, sonst werd' ich verrückt"* [cf. L. Kollros, *Albert Einstein en Suisse. Souvenirs*, in *Fünfzig Jahre Relativitätstheorie*, Helvetica Physica Acta, Supplementum IV, Birkhäuser Verlag, Basel 1956, pp. 271–281.]

[10] G. H. Hardy, *A Mathematician's Apology*, Cambridge University Press, 1940; new printing with a foreword by C. P. Snow, pp. 139–140.

[11] "Endlich vor ein Paar Tagen ist's gelungen – aber nicht meinem mühsamen Suchen, sondern bloß durch die Gnade Gottes möchte ich sagen. Wie der Blitz einschlägt, hat sich das Räthsel gelöst; ich selbst wäre nicht im Stande, den leitenden Faden zwischen dem, was ich vorher wußte, dem, vomit ich die letzten Versuche gemacht hatte, – und dem, wodurch es gelang, nachzuweisen..." [In a letter to H. W. M. Olbers, September 3, 1805, *Gesammelte Werke* 10$_I$, p. 23.]

[12] K. Oka, *Ten Essays in Spring Evenings* (in Japanese). See K. Oka, *Collected Essays*, Gakushu Kenkya-sha, 1969. I thank T. Oda, of Sendai University, who drew my attention to them and translated some excerpts for me.

[13] "Die Herren Berliner spekulieren mit mir wie mit einem prämierten Leghuhn. I weiß nicht, ob ich noch Eier legen kann" [*loc. cit.*[9], pp. 123–124.]

[14] *loc. cit.*[10].

[15] R. Penrose, *The Emperor's New Mind*, Oxford University Press 1989, p. 96, reproduced here by permission of Oxford University Press.

[16] "Il existe, si je ne me trompe, tout un monde qui est l'ensemble des vérités mathématiques, dans lequel nous n'avons accès que par l'intelligence, comme il existe un monde des réalités physiques, l'un et l'autre indépendants de nous, tous deux de création divine." [G. Darboux, *La vie et l'Oeuvre de Charles Hermite*, Revue du Mois **10**, January 1906, p. 46.]

[17] For all this, see W. J. W. Dauben, *Georg Cantor. His Mathematics and Philosophy of the Infinite*, Harvard University Press, in particular pp. 132-146, 232–239.

[18] R. Kanigel, *The Man Who Knew Infinity*, C. Scribner's Sons, New-York 1991. *See* notably pp. 36, 64–67.

[19] The Paris Review N° 86, 1982, 127–157, see p. 132. The first paragraph is also reproduced in *The Writer's Chapbook*, G. Plimpton ed. Viking 1989, p. 194.

[20] As was pointed out, it is rather natural for a believer in God to conclude that mathematics preexists us, in the mind of God, but such a religious belief may also lead to the opposite view. At any rate, it did for R. Dedekind, who maintained in a letter to H. Weber (January 24, 1888) that we are of divine race and possess without any doubt a creative power, not only in material things (trains, telegraphs), but especially in spiritual things. Let me add, for the mathematician, that Dedekind made this statement in a discussion of the notion of number and of his definition

of irrational numbers by what became known as Dedekind cuts. Weber held the view that the irrational numbers were the cuts themselves, whereas for Dedekind they were something new, created by the mind, corresponding to a cut, but not to be identified with it:

"so möchtee ich doch rathen, unter der Zahl (Anzahl, Cardinalzahl) lieber nicht die Classe (das System aller einander ähnlichen endlichen Systeme) selbst zu verstehen, sondern etwas Neues (dieser Classe Entsprechendes), was der Geist erschafft. Wir sind göttlichen Geschlechtes und besitzen ohne jeden Zweifel schöpferische Kraft nicht blos in materiellen Dingen (Eisenbahnen, Telegraphen), sondern ganz besonders in geistigen Dingen. Es ist dies ganz dieselbe Frage, von der Du am Schlusse Deines Briefes bezüglich meiner Irrational-Theorie sprichst, wo Du sagst, die Irrationalzahl sei überhaupt Nichts anderes als der Schnitt selbst, während ich es vorziehe, etwas N e u e s (vom Schnitte Verschiedenes) zu erschaffen, was dem Schnitte entspricht, und wovon ich sage, daß es den Schnitt hervorbringe, erzeuge. Wir haben das Recht, uns eine solche Schöpfungskraft zuzusprechen, und außerdem ist es der Gleichartigkeit aller Zahlen wegen viel zweckmäßiger, so zu verfahren" [R. Dedekind, *Gesammelte Werke*, Vol. III, p.489, Viehweg and Sohn, Braunschweig 1932.]

(21) "... Alle Messungen der Welt wiegen nicht ein Theorem auf, wodurch die Wissenschaft der ewigen Wahrheiten wahrhaft weiter gebracht wird." [*Briefwechsel zwischen Gauss und Bessel*, Verlag W. Engelmann, Leipzig 1880, letter 143, March 14, 1824, p. 428.]

(22) "Denn je höher ich diesen Theil der Mathematik über alle andern setze und von jeher gesetzt habe, um so schmerzhafter ist es mir, dass – unmittelbar oder mittelbar durch die äussern Verhältnisse – ich so sehr von meiner Lieblingsbeschäftigung entfernt bin." [Letter to G. Lejeune Dirichlet, February 2, 1838, *Gauss' Werke* **12**, pp.309–312.]

(23) "Die Bedingung zur Aufstellung einer allgemeinen Arithmetik ist daher eine von aller Anschauung losgelöste, rein intellectuelle Mathematik, eine reine Formenlehre, in welcher nicht Quanta oder ihre Bilder, die Zahlen verknüpft werden, sondern intellectuelle Objecte, Gedankendinge, denen actuelle Objecte oder Relationen solcher entsprechen können, aber nicht müssen. [H. Hankel, *Vorlesungen ueber die complexen Zahlen und ihre Funktionen*, I. Teil, *Theorie der complexen Zahlensysteme*, L. Voss, Leipzig, 1867, p. 10.]

(24) G. Cantor, *Ueber unendlich lineare Punktmannigfaltigkeiten*, 5. Fortsetzung, Math. Annalen **21** (1883), 545–586; G.W. 165-208, 8.

(25) "Si l'on veut me permettre de poursuivre ma comparaison avec les beaux-arts, le mathématicien pur qui oublierait l'existence du monde extérieur serait semblable à un peintre qui saurait harmonieusement combiner les couleurs et les formes, mais à qui les modèles feraient défaut. Sa puissance créatrice serait bientôt tarie."

(26) J. A. McNeill Whistler, *The Gentle Art of Making Enemies*, G. P. Putnam and Sons, New-York 1890, see p. 126 (May 22, 1878).

(27) "Se rappeler qu'un tableau – avant d'être un cheval de bataille, une femme nue, ou une quelconque anecdote est essentiellement une surface plane recouverte de couleurs en un certain ordre assemblées. [Art et Critique, Paris 1890; also in Théories (1890–1910), Hermann, Paris 1964. Translated in H. B. Chipp, *Theories of Modern Art*, University of California Press 1968.]

(28) A. Endell, *Formenschönheit und dekorative Kunst*, Dekorative Kunst I, N° 2, Nov. 1897, p.75, see also P. Weiss, *Kandinsky in Munich*, Princeton University Press 1979, p. 34.

(29) I have limited myself to some signposts, without any attempt at comprehensiveness. They point to a trend towards "Art for Art's Sake", an idea which had also been entertained, in various forms, by some writers in the first half of the nineteenth century. In fact, that expression itself seems to appear first in the diary of Benjamin Constant, who wrote on February 10, 1804, during a visit in Weimar:
    "L'art pour l'art et sans but, tout but dénature l'art. Mais l'art atteint un but qu'il n'a pas."
    (Art for art's sake and without a goal, every goal falsifies art. But art achieves a goal it does not have.) a statement which strikes me as not devoid of similarity with the one of J. v. Neumann quoted earlier.

(30) F. X. Niemetschek, *Lebensbeschreibung des k.k. Kappellmeisters Wolfgang Amadeus Mozart*, 1798. Reprint VEB Deutscher Verlag für Musik, p. 54. Translation by H. Mautner: *Life of Mozart*, L. Hyman, London 1956, p. 45.

(31) *Vom Musikalisch-Schönen*, 1854, translated by G. Cohen, *The Beautiful in Music*, Ewer and Co. N.Y. 1891, reprinted by Da Capo Press, N.Y.

(32) The full aphorism reads in the original: "Als getrennt muß sich darstellen: Physik von Mathematik. Jene muß in einer entschiedenen Unabhängigkeit bestehen und mit allen liebenden, verehrenden, frommen Kräften in die Natur und das heilige Leben derselben einzudringen suchen, ganz unbekümmert, was die Mathematik von ihrer Seite leistet und tut. Diese muß sich dagegen unabhängig von allem Äußern erklären, ihren eigenen großen Geistesgang gehen und sich selber reiner ausbilden, als es geschehen kann, wenn sie wie bisher sich mit dem Vorhandenen abgibt und diesem etwas abzugewinnen oder anzupassen trachtet." [See: *Betrachtungen in Sinne der Wanderer*, Goethes Werke, Propyläen Ausgabe, Berlin, Vol. **41** p. 380.]
    This collection of aphorisms (the one quoted here being number 134) was published in 1829, as a supplement to "*Wilhem Meisters Wanderjahre*", Bd. 2, but is likely to have been written earlier.
    Goethe's advocacy of a strict separation between mathematics and physics may not be unrelated to the conflict between his theory of colors and that of Newton. In his opinion, the latter was mistaken, and it had gained support chiefly because of Newton's prestige as a mathematician. *See Ferneres ueber Mathematik und Mathematiker*, ibid. Vol. **39**, pp. 89 and 439.

(33) notably *The Mathematical Experience*, by P. J. Davis and R. Hersch, Birkhäuser, Boston 1980.

Mathematical Shell Theory: Recent Developments and Open Problems

Philippe G. Ciarlet

1. Introduction

Shells and their assemblages constitute, or are found in, a wide variety of elastic structures of considerable interest in contemporary engineering, such as:

- Blades of a rotor (Fig. 1);
- cylindrical tanks as found in an oil refinery (Fig. 2);
- cooling tower of a nuclear plant (Fig. 3);
- parts of a car body: doors, fenders, hood, windshield, etc.;
- hull of a vessel;
- dams;
- sails of a sailboat;
- parts of an aircraft: wings, tail, etc.;
- balloons;
- etc.

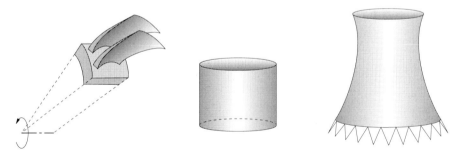

Fig. 1. A *rotor and its blades* is an example of an *elastic multi-structure* composed of a three-dimensional substructure (the rotor), and "two-dimensional" substructures (the blades). The *blades* are often modeled as *nonlinearly elastic shells*

Fig. 2. A *cylindrical tank* is another instance of an *elastic multi-structure* composed of "two-dimensional" substructures: a *cylindrical shell* and a *circular plate*

Fig. 3. A *cooling tower* is a *shell*, whose middle surface is hyperbolic. Together with its supporting rods, it constitutes an *elastic multi-structure* composed of a "two-dimensional" substructure (the shell) and "one-dimensional" substructures (the rods)

If anyone of these structures is viewed as a three-dimensional elastic body, the situation is on firm grounds, especially as regards its mathematical modeling or its mathematical analysis (see Ciarlet [1988]). After the fundamental ideas set forth by Ball [1977], there indeed remain various unresolved mathematical problems in the nonlinearly elastic case, but there are otherwise well developed constitutive and existence theories.

The numerical analysis, i.e., the conception and mathematical analysis of convergent approximation schemes, is likewise well developed in three-dimensional elasticity, especially in the linear case (Ciarlet [1991]), but also in the nonlinear case, where substantial progress has recently been made (see Le Tallec [1993] for an overview). These is nevertheless a strong *proviso: these numerical schemes fail when they are applied to elastic structures that have a "small" thickness,* such as *plates, shells, rods,* and their *assemblages.*

The *"small" thickness* of a shell (or of a plate for that matter) makes it natural to "replace" the genuine *three-dimensional model* by a "simpler" *two-dimensional model,* i.e., that is *posed over the middle surface of the shell.* Not only is this replacement natural from a theoretical viewpoint, but it becomes a *necessity* when *numerical methods* must be devised for computing (approximate) displacements and stresses: Any reasonably accurate three-dimensional discretization necessarily involves an astronomical number of unknows, which renders it prohibitively expensive. For instance, the ingenious three-dimensional computations reported by Babuška and Li [1992] already require about three hours on a CRAY supercomputer; yet they were devised for the very special case of a rectangular plate. Think of what this means for a "general" shell!

By contrast, the situation is on fairly safe grounds, at least on the theoretical side, as regards the application of finite element methods to two-dimensional shell models (see in this respect Bernadou [1993] for the "state of the art").

This is why *two-dimensional shell models are by and large preferred.* But nevertheless, *three major questions naturally arise*:

(i) *Can a known two-dimensional shell model be derived in a systematic and rational manner from three-dimensional elasticity?*

(ii) *Has the mathematical analysis* (existence, uniqueness, regularity, etc., of solutions) *of any known two-dimensional shell model reached a satisfactory stage?*

(iii) In a given physical situation, *how to choose among the various available shell models* (e.g., those described in Sects. 3–5), in order that the chosen one be "as good as possible" an approximation of the three-dimensional model it "replaces"? This last question is of paramount *practical* importance; note that it makes no sense to devise accurate methods for approximating the solution of a "wrong" model!

In view of the old age of the subject, and also of the ever increasing number of actual structures that comprise shells, an apparently safe guess would be that these questions have by now received satisfactory and complete answers. To show that this is far from being the case is the main purpose of this paper; a special emphasis will be also placed on the analogies and (mostly!) on differences with two-dimensional *plate* theories.

Note that, for simplicity, we will mostly confine our discussions to *linearly elastic* shells, and that no attempt has been made to compile an exhaustive bibliography.

2. The Three-Dimensional Equations of a Linearly Elastic Shell

In what follows, Greek indices or exponents (except ε) vary in $\{1, 2\}$, Latin indices or exponents vary in $\{1, 2, 3\}$, and the summation convention with respect to repeated indices and exponents is used; the Euclidean scalar and vector products of two vectors $\mathbf{a}, \mathbf{b} \in \mathbf{R}$ are denoted $\mathbf{a} \cdot \mathbf{b}$ and $\mathbf{a} \times \mathbf{b}$ respectively; the Euclidean norm in denoted $| \cdot |$.

Let ω be *domain* in $\mathbf{R}^2$, i.e., a bounded, open, connected subset with a Lipschitz-continuous boundary γ, and let $\gamma_0 \subset \gamma$ be given such that *length* $\gamma_0 > 0$. For each $\varepsilon > 0$, we let $\Omega^\varepsilon = \omega \times] - \varepsilon, \varepsilon[$ and $\Gamma_0^\varepsilon = \gamma_0 \times] - \varepsilon, \varepsilon[$. Let $\varphi : \bar{\omega} \to \mathbf{R}^3$ be a given, smooth enough, injective mapping, and let $\boldsymbol{\Phi} : \bar{\Omega}^\varepsilon \to \mathbf{R}^3$ be defined by

$$\boldsymbol{\Phi}\left(x_1, x_2, x_3^\varepsilon\right) = \varphi\left(x_1, x_2\right) + x_3^\varepsilon \mathbf{a}_3\left(x_1, x_2\right) \quad \text{for all} \quad \left(x_1, x_2, x_3^\varepsilon\right) \in \bar{\Omega}^\varepsilon,$$

where $\mathbf{a}_3$ is a continuously varying unit vector, normal to the surface $S := \varphi(\bar{\omega})$. Let $\partial_\alpha^\varepsilon = \partial_\alpha := \partial/\partial x_\alpha$ and $\partial_3^\varepsilon := \partial/\partial x_3^\varepsilon$. We assume that the three vectors $\mathbf{g}_j := \partial_j^\varepsilon \boldsymbol{\Phi}$ are linearly independent at all points of $\bar{\Omega}^\varepsilon$ and that $\boldsymbol{\Phi} : \bar{\Omega}^\varepsilon \to \boldsymbol{\Phi}(\bar{\Omega}^\varepsilon)$ is a C^2-diffeomorphism.

At each point of the set $\boldsymbol{\Phi}(\bar{\Omega}^\varepsilon)$, the *contravariant basis* $(\mathbf{g}^i)$ is derived from the *covariant basis* $(\mathbf{g}_j)$ by setting $\mathbf{g}^i \cdot \mathbf{g}_j = \delta_j^i$. We then let $g_{ij} := \mathbf{g}_i \cdot \mathbf{g}_j$ and $g^{ij} := \mathbf{g}^i \cdot \mathbf{g}^j$ denote the covariant and contravariant components of the *metric tensor*, and we set $g = \det(g_{ij})$. Finally, we let $\Gamma_{ij}^p := \mathbf{g}^p \cdot \partial_j^\varepsilon \mathbf{g}_i$ denote the *Christoffel symbols*.

The set $\boldsymbol{\Phi}(\bar{\Omega}^\varepsilon)$ is the reference configuration of an *elastic shell*, with *middle surface* S and *thickness* 2ε (Fig. 4). We assume that the reference configuration is a natural state, and that the material constituting the shell is homogeneous and isotropic. The *unknowns* of the *three-dimensional shell problem* are the *three covariant components* $u_i^\varepsilon : \bar{\Omega}^\varepsilon \to \mathbf{R}$ *of the displacement* $u_i^\varepsilon \mathbf{g}^i$ of the points of the shell (Fig. 4). We also assume that the displacement vanishes on the portion $\boldsymbol{\Phi}(\Gamma_0^\varepsilon)$ of the lateral surface of the shell.

In *linearized elasticity*, the unknown $\mathbf{u}^\varepsilon := (u_i^\varepsilon)$ solves the following *three-dimensional variational problem*:

(1) $\mathbf{u}^\varepsilon \in \mathbf{V}(\Omega^\varepsilon) := \left\{\mathbf{v} = (v_i) \in \mathbf{H}^1(\Omega^\varepsilon); \mathbf{v} = \mathbf{0} \text{ on } \Gamma_0^\varepsilon\right\},$

$$\int_{\Omega^\varepsilon} A^{ijkl} \left(e_{ij}^\varepsilon\left(\mathbf{u}^\varepsilon\right) - \Gamma_{ij}^p u_p^\varepsilon\right) \left(e_{kl}\left(\mathbf{v}\right) - \Gamma_{kl}^q v_q\right) \sqrt{g} \, dx^\varepsilon$$

(2) $$= \int_{\Omega^\varepsilon} f^{i,\varepsilon} v_i \sqrt{g} \, dx^\varepsilon \quad \text{for all} \quad \mathbf{v} \in \mathbf{V}(\Omega^\varepsilon),$$

where ($\lambda > 0$ and $\mu > 0$ are the Lamé constants of the material):

162   Philippe G. Ciarlet

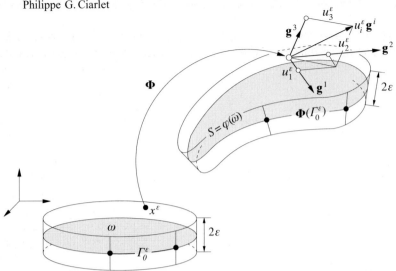

Fig. 4. When a shell is modeled by *three-dimensional equations*, the three unknowns are the covariant components $u_i^\varepsilon : \bar{\Omega}^\varepsilon \to \mathbf{R}$ of the displacement $u_i^\varepsilon \mathbf{g}^i$ of the points of the reference configuration $\Phi(\bar{\Omega}^\varepsilon)$ of the shell, where $\Omega^\varepsilon = \omega \times]-\varepsilon, \varepsilon[$ and $(\mathbf{g}^i)$ denotes the covariant basis at each point of the shell

$$A^{ijkl} := \lambda g^{ij} g^{kl} + \mu(g^{ik} g^{jl} + g^{il} g^{jk}),$$

$$e_{ij}^\varepsilon(\mathbf{v}) := \frac{1}{2}(\partial_j^\varepsilon v_i + \partial_i^\varepsilon v_j),$$

and $f^{i,\varepsilon} \mathbf{g}_i$, with $f^{i,\varepsilon} \in L^2(\Omega^\varepsilon)$, denotes the body force density (for simplicity, we assume that there are no surface forces). Problem (1)–(2) possesses a unique solution. As observed in Ciarlet [1993], this is a consequence of a *generalized Korn inequality*, valid for all $\mathbf{v} \in \mathbf{H}^1(\Omega^\varepsilon)$:

$$\|\mathbf{v}\|_{\mathbf{H}^1(\Omega^\varepsilon)} \le c(\Omega^\varepsilon) \left\{ \sum_{i,j} \|e_{ij}^\varepsilon(\mathbf{v}) - \Gamma_{ij}^p v_p\|_{L^2(\Omega^\varepsilon)}^2 + \sum_i \|v_i\|_{L^2(\Omega^\varepsilon)}^2 \right\}^{1/2},$$

itself a corollary of a *lemma of J. L. Lions* (see footnote (27), p. 320, in Magenes and Stampacchia [1958], or Duvaut and Lions [1972, Th. 3.2, p. 111]), which states that if Ω is a domain in $\mathbf{R}^n$ with a smooth boundary, $w \in H^{-1}(\Omega)$ and $\partial_i w \in H^{-1}(\Omega)$, $1 \le i \le n$, imply that $w \in L^2(\Omega)$. This result has been recently extended to domains whose boundaries are, as here, "only" Lipschitz-continuous; see Amrouche and Girault [1990], or Borchers and Sohr [1990].

Remark. The existence and uniqueness of the solution of the variational problem (1)–(2) also follows from the existence and uniqueness of the variational solution of the three-dimensional system of linearized elasticity expressed in Cartesian coordinates.   □

3. Asymptotic Analysis as the Thickness Approaches Zero: The Two-Dimensional "Bending" and "Membrane" Models

A *two-dimensional shell model* is one which is "posed over the middle surface of the shell". More specifically, at each point of the surface $S = \varphi(w)$, the *covariant basis* $(\mathbf{a}_\beta)$ and the *contravariant basis* $(\mathbf{a}^\alpha)$ of the tangent plane are defined by letting $\mathbf{a}^\beta := \partial_\beta \varphi$ and $\mathbf{a}^\alpha \cdot \mathbf{a}_\beta = \delta_\beta^\alpha$; one also defines the unit normal vector $\mathbf{a}_3 = \mathbf{a}^3 = (\mathbf{a}_1 \times \mathbf{a}_2)/|\mathbf{a}_1 \times \mathbf{a}_2|$. The *Christoffel symbols* of S are then defined as $\Gamma_{\alpha\beta}^\rho := \mathbf{a}^\rho \cdot \partial_\beta \mathbf{a}_\alpha$, and the *covariant components of the first, second,* and *third, fundamental forms* of S are given by

$$a_{\alpha\beta} = \mathbf{a}_\alpha \cdot \mathbf{a}_\beta , \qquad b_{\alpha\beta} = -\mathbf{a}_\alpha \cdot \partial_\beta \mathbf{a}_3 , \qquad c_{\alpha\beta} = b_\alpha^\rho b_{\rho\beta} ,$$

where

$$b_\alpha^\beta = a^{\beta\rho} b_{\rho\alpha} \quad \text{and} \quad a^{\alpha\beta} = \mathbf{a}^\alpha \cdot \mathbf{a}^\beta .$$

Finally, one sets $a = \det(a_{\alpha\beta})$.

Then, typically, the *unknowns* of a *two-dimensional shell model* are the *three covariant components* $\zeta_i^\varepsilon : \bar{\omega} \to \mathbf{R}$ *of the displacement* $\zeta_i^\varepsilon \mathbf{a}^i$ of the points of the shell (Fig. 5).

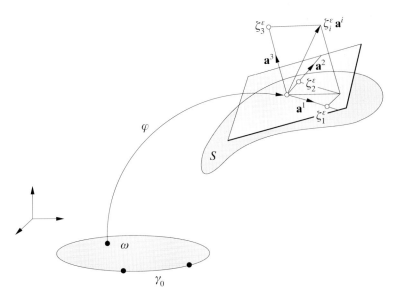

Fig. 5. When a shell is modeled by *two-dimensional equations*, the three unknowns are the covariant components $\zeta_i^\varepsilon : \bar{\omega} \to \mathbf{R}$ of the displacement $\zeta_i^\varepsilon \mathbf{a}^i$ of the points of the middle surface $\varphi(\bar{\omega})$ of the shell, where $(\mathbf{a}^\alpha)$ denotes the covariant basis of the tangent plane, and $\mathbf{a}^3 = (\mathbf{a}_1 \times \mathbf{a}_2)/(|\mathbf{a}_1 \times \mathbf{a}_2|)$, at each point of S

Remark. In some two-dimensional shell models, there may be additional unknowns. This is the case for instance of *Naghdi's model*, described in Sect. 5. □

In order to derive in a mathematically acceptable fashion a two-dimensional model from the three-dimensional equations (1)–(2) of a linearly elastic shell, a natural idea consists in studying the *asymptotic behavior, as $\varepsilon \to 0$, of the solution $\mathbf{u}^\varepsilon$ of problem* (1)–(2). To this end, we use the same *preliminaries* as those set forth by Ciarlet and Destuynder [1979 a] for the asymptotic analysis of a *plate* (see also the monographs Destuynder [1986] and Ciarlet [1990] for an overview): *Passage to a fixed domain, scalings of the* (covariant) *components of the displacement, assumptions on the data:*

Let $\Omega = \omega \times]-1, 1[$, and let $x^\varepsilon = (x_1, x_2, \varepsilon x_3)$ be the point in the set $\bar{\Omega}^\varepsilon$ that corresponds to an arbitrary point $x = (x_1, x_2, x_3)$ in the set $\bar{\Omega}$; let $\partial_{\alpha\beta} = \partial^2 / \partial x_\alpha \partial x_\beta$, and let ∂_ν denote the outer normal derivative operator along γ. We then define the *scaled unknown* $\mathbf{u}(\varepsilon) = (u_i(\varepsilon)) : \bar{\Omega} \to \mathbf{R}^3$ by letting

$$(3) \qquad u_i^\varepsilon(x^\varepsilon) = u_i(\varepsilon)(x) \quad \text{for all } x^\varepsilon \in \bar{\Omega}^\varepsilon ,$$

where $\mathbf{u}^\varepsilon = (u_i^\varepsilon) : \bar{\Omega}^\varepsilon \to \mathbf{R}^3$ is the solution of problem (1)–(2). Finally, we *assume* that there exist functions $f^i \in L^2(\Omega)$ *independent* of ε such that

$$(4) \qquad f^{i,\varepsilon}(x^\varepsilon) = \varepsilon^2 f^i(x) \quad \text{for all } x^\varepsilon \in \bar{\Omega}^\varepsilon .$$

Remark. Both relations (3) and (4) are in sharp contrast with the analogous ones for *plates*: then the scalings and assumptions are *different* for the "horizontal" and "vertical" components; for more details, see Ciarlet [1990, 1992 b], Miara [1993 a, 1993 b].   □

We next define the *space of inextensional displacements* (to within the first order; we recall that we only consider linearized theories) as

$$\mathbf{V}_0(\omega) := \{\boldsymbol{\eta} = (\eta_i) \in H^1(\omega) \times H^1(\omega) \times H^2(\omega) ;$$
$$(5) \qquad \eta_i = \partial_\nu \eta_3 = 0 \text{ on } \gamma_0 , \ \gamma_{\alpha\beta}(\boldsymbol{\eta}) = 0 \text{ in } \omega\} ,$$

where

$$(6) \qquad \gamma_{\alpha\beta}(\boldsymbol{\eta}) := \frac{1}{2}(\partial_\beta \eta_\alpha + \partial_\alpha \eta_\beta) - \Gamma_{\alpha\beta}^\rho \eta_\rho - b_{\alpha\beta} \eta_3$$

denote the *covariant components of the (linearized) strain tensor*. We also let

$$\Upsilon_{\alpha\beta}(\boldsymbol{\eta}) := \partial_{\alpha\beta} \eta_3 - \Gamma_{\alpha\beta}^\rho \partial_\beta \eta_3 - c_{\alpha\beta} \eta_3 + b_\beta^\rho(\partial_\alpha \eta_\rho - \Gamma_{\alpha\rho}^\sigma \eta_\sigma)$$
$$(7) \qquad \qquad + b_\alpha^\rho(\partial_\beta \eta_\rho - \Gamma_{\rho\beta}^\sigma \eta_\sigma) + (\partial_\alpha b_\beta^\rho + \Gamma_{\alpha\sigma}^\rho b_\beta^\sigma - \Gamma_{\alpha\beta}^\sigma b_\sigma^\rho) \eta_\rho ,$$

denote the *covariant components of the (linearized) change of curvature tensor*.

Sanchez-Palencia [1990] has recently established the following fundamental result: (i) *If* $\mathbf{V}_0(\omega) \neq \{\mathbf{0}\}$, *the first nonzero term of a formal asymptotic expansion of the scaled unknown* $\mathbf{u}(\varepsilon)$ *(as a power series in terms of ε) is of order* 0, i.e.,

$$(8) \qquad \mathbf{u}(\varepsilon) = \boldsymbol{\varsigma} + \varepsilon \mathbf{u}^1 + \varepsilon^2 \mathbf{u}^2 + \dots .$$

Furthermore, ς is independent of x_3, and the "de-scaled" field $\varsigma^\varepsilon := \varsigma$ solves the following two-dimensional variational problem:

$$(9) \qquad\qquad\qquad \varsigma^\varepsilon \in \mathbf{V}_0(\omega),$$

$$(10) \quad \int_\omega \frac{\varepsilon^3}{3} a^{\alpha\beta\rho\sigma} \Upsilon_{\alpha\beta}(\varsigma^\varepsilon) \Upsilon_{\rho\sigma}(\boldsymbol{\eta}) \sqrt{a}\, d\omega = \int_\omega p^{i,\varepsilon} \eta_i \sqrt{a}\, d\omega \quad \text{for all } \boldsymbol{\eta} \in \mathbf{V}_0(\omega),$$

where the functions $\Upsilon_{\alpha\beta}(\boldsymbol{\eta})$ are defined as in (7), and

$$(11) \qquad\qquad a^{\alpha\beta\rho\sigma} := \frac{4\lambda\mu}{(\lambda + 2\mu)} a^{\alpha\beta} a^{\rho\sigma} + 2\mu(a^{\alpha\rho} a^{\beta\sigma} + a^{\alpha\sigma} a^{\beta\rho}),$$

$$(12) \qquad\qquad\qquad p^{i,\varepsilon} := \int_{-\varepsilon}^\varepsilon f^{i,\varepsilon}\, dx_3^\varepsilon.$$

(ii) *If $\mathbf{V}_0(\omega) = \{\mathbf{0}\}$, the first nonzero term of a formal asymptotic expansion of the scaled unknown $\mathbf{u}(\varepsilon)$ is of order 2, i.e.,*

$$(13) \qquad\qquad \mathbf{u}(\varepsilon) = \varepsilon^2 \varsigma^2 + \varepsilon^3 \mathbf{u}^3 + \varepsilon^4 \mathbf{u}^4 + \dots$$

Furthermore, ς is independent of x_3, and the "de-scaled" field $\varsigma^\varepsilon := \varepsilon^2 \varsigma^2$ solves the following two-dimensional variational problem:

$$(14) \quad \varsigma^\varepsilon \in \mathbf{V}_1(\omega) := \{\boldsymbol{\eta} = (\eta_i) \in H^1(\omega) \times H^1(\omega) \times L^2(\omega)\,;\ \eta_\alpha = 0 \text{ on } \gamma_0\},$$

$$(15) \quad \int_\omega \varepsilon a^{\alpha\beta\rho\sigma} \gamma_{\alpha\beta}(\varsigma^\varepsilon) \gamma_{\rho\sigma}(\boldsymbol{\eta}) \sqrt{a}\, d\omega = \int_\omega p^{i,\varepsilon} \eta_i \sqrt{a}\, d\omega \quad \text{for all } \boldsymbol{\eta} \in \mathbf{V}_1(\omega),$$

where the functions $\gamma_{\alpha\beta}(\boldsymbol{\eta})$, $a^{\alpha\beta\rho\sigma}$, $p^{i,\varepsilon}$ are defined as in (6), (11), (12).

We thus note that the *limit problem* (i.e., as $\varepsilon \to 0$) is *either* a "bending" model (cf. (9)–(10)) *if* $\mathbf{V}_0(\omega) \neq \{\mathbf{0}\}$, *or* a "membrane" model (cf. (14)–(15)) *if* $\mathbf{V}_0(\omega) = \{\mathbf{0}\}$. Both limit problems are *two-dimensional*, and both involve the *three* unknowns ζ_i^ε and the *three* functions $p^{i,\varepsilon}$.

Note that whether the spaces $\mathbf{V}_0(\omega)$ of (5) reduces to $\{\mathbf{0}\}$ depends only on the "*geometry*" of the shell (i.e., on the mapping φ), and on the *boundary conditions* (i.e., on the set γ_0; recall that the three-dimensional boundary conditions is imposed on the set $\Phi(\Gamma_0^\varepsilon)$, with $\Gamma_0^\varepsilon = \gamma_0 \times]-\varepsilon, \varepsilon[$).

Remarks. (1) The idea of expanding the three-dimensional solution of a shell problem as a formal series of powers of the thickness seems to be due to Goldenveizer [1963].

(2) That the membrane model corresponds to $\mathbf{V}_0(\omega) = \{\mathbf{0}\}$ was already noted by Destuynder [1980, 1985].

(3) Again, the above conclusions are in sharp contrast with those corresponding to *plates*: The "bending" model for a plate only involves the unknown ζ_3^ε, and the

"membrane" model for a plate only involves the unknowns ζ_α^ε (at least in the linearized case considered here; see e. g. Ciarlet [1990] for more details).

(4) A comparison between various shell models with special geometries is made in Piila and Pitkäranta [1991, 1992]. □

A fundamental problem then consists in establishing the *convergence* (in an appropriate space) of the scaled unknown $\mathbf{u}(\varepsilon)$ to the leading term of either formal asymptotic expansions (8) and (13), i. e., in establishing that $\mathbf{u}(\varepsilon) \to \boldsymbol{\varsigma}$ if $\mathbf{V}_0(\omega) \neq \{\mathbf{0}\}$, and $\mathbf{u}(\varepsilon) \to \varepsilon^2\boldsymbol{\varsigma}^2$ if $\mathbf{V}_0(\omega) = \{\mathbf{0}\}$), with respect to some norm.

One approach consists in applying the general method set forth by Lions [1973] for analyzing *singular perturbation problems posed in variational form*. This method has been successfully applied to *plates* (cf. Destuynder [1981] and Ciarlet and Kesavan [1981]; see also Ciarlet [1990, Th. 3.3–1]) and to *shallow shells* (Ciarlet and Miara [1992b]).

Remark. Shallow shells constitute an instance where the mapping that defines the middle surface of the shell *also depends on* ε (in a well specified manner). Other such instances have been considered in the pioneering works of Destuynder [1980, 1985], who could in this fashion justify the two-dimensional shell model of Novozhilov [1959]. □

Another approach consists in applying techniques from Γ-*convergence theory*. It has already been successfully applied, to *plates*, by Anzellotti, Baldo and Percivale [1993], Bourquin, Ciarlet, Geymonat and Raoult [1992], and also to *general shells*, by Acerbi, Buttazzo and Percivale [1988] (in a setting that is however different from the one considered here).

Another important problem consists in carrying out a formal *asymptotic expansion method* that extends the results of Sanchez-Palencia [1990] to *nonlinearly elastic shells*. This approach has already been applied to nonlinearly elastic plates (cf. Ciarlet and Destuynder [1979 b] and Ciarlet [1980]), to *nonlinearly elastic shallow shells* (Ciarlet and Paumier [1986]), and to nonlinearly elastic plates undergoing *"large deformations"* (Fox, Raoult and Simo [1993]). In this respect, see in particular Rao [1989] and de Figueiredo [1990 a] for "special" shells, Muttin [1991] for *sails* viewed as nonlinear membrane models, Basar and Krätzig [1988] for shells undergoing "large" deformations.

It should also be noted that, even in the *nonlinear* case, *convergence* is likely to be "at hand": This hope is motivated by convergence results recently obtained by Acerbi, Buttazzo and Percivale [1991] for *nonlinearly elastic strings*, and by Le Dret and Raoult [1993] for *nonlinearly elastic plates*; in both instances, the proofs again rely on Γ-*convergence theory*.

Remarks. (1) This approach (formal asymptotic expansion method, supplemented by a convergence analysis whenever it is possible) should be extended to *eigenvalue problems*, as in Ciarlet and Kesavan [1981] for *plates*, and to *time-dependent problems*, as in Raoult [1985, 1988] for *plates*.

(2) Another problem consists in computing *"higher-order" terms in the formal asymptotic expansion* of the scaled solution, as in Raoult [1985] for *plates*. This is probably related to the *spectral method*, where one seeks the best "projection" on a specific space of "polynomial functions"; see Miara [1989] for *plates*, Miara and Trabucho [1992] for *beams*, Figueiredo and Trabucho [1992] for *shallow shells*. □

4. Koiter's Two-Dimensional Model

The fundamental work of John [1965] has led Koiter [1970] to propose two-dimensional shell models in both the linear and nonlinear cases, which are very commonly used, especially in Computational Mechanics. For a *linearly elastic shell, Koiter's model* consists in finding a field ς^ε that solves the following two-dimensional variational problem:

(16) $\varsigma^\varepsilon \in \mathbf{V}(\omega) := \{\boldsymbol{\eta} = (\eta_i) \in H^1(\omega) \times H^1(\omega) \times H^2(\omega) ; \ \eta_i = \partial_\nu \eta_3 \text{ on } \gamma_0\}$,

$$\int_\omega \left\{ \frac{\varepsilon^3}{3} a^{\alpha\beta\rho\sigma} \Upsilon_{\alpha\beta}(\varsigma^\varepsilon) \Upsilon_{\rho\sigma}(\boldsymbol{\eta}) + \varepsilon a^{\alpha\beta\rho\sigma} \gamma_{\alpha\beta}(\varsigma^\varepsilon) \gamma_{\rho\sigma}(\boldsymbol{\eta}) \right\} \sqrt{a} \, d\omega$$

(17) $$= \int_\omega p^{i,\varepsilon} \eta_i \sqrt{a} \, d\omega \quad \text{for all } \boldsymbol{\eta} \in \mathbf{V}(\omega),$$

where the tensors $(\gamma_{\alpha\beta}(\varsigma))$, $(\Upsilon_{\alpha\beta}(\varsigma))$, and $(a^{\alpha\beta\rho\sigma})$ are defined as in (6), (7), and (11), and the field $p^{i,\varepsilon} \mathbf{a}_i$, with $p^{i,\varepsilon} \in L^2(\omega)$, represents the resultant of the applied forces along the middle surface S.

A simple inspection of eqs. (17) thus reveals that their left-hand side is nothing but the *sum* of the left-hand side found in the eqs. (10) of the "bending" model, and of the left-hand side found in the eqs. (15) of the "membrane" model!

The justification of this "addition", i.e., a rigorous mathematical justification of Koiter's model from three-dimensional elasticity, remains to be seen! Clearly, Koiter's model is *not* a "limit" problem, i.e., one that can be obtained in some rational fashion as a *limit* when $\varepsilon \to 0$. As observed in Ciarlet [1992 b], this can be easily checked by considering the orders (with respect to ε) of the different terms found in eqs. (17). Another convincing argument was given by Sanchez-Palencia [1989 a, 1989 b] who showed that the solution of Koiter's model itself converges as $\varepsilon \to 0$ *either* to that of the *bending problem* (9)–(10) *if* $\mathbf{V}_0(\omega) \neq \{\mathbf{0}\}$ *or* to that of the *membrane problem* (14)–(15) *if* $\mathbf{V}_0(\omega) = \{\mathbf{0}\}$.

It is nevertheless plausible, although it remains to be proven, that, for "moderately small' values of ε, Koiter's model is a "better" model than the limit problem found by the asymptotic analysis. This would explain why it has gained so much popularity among the Engineering community.

Remark. A "partial" justification of Koiter's model from three-dimensional elasticity has been provided by Destuynder [1985], through ad'hoc *a priori* assumptions, a cruder variant of which leads to the two-dimensional model proposed by Budiansky and Sanders [1967].   □

5. Naghdi's Two-Dimensional Model

Naghdi [1963] (see also Naghdi [1972]) has proposed yet another two-dimensional linear shell model, whose unknowns are the three covariant components $\zeta_i^\varepsilon : \bar{\omega} \to \mathbf{R}$ of the displacement of the points of the middle surface S as before, *and* the two (linearized) components r_α^ε of the rotation of the normal vector $\mathbf{a}_3$. The unknowns $\boldsymbol{\varsigma}^\varepsilon = (\zeta_i^\varepsilon)$ and $\mathbf{r}^\varepsilon = (r_\alpha^\varepsilon)$ solve the following two-dimensional variational problem:

$$(18) \quad (\boldsymbol{\varsigma}^\varepsilon, \mathbf{r}^\varepsilon) \in \mathbf{W}(\omega) := \{(\boldsymbol{\eta}, \mathbf{s}) = ((\eta_i), (s_\alpha)) \in \mathbf{H}^1(\omega) \,; \ \eta_i = s_\alpha = 0 \text{ on } \gamma_0\},$$

$$
(19) \quad \int_\omega \left\{ \frac{\varepsilon^3}{3} a^{\alpha\beta\rho\sigma} \chi_{\alpha\beta}(\boldsymbol{\varsigma}^\varepsilon, \mathbf{r}^\varepsilon) \chi_{\rho\sigma}(\boldsymbol{\eta}, \mathbf{s}) + \varepsilon a^{\alpha\beta\rho\sigma} \gamma_{\alpha\beta}(\boldsymbol{\varsigma}^\varepsilon) \gamma_{\rho\sigma}(\boldsymbol{\eta}) \right.
$$
$$
\left. + 2\mu\varepsilon a^{\alpha\beta}(\theta_\alpha(\boldsymbol{\varsigma}^\varepsilon) + r_\alpha^\varepsilon)(\theta_\beta(\boldsymbol{\eta}) + s_\beta) \right\} \sqrt{a}\, d\omega = L(\boldsymbol{\eta}, \mathbf{s})
$$

for all $(\boldsymbol{\eta}, \mathbf{s}) \in \mathbf{W}(\omega)$, where the tensors $(\gamma_{\alpha\beta}(\boldsymbol{\eta}))$ and $(a^{\alpha\beta\rho\sigma})$ are defined as in (6) and (11),

$$
\chi_{\alpha\beta}(\boldsymbol{\eta}, \mathbf{s}) := \frac{1}{2}(\partial_\beta s_\alpha + \partial_\alpha s_\beta) - \Gamma_{\alpha\beta}^\rho s_\rho - \frac{1}{2} b_\alpha^\rho (\partial_\beta \eta_\rho - \Gamma_{\rho\beta}^\sigma \eta_\sigma)
$$
$$
- \frac{1}{2} b_\beta^\rho (\partial_\alpha \eta_\rho - \Gamma_{\rho\alpha}^\sigma \eta_\sigma) + \frac{1}{2}(b_\alpha^\rho b_{\rho\beta} + b_\beta^\rho b_{\rho\alpha})\eta_3 \,,
$$

$$
\theta_\alpha(\boldsymbol{\eta}) := \partial_\alpha \eta_3 + b_\alpha^\rho \eta_\rho \,,
$$

and $L(\cdot)$ is a linear form, continuous over $\mathbf{W}(\omega)$, that takes into account the applied forces.

Finding a mathematical justification of Naghdi's model from three-dimensional elasticity is probably a challenging problem, since its "plate counterpart", i. e., that of justifying the *Reissner-Mindlin model* has not yet received a fully satisfactory answer. In this direction, see however Babuška, d'Harcourt and Schwab [1991], Babuška and Li [1992], Schwab [1989].

6. Existence Theory

A major difficulty in showing the existence of a solution to anyone of the linear two-dimensional shell models described above lies of course in checking whether the bilinear form appearing in the left-hand side of the variational equations is *elliptic*, or equivalently, coercive over the space where the unknown is sought.

The $\mathbf{V}(\omega)$-ellipticity of (the bilinear form found in) *Koiter's model* (16)–(17) was first established by Bernadou and Ciarlet [1976]. The proof relied on various equivalences of norms involving covariant derivatives (some due to Rougée [1969]), on a (linearized) "rigid displacement lemma", and on "technical" inequalities combined with weak lower semi-continuity properties of the associated quadratic functional. A simpler proof was recently proposed by Ciarlet and Miara [1992 a]; see also Bernadou, Ciarlet and Miara [1993]. This proof still uses the same "rigid displacement lemma", but it otherwise relies on the same crucial *lemma of J. L. Lions* that was used in proving the existence of a solution to problem (1)–(2). As shown in Bernadou, Ciarlet and Miara [1993], this "new" proof applies to *Naghdi's model* (18)–(19); The corresponding "rigid displacement lemma" is due to Coutris [1978].

While the $\mathbf{V}_0(\omega)$-ellipticity of the *bending model* (9)–(10) is an easy corollary of the $\mathbf{V}(\omega)$-ellipticity of *Koiter's model* (16)–(17), proving the $\mathbf{V}_1(\omega)$-ellipticity of the *membrane model* (14)–(15) is quite another story. To understand why, observe that the third unknown ζ_3^ε is only required to lie in the space $L^2(\omega)$ (see the definition (14) of the space $\mathbf{V}_1(\omega)$). This is also reflected by the equivalence (at least formally) with a system of partial differential equations which is of the *second* order with respect to the unknowns ζ_α^ε, but only of the *first* order with respect to the unknown ζ_3^ε. More specifically, this system reads (together with appropriate boundary conditions):

$$\begin{cases} \dfrac{-\varepsilon}{\sqrt{a}}\partial_\beta[a^{\alpha\beta\rho\sigma}\gamma_{\rho\sigma}(\varsigma^\varepsilon)] - a^{\tau\beta\rho\sigma}\Gamma_{\tau\beta}^\alpha\gamma_{\rho\sigma}(\varsigma^\varepsilon) = p^{\alpha,\varepsilon} \text{ in } \omega\,, \\[2mm] \qquad\qquad - \varepsilon a^{\alpha\beta\rho\sigma}b_{\alpha\beta}\gamma_{\rho\sigma}(\varsigma^\varepsilon) = p^{3,\varepsilon} \text{ in } \omega\,. \end{cases}$$

It has been recently shown (see Ciarlet and Sanchez-Palencia [1993], Ciarlet and Lods [1993]) and also the earlier works of Sanchez-Palencia [1989 a, 1989 b], 1990] and Geymonat and Sanchez-Palencia [1991]) that, if the middle surface S of the shell is *uniformly elliptic* and if all the data are *analytic*, the membrane model (14)–(15) has one and only one solution in the space $\mathbf{V}_1(\omega)$ and the space $\mathbf{V}_0(\omega)$ of (5) reduces to $\{\mathbf{0}\}$ (as it should, since the membrane model is derived precisely under this assumption; cf. Sect. 3). We recall that a smooth surface is "uniformly elliptic" if the two principal radii of curvature $R_1(s)$ and $R_2(s)$ are of the same sign for all $s \in S$ and there exists a constant $c > 0$ such that $c^{-1} \leq \frac{1}{|R_\alpha(s)|} \leq c$ for all $s \in S$, $\alpha = 1, 2$.

The problems of determining whether the space $\mathbf{V}_0(\omega)$ reduces to $\{\mathbf{0}\}$ or not and of finding an existence theory for other "general" shells and boundary conditions remain however basically open. In this direction, see however Sanchez-Palencia [1993] where membrane models in the "hyperbolic" case are considered.

The linear two-dimensional shell models mentioned in Sects. 3,4,5 are in fact linearized versions of more general *nonlinear two-dimensional models*, different nonlinear models possibly giving rise to the same linear one. For instance, a *nonlinear membrane model* is described in Destuynder [1982], a *nonlinear Koiter's model* is described in Koiter [1966], etc. There then remains to carry out a mathematical analysis of the *existence, regularity, and possible non-uniqueness* (bifurcation) of solutions of each one of these available two-dimensional models for nonlinearly elastic shells. Partial results are already available, especially for *shallow shells*: See Bernadou and Oden [1981], Rupprecht [1981], Destuynder [1983], Kesavan and Srikanth [1983], Rao [1989], Paumier and Rao [1990], de Figueiredo [1990 b].

Remark. By contrast, a wealth of results are available concerning two-dimensional models of *nonlinearly elastic plates*. See in particular Berger [1967], Knightly [1967], Lions [1969], Duvaut and Lions [1974 a, 1974 b], Berger [1977], Golubitsky and Shaeffer [1979], Ciarlet and Rabier [1980].   □

7. What Happens When a Shell Becomes a Plate?

Let us imbed the mapping $\varphi : \bar{\omega} \to \mathbb{R}^3$ that defines the middle surface of a given shell into a one-parameter family of mappings $\varphi(t) : \bar{\omega} \to \mathbb{R}^3$, by letting

$$\varphi(t) = \begin{pmatrix} (1-t)x_1 + t\varphi_1(x_1, x_2) \\ (1-t)x_2 + t\varphi_2(x_1, x_2) \\ t\varphi_3(x_1, x_2) \end{pmatrix}, \quad 0 \le t \le 1,$$

so that $\varphi(1) = \varphi$ corresponds to the given *shell*, while $\varphi(0)$ corresponds to a *plate*, with $\bar{\omega}$ as its middle surface. We say that *"the shell becomes a plate"* it $t \to 0$.

As shown in Ciarlet [1992 a], *for a fixed $\varepsilon > 0$, the solution* $\mathbf{u}^\varepsilon(t)$ *of the three-dimensional shell equations* (1)–(2), with coefficients $A^{ijkl}(t)$, $\Gamma_{ij}^p(t)$, $g(t)$ (now depending on t in an obvious manner) *converges as $t \to 0$ in the space* $\mathbf{H}^1(\Omega^\varepsilon)$, where $\Omega^\varepsilon = \omega \times] - \varepsilon, \varepsilon[$, *to the solution* $\mathbf{u}^\varepsilon(0)$ *of the three-dimensional plate equations*

$$(20) \qquad \mathbf{u}^\varepsilon(0) \in \mathbf{V}(\Omega^\varepsilon) = \{\mathbf{v} = (v_i) \in \mathbf{H}^1(\Omega^\varepsilon);\ \mathbf{v} = \mathbf{0} \text{ on } \Gamma_0^\varepsilon\},$$

$$(21) \qquad \int_{\Omega^\varepsilon} B^{ijkl} e_{ij}(\mathbf{u}^\varepsilon(0)) e_{kl}(\mathbf{v})\, dx^\varepsilon = \int_{\Omega^\varepsilon} f^{i,\varepsilon} v_i\, dx^\varepsilon \quad \text{for all } \mathbf{v} \in \mathbf{V}(\Omega^\varepsilon),$$

where (δ^{ij} denotes the Kronecker's symbol)

$$B^{ijkl} := \lambda \delta^{ij}\delta^{kl} + \mu(\delta^{ik}\delta^{jl} + \delta^{il}\delta^{jk}).$$

As shown in Ciarlet [1990, Theorem 3.3–1], the appropriately "scaled" solution of the three-dimensional plate equations (20)–(21) *converges as $\varepsilon \to 0$ with respect to the norm* $\|\cdot\|_{\mathbf{H}^1(\Omega)}$, $\Omega = \omega \times] - 1, 1[$, *towards the solution of a two-dimensional plate model*, where *both* the *"bending"* and *"membrane"* equations *simultaneously appear*.

Suppose instead that we now *"interchange the orders of the limits"*: This time, the thickness goes to zero first and secondly, "the shell becomes a plate". Then, according to the asymptotic analysis described in Sect. 3, the limit problem as $\varepsilon \to 0$ is *either* the two-dimensional "bending model" (9)–(10) *or* the two-dimensional "membrane model" (14)–(15), according to the geometry of the middle surface of the shell and the boundary conditions. When "the shell becomes a plate", each one of these two-dimensional shell model "formally converges" to the corresponding two-dimensional plate model. In other words, it is no longer possible to *simultaneously* obtain both the "bending" and "membrane" equations of a plate (cf.

Ciarlet [1992 b] and Ciarlet and Lods [1994] for a more detailed analysis of this phenomenon).

Although *Koiter's model* cannot be viewed as a limit model as $\varepsilon \to 0$ as was pointed out in Sect. 4, it nevertheless possesses a precious virtue, which is not shared by either the bending or the membrane models: As "the shell becomes a plate", i.e., as $t \to 0$, its solution $\varsigma^\varepsilon(t) = (\zeta_i^\varepsilon(t))$ (now depending on t in an obvious manner) converges in the space $H^1(\omega) \times H^1(\omega) \times H^2(\omega)$ to the solution of the "complete" two-dimensional plate model, i.e., that includes both "bending" and "membrane" equations (cf. Ciarlet [1992 b]).

8. Elastic Multi-Structures That Comprise Shells

One of the simplest *elastic multi-structures* (this definition was proposed by Ciarlet [1990] that comprises a shell is a *rotor and its blades* (Fig. 1). Then the asymptotic analysis developed by Ciarlet, Le Dret and Nzengwa [1989] (see also Ciarlet [1990] and Le Dret [1991] for an overview) for modeling the *"canonical" elastic multi-structure*, composed of a three-dimensional substructure and a plate, likewise allows to model the rotor (the three-dimensional substructure) together with one blade (the "two-dimensional" substructure, either viewed as a "shallow shell" or as a "general" shell) as an elastic multi-structure, both in the linearly and nonlinearly elastic cases; see Rodriguez [1993]. The corresponding *eigenvalue, time-dependent*, and *nonlinearly elastic, problems* can be likewise modeled by the techniques of Bourquin and Ciarlet [1989], Raoult [1992] and Aufranc [1993], respectively, again developed for the "canonical" elastic multi-structure.

Using the (closely related) asymptotic analysis that Le Dret [1989, 1990] developed for handling elastic multi-structures composed of *plates*, one should be able to likewise model a *cylindrical tank* (Fig. 2) as an instance of a multi-structure comprising a cylindrical shell and a plate, or a *multi-structure comprising two cylindrical shells*, as shown in Figs. 6 and 7. Such structures are commonly found, for instance in offshore platforms for oil extraction; their appropriate modeling is thus a question of paramount practical importance.

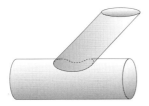
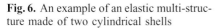

Fig. 6. An example of an elastic multi-structure made of two cylindrical shells

Fig. 7. Another example of an elastic multi-structure made of two cylindrical shells

Note in passing that elastic multi-structures consitute an ideal field of investigation and testing of the performances of *domain decomposition methods*: For once, the "subdomains" are clearly identified! See in this direction d'Hennezel [1993].

A related problem consists in studying the possible *rigidification of the folds* that are found when *shells are assembled in an elastic multi-structure*. This problem has been completely solved for junctions between *plates*, by Le Dret [1989, 1990], who has shown that such folds "become" *rigid* as the thickness of the plates approaches zero. The situation is far less clear for shells, and seems quite challenging (see in this direction Geymonat and Sanchez-Palencia [1991], Akian and Sanchez-Palencia [1992], Percivale [1990]).

A shell with *stiffeners* is another instance of an elastic multi-structure, this time comprising a "two-dimensional" substructure (the shell) and "one-dimensional" substructures (the stiffeners). In the corresponding asymptotic analysis, the stiffeners are to be modeled as *curved beams* (see Trabucho and Viaño [1994] for an overview of the asymptotic analysis applied to "one-dimensional" bodies, such as beams and rods).

Another multi-structure likewise comprising a shell and "one-dimensional" substructures, differently assembled however, is a *cooling tower*, together with its supporting *rods* (Fig. 3); in this respect, see Aufranc [1990, 1991] for plates with stiffeners, Gruais [1993 a, 1993 b] for plates and rods.

It should be clear that the modeling and mathematical analysis of the multi-structures described in this section pose extremely challenging, and thus fascinating, mathematical problems.

Acknowledgement. During the past 15 years I have had the good fortune of participating to several conferences held at the Oberwolfach Institute. I will always remember with special fondness the serene and congenial atmosphere there, the perfect accomodations, and the endless conversations, whether they be leisurely after-dinner chats in the lounge, or lively mathematical discussions during long walks in the surrounding hills. I am grateful that I could benefit from such a unique environment; a conference at the Institute stands as the paradigm of an ideal gathering of mathematicians.

Reference

Acerbi, E., Buttazzo, G., Percivale, D. (1988): Thin inclusions in linear elasticity: a variational approach. J. Reine Angew. Math. **386**, 99–115

Acerbi, E., Buttazzo, G., Percivale, D. (1991): A variational definition of the strain energy for an elastic string. J. Elasticity **25**, 137–148

Akian, J.L., Sanchez-Palencia, E. (1992): Approximation de coques élastiques minces par facettes planes: Phénomène de blocage membranaire. C.R. Acad. Sci. Paris, Sér. I **315**, 363–369

Amrouche, E., Girault, V. (1990): Propriétés fonctionnelles d'opérateurs; applications au problème de Stokes en dimension quelconque. Rapport R90025, Laboratoire d'Analyse Numérique, Université Pierre et Marie Curie, Paris

Anzellotti, G., Baldo, S., Percivale, D. (1993): Dimension reduction in variational problems, asymptotic development in Γ-convergence and thin structures in elasticity. Asymptotic Anal. (To appear)

Aufranc, M. (1990): Plaques raidies par des poutres. C.R. Acad. Sci. Paris, Ser. I **311**, 835–838

Aufranc, M. (1991): Junctions between three-dimensional and two-dimensional nonlinearly elastic structures. Asymptotic Anal. **4**, 319–338

Babuška, I., d'Harcourt, J.M., Schwab, C. (1991): Optimal shear correction factors in hierarchic plate modeling. Mathematical Modelling & Sc. Computing **1**, 1–30

Babuška, I., Li, L. (1992): The problem of plate modeling: Theoretical and computational results. Comput. Meth. Appl. Mech. Engrg. **100**, 249–273

Ball, J.M. (1997): Convexity conditions and existence theorems in nonlinear elasticity. Arch. Rational Mech. Anal. **63**, 337–403

Başar, Y., Krätzig, W.B. (1988): A consistent shell theory for finite deformations. Acta Mech. **76**, 73–87

Berger, M.S. (1967): On the von Kármán equations and the buckling of a thin elastic plate, I. The clamped plate. Comm. Pure Appl. Math. **20**, 687–719

Berger, M.S. (1977): Nonlinearity and functional analysis. Academic Press, New York

Bernadou, M. (1993): Méthodes d'eléments finis pour les coques minces. Masson, Paris

Bernadou, M., Ciarlet, P.G. (1976): Sur l'ellipticité du modèle linéaire de coques de W.T. Koiter. In: Computing methods in applied sciences and engineering, R. Glowinski and J.-L. Lions (eds.) (Lecture Notes in Economics and Mathematical Systems, Vol. 134, pp. 89–136) Springer, Berlin

Bernadou, M., Ciarlet, P.G., Miara, B. (1993): Existence theorems for two-dimensional linear shell theories. J. Elasticity (To appear)

Bernadou, M., Oden, J.T. (1981): An existence theorem for a class of nonlinear shallow shell problems. J. Math. Pures Appl. **60**, 285–308

Borchers, W., Sohr, H. (1990): On the equations rot $v = g$ and div $u = f$ with zero boundary conditions. Hokkaido Math. J. **19**, 67–87

Bourquin, F., Ciarlet, P.G. (1989): Modeling and justification of eigenvalue problems for junctions between elastic structures. J. Funct. Anal. **87**, 392–427

Bourquin, F., Ciarlet, P.G., Geymonat, G., Raoult, A. (1992): Γ-convergence et analyse asymptotique des plaques minces. C.R. Acad. Sci. Paris, Sér. I **315**, 1017–1024

Budiansky, B., Sanders, J.L. (1967): On the "best" first-order linear shell theory. In: Progress in Applied Mechanics (W. Prager Anniversary Volume, pp. 129–140). Macmillan, New York

Ciarlet, P.G. (1980): A justification of the von Kármán equations. Arch. Rational Mech. Anal. **73**, 349–389

Ciarlet, P.G. (1988): Mathematical elasticity, Vol. I: Three-dimensional elasticity. North-Holland, Amsterdam

Ciarlet, P.G. (1990): Plates and junctions in elastic multi-structures: an asymptotic analysis. Masson, Paris

Ciarlet, P.G. (1991): Basic error estimates for elliptic problems. In: Handbook of Numerical Analysis (P.G. Ciarlet and J.-L. Lions (eds.)), Vol. II: Finite element methods, Part 1, pp. 17–351. North-Holland, Amsterdam

Ciarlet, P.G. (1992a): Echange de limites en théorie asymptotique de coques. I. En premier lieu, "la coque devient une plaque". C.R. Acad. Sci. Paris, Sér. I **315**, 107–111

Ciarlet, P.G. (1992b): Echange de limites en théorie asymptotique de coques. II. En premier lieu, l'épaisseur tend vers zéro. C.R. Acad. Sci. Paris, Sér. I **315**, 227–233

Ciarlet, P.G. (1993): Modèles bi-dimensionals de coques: analyse asymptotique et théorèmes d'existence. (To appear)

Ciarlet, P.G., Destuynder, P. (1979a): A justification of the two-dimensional plate model. J. Mécanique **18**, 315–344

Ciarlet, P.G., Destuynder, P. (1979b): A justification of a nonlinear model in plate theory. Comput. Methods Appl. Mech. Engrg. **17–18**, 227–258

Ciarlet, P.G., Kesavan, S. (1981): Two-dimensional approximation of three-dimensional eigenvalue problems in plate theory. Comput. Methods Appl. Mech. Engrg. **26**, 149–172

Ciarlet, P.G., Le Dret, H., Nzengwa, R. (1989): Junctions between three-dimensional and two-dimensional linearly elastic structures. J. Math. Pures Appl. **68**, 261–295

Ciarlet, P.G., Lods, V. (1993): On the ellipticity of linear membrance shell equations. (To appear)

Ciarlet, P.G., Lods, V. (1994): Limites des modèles d'une coque "en flexion" et "en membrane" lorsque "la coque devient une plaque". (To appear)

Ciarlet, P.G., Miara B. (1992a): On the ellipticity of linear shell models. Z. angew. Math. Phys. **43**, 243–253

Ciarlet, P.G., Miara, B. (1992b): Justification of the two-dimensional equations of a linearly elastic shallow shell. Comm. Pure Appl. Math. **XLV**, 327–360

Ciarlet, P.G., Paumier, J.C. (1986): A justification of the Marguerre-von Kármán equations. Computational Mechanics **1**, 177–202

Ciarlet, P.G., Rabier, P. (1980): Les equations de von Kármán. Lecture Notes in Mathematics, Vol. 826. Springer, Berlin

Ciarlet, P.G., Sanchez-Palencia, E. (1993): Existence theory for two-dimensional membrane shell equations. (To appear)

Coutris, N. (1978): Théorème d'existence et d'unicité pour un problème de coque élastique dans le cas d'un modèle linéaire de P.M. Naghdi. RAIRO Analyse Numérique **12**, 51–57

Destuynder, P. (1980): Sur une justification des modèles de plaques et de coques par les méthodes asymptotiques. Thèse d'Etat, Université Pierre et Marie Curie

Destuynder, P. (1981): Comparison entre les modèles tri-dimensionnels et bi-dimensionnels de plaques en élasticité. RAIRO Analyse Numérique **15**, 331–369

Destuynder, P. (1982): On nonlinear membrane theory. Comput. Methods Appl. Mech. Engrg. **32**, 377–399

Destuynder, P. (1983): An existence theorem for a nonlinear shell model in large displacement analysis. Math. Meth. Applied Sci. **5**, 68–83

Destuynder, P. (1985): A classification of thin shell theories. Acta Applicandæ Mathematicæ **4**, 15–63

Destuynder, P. (1986): Une théorie asymptotique des plaques minces en élasticité linéaire. Masson, Paris

Duvaut, G., Lions, J.-L. (1972): Les inéquations en mécanique et en physique. Dunod, Paris

Duvaut, G., Lions, J.-L. (1974a): Problèmes unilatéraux dans la théorie de la flexion forte des plaques. J. Mécanique **13**, 51–74

Duvaut, G., Lions, J.-L. (1974b): Problèmes unilatéraux dans la théorie de la flexion forte des plaques, II: le cas d'évolution. J. Mécanique **13**, 245–266

de Figueiredo, I.N. (1990a): A justification of the Donnell-Mushtari-Vlasov model by the asymptotic expansion method. Asymptotic Anal. **36**, 221–234

de Figueiredo, I.N. (1990b): Local existence and regularity of the solution of the nonlinear thin shell model of Donnell-Mushtari-Vlasov. Applicable Anal. **36**, 221–234

de Figueiredo, I.N., Trabucho, L. (1992): A Galerkin approximation for linear elastic shallow shells. Computational Mechanics **10**, 107–119

Fox, D.D., Raoult, A., Simo, J.C. (1993): A justification of nonlinear properly invariant plate theories. Arch. Rational Mech. Anal. (To appear)

Geymonat, G., Sanchez-Palencia, E. (1991): Remarques sur la rigidité infinitésimale de certaines surfaces elliptiques non régulières, non convexes et applications. C.R. Acad. Sci. Paris, Sér. I **313**, 645–651

Goldenveizer, A.L. (1963): The construction of an approximate theory of shells by means of asymptotic integration of elasticity equations. Prikl. Math. Mekh. **27**, 594–608

Golubitsky, M., Shaeffer, D. (1979): Boundary conditions and mode jumping in the buckling of a rectangular plate. Comm. Math. Phys. **69**, 209–236

Gruais, I. (1993a): Modélisation de la jonction entre une plaque et une poutre en élasticité linéarisée. Modél. Math. Anal. Numér. **27**, 77–105

Gruais, I. (1993b): Modeling of the junction between a plate and a rod in nonlinear elasticity. Asymptotic Anal. **7**, 179–194

d'Hennezel, F. (1993): Domain decomposition method and elastic multi-structures: the stiffened plate problem. Numer. Math. (To appear)

John, F. (1965): Estimates for the derivatives on the stresses in a thin shell and interior shell equations. Comm. Pure Appl. Math. **18**, 235–267

Kesavan, S., Srikanth, P.N. (1983): On the Dirichlet problem for the Marguerre equations. Nonlinear Anal. **7**, 209–216

Knightly, G.H. (1967): An existence theorem for the von Kármán equations. Arch. Rational Mech. Anal. **27**, 233–242

Koiter, W.T. (1966): On the nonlinear theory of thin elastic shells. Proc. Kon. Nederl. Akad. Wetensch. **B69**, 1–54

Koiter, W.T. (1970): On the foundation of the linear theory of thin elastic shells. Proc. Kon. Nederl. Akad. Wetensch. **B73**, 169–195

Le Dret, H. (1989): Folded plates revisited. Comput. Mech. **5**, 345–365

Le Dret, H. (1990): Modeling of a folded plate. Comput. Mech. **5**, 401–416

Le Dret, H. (1991): Problèmes variationnels dans les multi-domaines-Modélisation des jonctions et applications. Masson, Paris

Le Dret, H., Raoult, A. (1993): Le modèle de membrane non linéaire comme limite variationnelle de l'élasticité non linéaire tridimensionnelle. C.R. Acad. Sci. Paris, Sér. I. **317**, 221–226

Le Tallec, P. (1993): Numerical methods for nonlinear three-dimensional elasticity. In: Handbook of Numerical Analysis (P.G. Ciarlet and J.-L. Lions (eds.)), Vol. III. North-Holland, Amsterdam (To appear)

Lions, J.-L. (1969): Quelques méthodes de résolution des problèmes aux limites non linéaires. Dunod, Paris

Lions, J.-L. (1973): Perturbations singulières dans les problèmes aux limites et en contrôle optimal. Lecture Notes in Mathematics, Vol. 323. Springer, Heidelberg

Magenes, E., Stampacchia, G. (1958): I problemi al contorno per le equazioni differenziali di tipo ellitico. Ann. Scuola Norm. Sup. Pisa **12**, 247–358

Miara, B. (1989): Optimal spectral approximation in linearized plate theory. Applicable Anal. **31**, 291–307

Miara, B. (1993a): Justification of the asymptotic analysis of elastic plate, I: The linear case. Asymptotic Anal. (To appear)

Miara, B. (1993b): Justification of the asymptotic analysis of elastic plates, II: The nonlinear case. Asymptotic Anal. (To appear)

Miara, B., Trabucho, L. (1992): A Galerkin spectral approximation in linearized beam theory. Modél. Math. Anal. Numér. **26**, 425–446

Muttin, F. (1991): Structural analysis of sails. European J. Mech. A, Solids **10**, 517–534

Naghdi, P.M. (1963): Foundations of elastic shell theory. In: Progress in Solid Mechanics, Vol. 4, pp. 1–90. North-Holland, Amsterdam

Naghdi, P.M. (1972): The theory of shells and plates. In: Handbuch der Physik, Vol. VI a.2, pp. 425–640. Springer, Berlin

Novozhilov, V.V. (1959): Thin shell theory. Wolters-Noordhoff, Groningen

Paumier, J.C., Rao, B. (1989): Qualitative and quantitative analysis of buckling of shallow shells. European J. Mech. A, Solids **8**, 461–489

Percivale, D. (1990): Folded shells: a variational approach. Report SISSA, Trieste

Piila, J., Pitkäranta, J. (1991): Energy estimates relating different linear elastic models of a thin cylindrical shell. (To appear)

Piila, J., Pitkäranta, J. (1992): Characterization of the membrane theory of a thin clamped shell of resolution; the parabolic case. (To appear)

Rao, B. (1989): Sur quelques questions d'élasticité non linéaire relatives aux coques minces. Doctoral dissertation, Université Pierre et Marie Curie, Paris

Raoult, A. (1985): Construction d'un modèle d'évolution de plaques avec terme d'inertie de rotation. Ann. Mat. Pura Appl. **139**, 361–400

Raoult, A. (1988): Analyse mathématique de quelques modèles de plaques et de poutres élastiques ou élasto-plastiques. Doctoral dissertation, Université Pierre et Marie Curie, Paris

Raoult, A. (1992): Asymptotic modeling of the elastodynamics of a multi-structure. Asymptotic Anal. **6**, 73–108

Rodriguez, J.M. (1993): Mathematical modeling of a rotor and its blades. (To appear)

Rougée, P. (1969): Equilibre des coques élastiques minces inhomogènes en théorie non linéaire. Thèse d'Etat, Université de Paris

Rupprecht, G. (1981): A singular perturbation approach to nonlinear shell theory. Rocky Mountains J. Math. **11**, 75–98

Sanchez-Palencia, E. (1989a): Statique et dynamique des coques minces. I. Cas de flexion pure non inhibée. C.R. Acad. Sci. Paris, Sér. I **309**, 411–417

Sanchez-Palencia, E. (1989b): Statique et dynamique des coques minces. II. Cas de flexion pure inhibée. C.R. Acad. Sci. Paris, Sér. I **309**, 531–537

Sanchez-Palencia, E. (1990): Passage à la limite de l'élasticité tri-dimensionnelle à la théorie asymptotique des coques minces. C.R. Acad. Sci. Paris, Sér. II **311**, 909–916

Sanchez-Palencia, E. (1993): On the membrane approximation for thin elastic shells in the hyperbolic case. Revista Matematica de la Universidad Complutense de Madrid (To appear)

Schwab, C. (1989): The dimensional reduction method. Ph.D. Thesis, University of Maryland, College Park

Trabucho, L., Viaño, J.M. (1994): Mathematical modeling of rods. In: Handbook of Numerical Analysis (P.G. Ciarlet and J.-L. Lions (eds.)), Vol. IV. North-Holland, Amsterdam (To appear)

On Algebras, Wild and Tame

José A. de la Peña and Peter Gabriel

Ein Liedlein will ich singen
von Honigvögelein,
die hin und her sich schwingen,
wo bunte Blumen sein.
Das Völklein in dem Grünen,
es schmauset auf der Weid,
ich singe von den Bienen
auf dieser freien Heid.

Not 'big cats', small-scale 'beasts' only, finite-dimensional algebras and their representations, are to be described in these lines. Serviceable though they seem, our denizens still stay on the skirts of the well-established realms. Requiring sustained attention, they have attracted but small groups of 'hunters', now in the second generation. The present work of these searchers is to be surveyed here.

1. Some Rudiments

1.1 Indecomposable Modules

For the sake of simplicity, we fix an algebraically closed ground field k and suppose, unless stated otherwise, that our (left) modules and (associative) algebras (with 1) have finite dimensions over k. Then each module is a finite direct sum of *indecomposable* submodules, i.e. of non-zero submodules which are not direct sums of two non-zero submodules. The *iso(morphism)classes* of the indecomposables appearing in such a finite direct sum, together with the multiplicities of their occurrences, are independent of the considered decomposition. They only depend on the isoclass of the given module. For a given algebra, the classification of its modules (up to isomorphism) thus amounts to classifying the indecomposables [18].

1.2 Spectroids

We denote by mod_A the category of all modules over an algebra A, by ind_A the full subcategory formed by chosen representatives of the isoclasses of indecomposable A-modules. The category ind_A is a *spectroid* over k, i.e. a category $\mathcal{S}$ whose objects are pairwise not isomorphic, whose morphism-sets $\mathcal{S}(U, V)$, $U, V \in \mathcal{S}$, are finite-dimensional vector spaces, whose composition maps are bilinear and whose endomorphism-algebras $B = \mathcal{S}(U, U)$ are *local* ($B = k \, 1_B \oplus \mathcal{R}_B$, where $\mathcal{R}_B$ is a nilpotent ideal, the *radical* of B).

In this survey, we admit spectroids $\mathcal{S}$ which are not isomorphic to some ind_A. We call $\mathcal{S}$ *finite* if the number of its *'points'* (=objects) is finite. A typical example, with two points e and g, is the following: $\mathcal{S}(g, g) = k[\varepsilon]/\varepsilon^3$, $\mathcal{S}(g, e) = k\beta \oplus k\delta\beta$, $\mathcal{S}(e, g) = k\mu \oplus k\mu\delta$, $\mathcal{S}(e, e) = k[\delta, \eta]/(\delta^2, \eta^2)$ with $\beta\varepsilon = 0$, $\eta\beta = 0$, $\varepsilon\mu = 0$, $\mu\eta = 0$, $\beta\mu = \eta$ modulo η^2 and $\mu\beta = \mu\delta\beta = \varepsilon^2$ modulo ε^3.

By a *module* on a spectroid $\mathcal{S}$ we mean a k-linear functor from $\mathcal{S}$ to mod_k, by $\mathrm{mod}_\mathcal{S}$ we denote the category formed by these modules. For instance, each *'point'* x of $\mathcal{S}$ gives rise to an indecomposable *projective* module $^\mathrm{P}x$ such that $^\mathrm{P}x(y) = \mathcal{S}(x, y)$, to an indecomposable *injective* module $^\mathrm{I}x$ such that $^\mathrm{I}x(y) = \mathcal{S}(y, x)^\top$ and to a *simple* module $^\mathrm{S}x$ such that $^\mathrm{S}x(x) = k$ and $^\mathrm{S}x(y) = 0$ if $y \neq x$. ($V^\top$ denotes the *dual* of a vector space V.)

1.3 Spectroids and Algebras

Let $\mathcal{S}$ be a finite spectroid and ϖ a function which assigns a natural number ≥ 1 to each point of $\mathcal{S}$. We then denote by $\mathcal{S}^{\varpi \times \varpi}$ the algebra formed by all 'matrices' B whose entries B_{ij} are 'blocks' indexed by the pairs (i, j) of points of $\mathcal{S}$; each block B_{ij} itself is a matrix with $\varpi(i)$ rows, $\varpi(j)$ columns and entries in $\mathcal{S}(j, i)$. The addition of $\mathcal{S}^{\varpi \times \varpi}$ is performed blockwise; the multiplication is the usual blockwise multiplication of matrices. For instance, $\mathcal{S}^{\varpi \times \varpi}$ is isomorphic to the algebra $k[\Sigma_4]$ of the symmetric group Σ_4 if k has characteristic 2, if $\mathcal{S}$ is the spectroid with points e, g described above and if $\varpi(e) = 1$, $\varpi(g) = 2$ (Fig. 1).[1]

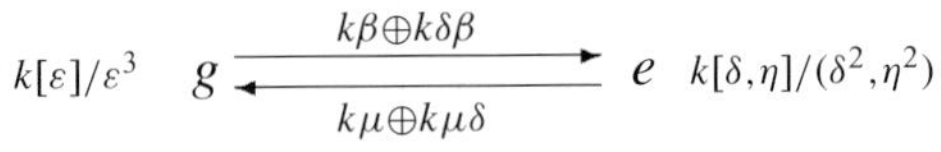

$$k[\varepsilon]/\varepsilon^3 \quad g \xleftarrow[{k\mu \oplus k\mu\delta}]{k\beta \oplus k\delta\beta} e \quad k[\delta, \eta]/(\delta^2, \eta^2)$$

Fig. 1

Similarly, each module M on $\mathcal{S}$ provides a module M^ϖ on $\mathcal{S}^{\varpi \times \varpi}$. The elements of M^ϖ are 'columns' x whose entries x_i, indexed by the points i of $\mathcal{S}$, run through $M(i)^{\varpi(i)}$. The algebra $\mathcal{S}^{\varpi \times \varpi}$ acts on M^ϖ from the left by ordinary matrix-multiplication, and the functor $M \mapsto M^\varpi$ is an equivalence $\mathrm{mod}_\mathcal{S} \xrightarrow{\sim} \mathrm{mod}_{\mathcal{S}^{\varpi \times \varpi}}$. Thus, when $\mathcal{S}$ stays fixed and ϖ varies, we obtain various non-isomorphic algebras $\mathcal{S}^{\varpi \times \varpi}$ whose categories of modules are equivalent.

In fact, each algebra A is isomorphic to $\mathcal{S}^{\varpi \times \varpi}$ for some $(\mathcal{S}, \varpi)$ which is determined up to isomorphism by A. To be precise, $\mathcal{S}$ must be (isomorphic to) the *spectroid of A*, i.e. (to) the full subspectroid $\mathcal{S}_A$ of $\mathrm{ind}_{A^{op}}$ formed by the projective indecomposables. And ϖ is defined by the formula $A \xrightarrow{\sim} \oplus_{P \in \mathcal{S}_A} P^{\varpi(P)}$.[2]

---

[1] See [18], p. 74, Example 3, where the lacking relation $\varepsilon^2 = \mu\delta\beta$ should be added.

[2] Since algebras and finite spectroids are 'equivalent' from the point of view of Representation Theory, we shall focus our report on spectroids, which provide a better geometric intuition.

1.4 Radical and Quivers

The non-invertible morphisms of an arbitrary spectroid S form a (two-sided) *ideal* $\mathcal{R}_S$ such that $\mathcal{R}_S(x, y) = S(x, y)$ if $x \neq y$ and $\mathcal{R}_S(x, x) \oplus k \mathbb{1}_x = S(x, x), \forall x \in S$ (1.2). This ideal is the *radical* of S. To each pair of points (x, y) it assigns a dimension

$$\delta(x, y) = \dim_k \mathcal{R}_S(x, y)/\mathcal{R}_S^2(x, y), \quad \text{where} \quad \mathcal{R}_S^2(x, y) = \sum_{z \in S} \mathcal{R}_S(z, y)\mathcal{R}_S(x, z),$$

thus giving rise to the *quiver* $\mathbf{Q}_S$ of S: The *vertices* of this quiver are the points of S, the number of *arrows* from a vertex x to a vertex y is $\delta(x, y)$. For instance, if S is the spectroid with two points e, g of 1.2, we have

$$(\mathcal{R}_S/\mathcal{R}_S^2)(e, e) = k\bar{\delta}, \qquad (\mathcal{R}_S/\mathcal{R}_S^2)(g, e) = k\bar{\beta}$$
$$(\mathcal{R}_S/\mathcal{R}_S^2)(e, g) = k\bar{\mu}, \qquad (\mathcal{R}_S/\mathcal{R}_S^2)(g, g) = k\bar{\varepsilon}ts.$$

Hence, $\mathbf{Q}_S$ is the quiver of Fig. 2.

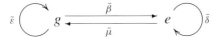

Fig. 2. Here we identify the arrows with residue classes of morphisms

If x, y are vertices of an arbitrary quiver Q, a *path* of *length* $n \geq 1$ from x to y is a formal composition $\alpha_n \ldots \alpha_2 \alpha_1$ of n arrows such that α_1 starts at x, α_2 starts where α_1 stops, $\ldots$, and α_n stops at y. Besides these paths, we also introduce *identical* paths $\mathbb{1}_x$ of length 0 from x to x, $\forall x \in Q$. Thus we are lead to the *linear path-category* kQ, whose objects are the vertices of Q, whose morphism-sets $kQ(x, y)$ are the vector spaces freely generated by the paths from x to y, whose (bilinear) composition is induced by the juxtaposition of paths.

In case $Q = \mathbf{Q}_S$, we can choose a basis in each space $(\mathcal{R}_S/\mathcal{R}_S^2)(x, y)$ and representatives of the basis vectors in $\mathcal{R}_S(x, y)$. Mapping the arrows of $\mathbf{Q}_S$ from x to y bijectively onto these representatives and extending the map in the obvious way to all morphisms of $k\mathbf{Q}_S$, we obtain a *display-functor*

$$\Phi : k\mathbf{Q}_S \to S$$

which is the identity on the objects and maps the ideal $k^+\mathbf{Q}_S$ of $k\mathbf{Q}_S$ generated by the paths of length ≥ 1 into $\mathcal{R}_S$. The *kernel* Ker Φ of Φ is contained in the square of the ideal $k^+\mathbf{Q}_S$.

In general, the maps $\Phi(x, y) : k\mathbf{Q}_S(x, y) \to S(x, y)$ are not surjective. If they are and if $(\rho_r)_{r \in R}$ is a family of morphisms generating the ideal Ker Φ, we say that the spectroid S *is defined by the quiver* $\mathbf{Q}_S$ *and the relations* $\rho_r = 0$. This happens whenever S is finite. For instance, the spectroid S with two points e, g of 1.2 is defined by the quiver of Fig. 2 and the relations $0 = \bar{\beta}\bar{\varepsilon} = \bar{\varepsilon}\bar{\mu} = \bar{\delta}^2 = \bar{\varepsilon}^2 - \bar{\mu}\bar{\beta} = \bar{\varepsilon}^2 - \bar{\mu}\bar{\delta}\bar{\beta} = \bar{\beta}\bar{\mu}\bar{\delta} - \bar{\delta}\bar{\beta}\bar{\mu}$.

2. Tame and Wild

2.1 The Varieties of Modules

Let S be a *finite* spectroid defined by the quiver $Q = \mathbf{Q}_S$ and the relations $\rho_r = 0$, $r \in R$. Let further D be a *dimension-function* which assigns a natural number to each point of S. The S-modules M such that $M(x) = k^{D(x)}$, $\forall x \in S$ are then determined by matrices $m_\alpha \in k^{D(h\alpha) \times D(t\alpha)}$ attached to the arrows $t\alpha \xrightarrow{\alpha} h\alpha$ of Q. These matrices are subjected to the relations[3] $m_{\rho_r} = 0$, $r \in R$, which we can interpret as algebraic equations in the entries $m_{\alpha i j}$ of the matrices m_α. The families

$$m := (m_\alpha)_\alpha \in \prod_\alpha k^{D(h\alpha) \times D(t\alpha)} =: k^{D \times D}$$

therefore range over an algebraic subvariety[4] $\mathbf{mod}_S^D \subset k^{D \times D}$ which is the *spectrum* of the 'algebra'[5]

$$\mathbf{A}_S^D = k[X_{\alpha i j}; \alpha i j]/(X_{\rho_r})_{r \in R} \, .$$

Conjointly with $\mathbf{A}_S^D$, S gives birth to an S-'module' $\mathbf{M}_S^D$ whose 'stalks' $\mathbf{M}_S^D(x) = (\mathbf{A}_S^D)^{D(x)}$ are free 'modules' over $\mathbf{A}_S^D$, whose matrices

$$\bar{X}_\alpha = [\bar{X}_{\alpha i j}] \in (\mathbf{A}_S^D)^{D(h\alpha) \times D(t\alpha)}$$

have their entries (= the residue-classes of the indeterminates $X_{\alpha i j}$) in $\mathbf{A}_S^D$, not in k! The 'module' $\mathbf{M}_S^D$ is *universal* in the sense that the map $\varphi \mapsto ([\varphi(\bar{X}_{\alpha i j})])_\alpha$ provides a bijection between the homomorphisms $\varphi : \mathbf{A}_S^D \to k$ and the points of $\mathbf{mod}_S^D$.

More generally, each homomorphism of 'algebras' $\varphi : \mathbf{A}_S^D \to B$ induces a functor[6]

$$\mathbf{F}_\varphi : \mathrm{mod}_B \to \mathrm{mod}_S , \quad N \mapsto \mathbf{M}_S^D \otimes_\varphi N \, .$$

We are interested in the case where $\mathbf{F}_\varphi$ *insets the indecomposables*[7]

---

[3] Replacing linear combinations and products of paths by linear combinations and products of matrices, we extend the map $\alpha \mapsto m_\alpha$ to all morphisms of the linear path-category kQ.

[4] In our terminology, an algebraic variety is the restriction of an algebraic scheme over k to the set of its closed points.

[5] The quotation marks hint at the fact that this 'algebra' is not finite-dimensional in general. The following notation should be self-explanatory.

[6] Though B may have infinite dimension, mod_B denotes the category of its finite-dimensional modules.

[7] i.e. maps indecomposables to indecomposables and non-isomorphic indecomposables to non-isomorphic ones. 'To inset': to insert a smaller map or picture within the border of a larger one.

   For the sake of brevity, we restrict 2.2 and the proof of 2.3 below to commutative 'algebras'. In the general case, $k[S, T]$ has to be replaced by the free associative 'algebra' $k\langle S, T \rangle$ and $\mathbf{A}_S^D$ by its non-commutative analogue. We thank B. Keller for reminding us in this point of our own teaching!

2.2 Example

Suppose that S has *one point* e with algebra of endomorphisms $S(e, e) = k[X, Y]/(X^3, X^2Y, XY^2, Y^3)$, and that $D(e) = 4$. Then $\mathbf{Q}_S$ has one vertex e and two 'loops' ξ, η, and S is defined by $\mathbf{Q}_S$ and the relations $0 = \xi^3 = \xi^2\eta = \xi\eta^2 = \eta^3$. Accordingly, $\mathbf{mod}_S^4 \subset k^{4\times 4} \times k^{4\times 4}$ consists of the pairs of 4×4-matrices x, y such that $0 = x^3 = x^2y = xy^2 = y^3$. In particular, $\mathbf{mod}_S^D$ contains the *affine plane* $P \subset k^{4\times 4} \times k^{4\times 4}$ formed by the pairs

$$
x(s) = \begin{bmatrix} 0 & 1 & 0 & 0 \\ 0 & 0 & 0 & s \\ 0 & 0 & 0 & 1 \\ 0 & 0 & 0 & 0 \end{bmatrix}, \quad
y(t) = \begin{bmatrix} 0 & 0 & 1 & 0 \\ 0 & 0 & 0 & 1 \\ 0 & 0 & 0 & t \\ 0 & 0 & 0 & 0 \end{bmatrix}.
$$

Now let $k[S, T]$ be the 'algebra' of polynomials in S and T. There is a homomorphism $\psi : \mathbf{A}_S^4 \to k[S, T]$ which maps the generators $\bar{X}_{\xi ij}$ and $\bar{X}_{\eta ij}$, $1 \le i, j \le 4$ to the corresponding entries of the matrices of $k[S, T]^{4\times 4}$ obtained from $x(s)$ and $y(t)$ by replacing $0, 1, s, t$ by $0, 1, S, T$ respectively. The induced functor $\mathbf{F}_\psi$ maps a $k[S, T]$-module V onto the S-module V^4, on which ξ and η act like the two matrices of $k[S, T]^{4\times 4}$ considered above.

$\mathbf{F}_\psi$ insets the indecomposables[7].

2.3

Proposition. *Let B be an 'algebra' generated by n elements $b_1, \ldots, b_n$ and S the spectroid with one point e and algebra of endomorphisms*

$$
S(e, e) = k[X, Y]/(X^3, X^2Y, XY^2, Y^3).
$$

Then there is a homomorphism $\varphi : \mathbf{A}_S^{4(n+1)} \to B$ such that the induced functor

$$
\mathbf{F}_\varphi : \mathbf{mod}_B \to \mathbf{mod}_S
$$

insets the indecomposables.[7]

Proof. We first consider the spectroid $\mathcal{T}$ with one point e and algebra of endomorphisms $\mathcal{T}(e, e) = k[S, T]/I$, where I is the ideal generated by all (non-commutative) monomials of degree $n + 1$ in S, T. The quiver $\mathbf{Q}_{\mathcal{T}}$ has two loops σ, τ, which are 'produced' by S and T. Accordingly, $\mathbf{mod}_{\mathcal{T}}^{n+1}$ contains an affine subspace of dimension n whose points are the pairs of matrices

$$
m_\sigma = \begin{bmatrix} 0 & 1 & 0 & \vdots & 0 & 0 \\ 0 & 0 & 1 & \vdots & 0 & 0 \\ 0 & 0 & 0 & \vdots & 0 & 0 \\ \multicolumn{6}{c}{\dotfill} \\ 0 & 0 & 0 & \vdots & 0 & 1 \\ 0 & 0 & 0 & \vdots & 0 & 0 \end{bmatrix}, \quad
m_\tau(u) = \begin{bmatrix} 0 & u_1 & 0 & \vdots & 0 & 0 \\ 0 & 0 & u_2 & \vdots & 0 & 0 \\ 0 & 0 & 0 & \vdots & 0 & 0 \\ \multicolumn{6}{c}{\dotfill} \\ 0 & 0 & 0 & \vdots & 0 & u_n \\ 0 & 0 & 0 & \vdots & 0 & 0 \end{bmatrix}.
$$

Now we have a homomorphism $\chi : A_T^{n+1} \rightarrow B$ which maps the generators $\bar{X}_{\sigma ij}, \bar{X}_{\tau ij}$ of A_T^{n+1} to the corresponding entries of the matrices of $B^{n+1 \times n+1}$ obtained from $m_\sigma, m_\tau(u)$ by replacing $0, 1, u_i$ by $0, 1, b_i$ respectively. The induced functor $\mathbf{F}_\chi$ maps a B-module V onto the T-module V^{n+1}, on which $\bar{S}, \bar{T}$ act like the two matrices of $B^{n+1 \times n+1}$ considered above. The functor $\mathbf{F}_\chi$ insets the indecomposables.

The functors $\mathbf{F}_\chi$ and $\mathbf{F}_\psi$ (2.2) provide a composed functor

$$\mathrm{mod}_B \xrightarrow{\mathbf{F}_\chi} \mathrm{mod}_T \xrightarrow{\mathrm{incl}} \mathrm{mod}_{k \langle S, T \rangle} \xrightarrow{\mathbf{F}_\psi} \mathrm{mod}_{k[X,Y]/(X^3, X^2 Y, X Y^2, Y^3)},$$

which insets the indecomposables. The composed functor equals $\mathbf{F}_\varphi$ for some $\varphi : A_S^{n+1} \rightarrow B$ whose determination should not be insuperable. $\checkmark$

2.4 The Definitions

The algebraic group $\mathbf{G}_S^D = \prod_{x \in S} \mathrm{GL}_{D(x)}$ acts on the algebraic variety $\mathbf{mod}_S^D$ according to the formula $(gm)_\alpha = g_{h\alpha} m_\alpha g_{t\alpha}^{-1}$. The orbits are the isoclasses of module-structures in $\mathbf{mod}_S^D$. They provide a *stratification* of $\mathbf{mod}_S^D$ by locally closed subsets

$$\mathbf{mod}_S^{D|e} = \{m \in \mathbf{mod}_S^D : \dim \mathrm{End}(m) = e\}.$$

The orbit $\mathbf{G}_S^D m$ of a point $m \in \mathbf{mod}_S^{D|e}$ has the dimension

$$\delta(e) = \dim \mathbf{G}_S^D - \dim \mathrm{End}(m) = \sum_{x \in S} D(x)^2 - e.$$

The number $\deg_S(D) = \max_e(\dim \mathbf{mod}_S^{D|e} - \delta(e))$ associated with this stratification is the *degree of freedom* of the S-modules with dimension-function D.

Within $\mathbf{mod}_S^D$ we also consider the *constructible* $\mathbf{G}_S^D$-stable subset indf_S^D formed by the *indecomposable* module-structures and stratified by the intersections

$$\mathrm{indf}_S^{D|e} = \mathrm{indf}_S^D \cap \mathbf{mod}_S^{D|e}.$$

The associated number $\deg_S^i(D) = \max_e(\dim \mathrm{indf}_S^{D|e} - \delta(e))$ is the *degree of freedom* of the indecomposables with dimension-function D.

Definition. A *finite* spectroid S is called *tame* (resp. *wild*) if $\deg_S^i(D) \leq 1$ (resp. > 1) for all dimension-functions D (resp. for some D).
An algebra is *tame* (resp. *wild*) if so is its spectroid.

In fact, the dichotomy 'Tame-Wild' is highly more spectacular than our definition. But the spectacle needs more preparations: We say that an affine (locally closed) subvariety $V = \mathrm{Spec} k[V]$ of $\mathbf{mod}_S^D$ is *reliable* if the functor $\mathbf{F}_\varphi : \mathrm{mod}_{k[V]} \rightarrow \mathrm{mod}_S$ attached to the restriction-map $\varphi : A_S^D \rightarrow k[V]$ insets the indecomposables. Under this condition, an S-module isomorphic to some $\mathbf{F}_\varphi(N)$ is said to be *produced* by V. Furthermore, we call *plane* (resp. *punched line*) of

$\mathbf{mod}_S^D$ a subvariety which is a translate of a vector subspace of dimension 2 (resp. a cofinite subvariety of a translate of a vector subspace of dimension 1) of $k^{D\times D}$.

Theorem. [8] *If S is wild, $\mathbf{mod}_S^D$ contains a reliable plane for some D. Hence, for each algebra B, there is a functor $\mathbf{F}_\psi : \mathrm{mod}_B \to \mathrm{mod}_S$ which insets the indecomposables.*

If S is tame, there exists a family of reliable punched lines $L_i \subset \mathbf{mod}_S^{D_i}, i \in I$, such that: For each D, the number of $i \in I$ satisfying $D = D_i$ is finite, and almost all orbits of indf_S^D consist of indecomposables produced by some L_i; moreover, if $i \neq j$, no indecomposable produced by L_i is produced by L_j.

2.5 Examples

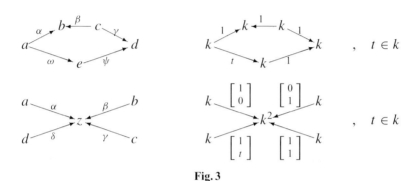

Fig. 3

Figure 3 depicts reliable lines of two spectroids defined by quivers without relation. The quivers are particular cases of *extended Dynkin* quivers. For each extended Dynkin quiver *one* reliable line suffices.

2.6 Example

Let Q be a finite quiver, L a set of loops of Q, Z a set of paths of lengths ≥ 2 whose arrows lie outside L, N a natural number ≥ 3 such that each path having N arrows outside L contains some subpath which lies in Z. Let further A be the quotient of the *quiver-'algebra*[9]

$$k[Q] = \oplus_{x,y\in Q} kQ(x, y)$$

by the ideal generated by all $\varepsilon^2 - \varepsilon, \varepsilon \in L$ and all paths of Z.

---

[8] See [17]. The statement goes back to a conjecture of P. Donovan and M.R. Freislich. A slightly weaker version was first proved by Ju.A. Drozd and W.W. Crawley-Boevey.

[9] An element of $k[Q]$ is a matrix whose entries are indexed by the pairs of vertices $x, y \in Q$ and lie in the morphism-spaces $kQ(x, y)$ of kQ (1.4). The multiplication of $k[Q]$ is the multiplication of matrices. Compare with the construction of $S^{\varpi \times \varpi}$.

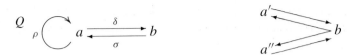

Fig. 4. This example, where $L = \{\rho\}$ and $Z = \{\delta\sigma, (\delta\rho\sigma)^n\}$, $n \geq 1$, is due to I.M. Gelfand. Notice that Q is different from the quiver (of the spectroid) of A, which is reproduced at the right of Q

Proposition. *The algebra A above is tame if the data satisfy the following two conditions:*

    a) At most two arrows start at each vertex, at most two stop.

    b) For each arrow $\alpha \notin L$, there is at most one arrow β such that $\beta\alpha \notin Z$ and at most one γ such that $\alpha\gamma \notin Z$.[10]

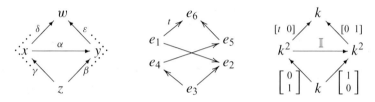

Fig. 5. $Z = \{\varepsilon\beta, \delta\gamma\}$, $\zeta = \zeta_n \cdots \zeta_1 = \bar{\delta}\varepsilon\alpha\gamma\bar{\beta}\alpha$. The reliable line attached to ζ consists of modules M_t. Each arrow ζ_i with tail x_i in $Q^{\pm}$ produces one basis vector e_i of $M_t(x_i)$. If ζ_i lies in Q, ζ_i maps e_i to e_{i+1} in case $i < n$, to te_1 in case $i = n$. If $\bar{\zeta}_i$ lies in Q, e_{i+1} is mapped to e_i in case $i < n$, and e_1 to te_n in case $i = n$

We describe a complete[11] family of reliable punched lines in the easy case[12] where $L = \emptyset$. For this sake, we extend Q to a quiver $Q^{\pm}$ by adding an arrow $\bar{\alpha} : y \to x$ to each arrow $\alpha : x \to y$ of Q. We further set $\bar{\bar{\alpha}} = \alpha$ for each arrow α of Q and consider the 'non-zero' and non-periodic *cycles*[13] $\zeta = \zeta_n \cdots \zeta_1$ of $Q^{\pm}$, i.e. the cycles with the following three properties:

    a) No power of ζ or of $\bar{\zeta} = \bar{\zeta}_1 \cdots \bar{\zeta}_n$ has a 'subpath' in Z.
    b) ζ is not a power of a proper subpath.
    c) ζ^2 contains no subpath of the form $\bar{\beta}\beta$, where β is an arrow of $Q^{\pm}$.

Defining a *rotation* of ζ as a cycle of the form $R^i = \zeta_i \cdots \zeta_1 \zeta_n \cdots \zeta_{i+1}$, we obtain rotation classes $\{\zeta, R\zeta, \ldots, R^{n-1}\zeta\}$, in each of which we choose *one* representative.

---

[10] The proposition and an adequate proof go back to L.A. Nazarova and A.V. Roiter. See [6, 12, 4].
[11] i.e. satisfying the statements of Theorem 2.4
[12] Q is then identified with the quiver of A.
[13] A *cycle* is a path of length ≥ 1 whose origin and terminus coincide.

Our reliable lines are indexed by these representatives. To explain what line is attached to a given representative, the example of Fig. 5 may suffice.

2.7 Polynomial Growth

In this subsection, we suppose that S is a tame finite spectroid and $(L_i)_{i \in I}$ a complete family of reliable punched lines.

For each natural number $d \geq 1$, we denote by $\rho_S(d)$ the number of lines $L_i \subset \mathbf{mod}_S^{D_i}$ whose elements have *total dimension*[14] $\sum_{x \in S} D_i(x) = d$. We thus obtain an *arithmetical function* which depends on S, but not on the choice of $(L_i)_{i \in I}$. For each $D \in \mathbb{N}^S$, we further consider the number of the irreducible components of indf_S^D which contain infinitely many orbits of $\mathbf{G}_S^D$. The sum of these numbers, when D is subjected to $d = \sum_{x \in S} D(x)$, is denoted by $\lambda_S(d)$. The function λ_S depends on ρ_S according to the formula

$$\lambda_S(d) = \sum_{c \mid d} \rho_S(c) \,.$$

If $\rho_S(d) = 0, \forall d$, i.e. if there is no reliable punched line at all, S is *representation-finite*, i.e. admits only finitely many isoclasses of indecomposables [5, 10]. A reference to the well-rounded store of knowledge about representation-finite algebras may suffice here[15].

If the number $\sum_{d \geq 1} \rho_S(d)$ of lines L_i is finite or, equivalently, if the function λ_S is bounded, S is called *domestic*. Algebras defined by an extended Dynkin quiver without relation have one line L_i (2.5) and are domestic. Figure 6 provides trivial examples of domestic spectroids, where the numbers of lines L_i are arbitrary. More substantial examples will be examined in Sect. 5.

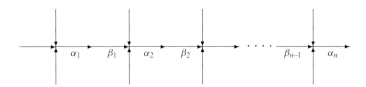

Fig. 6. The relations are $\beta_i \alpha_i = 0$

S is said to be tame *of polynomial growth* if there are constants $C \in \mathbb{N}$ and $c \in \mathbb{N}$ such that $\rho_S(d) \leq Cd^c, \forall d$.[16] Tame spectroids of polynomial growth are examined in Sect. 5.

The examples produced in 2.6 are *not* of polynomial growth in general. For instance, if S is defined by a quiver Q with one vertex, two arrows ξ, η and the

---

[14] The *total dimension* $\dim M$ of an S-module M is $\sum_{x \in S} \dim M(x)$.
[15] See [16] for a short survey, [18] for a more detailed one.
[16] We could equally well impose a bound on $\lambda_S(d)$. The notion comes from and is well understood in Representation Theory of partially ordered sets [30, 49, 42].

relations $\xi^3 = \xi\eta = \eta\xi = \eta^2 = 0$, there are, up to rotation, at least $2^p/3p$ non-zero and non-periodic cycles of $Q^\pm$ which contain a prime number p of terms equal to η.

2.8 Deformations of Tame Algebras

Suppose that the *finite* spectroid S is *tame*. Considering that the indecomposable summands of a point $m \in \mathbf{mod}_S^D$ 'depend on one parameter at most', that $\dim \mathrm{End}(m)$ is at least equal to the number of these summands and that $\rho_S(D) = \lambda_S(D) = 0$ if $|D| := \sum_{x \in S} D(x) = 1$, one proves that

$$\dim \mathbf{mod}_S^{D|e} \leq \min\{\tfrac{1}{2}|D|, e\} + \sum_{x \in S} D(x)^2 - e.$$

The inequality implies:

Proposition 1. *Let S be a tame finite spectroid. Then*

$$\dim \mathbf{mod}_S^D \leq \dim \mathbf{G}_S^D = \sum_{x \in S} D(x)^2, \quad \forall D.$$

Proposition 2 [33]. *If S is a finite spectroid, the following statements are equivalent:*

(i) S *is tame.*
(ii) $\dim \mathbf{mod}_S^{D|e} \leq \tfrac{1}{2}|D| + |D|^2 - e, \forall D, \forall e.$
(iii) $\deg_S(D) \leq \tfrac{1}{2}|D|, \forall D.$
(iv) $\limsup_{|D| \to \infty} \deg_S(D)/|D|^2 = 0.$

Let us now denote by $\mathbf{alg}^d$ the algebraic variety formed by the algebra structures on k^d. Two such structures belong to the same orbit under the natural action of the algebraic group GL_d if and only if they are isomorphic. We say that an algebra B of dimension d is a *deformation* of an algebra A of the same dimension if the orbit of $\mathbf{alg}^d$ associated with A lies in the Zariski-closure of the orbit associated with B.[17]

Proposition 3 [19]. *Deformations of tame algebras are tame.*

Consider for example the local algebras

$$A_t = k\langle X, Y \rangle/(X^2 - tYXY, \ Y^2 - tXYX, \ (XY)^2, \ (YX)^2),$$

whose dimension is 7. The algebra A_0 is tame by 2.6. In case $t \neq 0$, A_t is isomorphic to A_1. It follows that A_1 is tame[18] [19]. Unfortunately, the method provides no information on the structure of the indecomposables.

---

[17] A is then also called a *degeneration* of B.
[18] If $\mathrm{char}\,k = 2$ A_1 is isomorphic to the quotient of the group algebra of the quaternion group by its socle [7].

2.9 Tame Spectroids with One or Two Points

In general, it is a difficult problem to determine whether a given algebra is tame. Several techniques, such as quadratic forms, degenerations, Galois coverings or subspace problems, are used in practice. Some of them are to be considered later.

If S is a tame spectroid with one point e and if the characteristic of k is $\neq 2$, $S(e, e)$ is a quotient of one of the following algebras:

a) $k\langle X, Y\rangle/(YX - X^m, XY - Y^n, X^{m+1}, Y^{n+1})$, $m \geq 2, n \geq 3$.[19]
b) $k\langle X, Y\rangle/(YX - X^2, XY - \alpha Y^2)$, $0 \neq \alpha \neq 1$.[20]
c) $k\langle X, Y\rangle/(X^2 - YXY \cdots, Y^2 - XYX \cdots, X^4, Y^4)$, where $YXY \cdots$ and $XYX \cdots$ alternate $n \geq 3$ factors X and Y.[21]
d) $k\langle X, Y\rangle/(X^2 - YXY \cdots, Y^2, X^4)$, where $YXY \cdots$ is as in c).[21]

Among the local tame algebras thus described as quotients, only the truncated polynomial algebras $k[X]/X^n$ are representation-finite. The situation is rather different in general. Recently, Geiss established the impressive list of the tame spectroids with two points and finitely many ideals [20].[22] Let us here examine the trivial case of a spectroid S with quiver $a \xrightarrow{v} b\, \rho$, where ρ stands for a loop at b. The possible relations then are $\rho^m v = \rho^n = 0$ with $1 \leq m \leq n \neq 1$. The spectroid is representation-finite if $m = 1$ or if $2 = m \leq n \leq 5$. It still is tame if $2 = m$, $n = 6$. Wildness rules anywhere else.

2.10 Quadratic Forms

To exploit Prop.1 of 2.8, we need a lower bound for the dimension of $\mathbf{mod}_S^D$. For this sake, we return to our display-functor $\Phi : k\mathbf{Q}_S \to S$ of 1.4. Since S is finite, the ideal $\mathcal{K} = \mathrm{Ker}\,\Phi$ admits a *finite* family $\rho = (\rho_s)_{s \in R}$ of generators $\rho_s : a_s \to z_s$. The condition $m_{\rho_s} = 0$ attached to ρ_s subjects a point $m \in k^{D \times D}$ to $D(z_s)D(a_s)$ equations. Gathering up all these equations, we obtain the inequality

$$\dim \mathbf{mod}_S^D \geq \dim k^{D \times D} - \sum_{x, y \in S} e^2(x, y)D(x)D(y)$$

$$= \sum_{x, y \in S} e^1(x, y)D(x)D(y) - \sum_{x, y \in S} e^2(x, y)D(x)D(y),$$

---

[19] The dimension of this local algebra is $m + n + 1$. The algebra is a deformation of $k\langle X, Y\rangle/(YX, XY, X^{m+1}, Y^{n+1})$, which is tame by 2.6. A general reference for tame spectroids with one point is [38].

[20] The dimension here is 5. According to Geiss and de la Peña (unpublished), the indecomposables of the obvious Galois covering of S (see below) are supported by 3 points (Compare with [21]). The proof of the tameness is thus reduced to the quiver $\bullet \overset{\eta}{\underset{\eta}{\cdot}} \overset{\zeta}{\underset{\tau}{\cdot}} \bullet$ equipped with the relations $\tau\xi = \zeta\xi$, $\zeta\eta = \alpha\tau\eta$, which are 'equivalent' to $\tau\xi = \zeta\eta = 0$ and are 'tamed' by 2.6. In case $\alpha = 1$, the defined algebra has a wild quotient of dimension 5.

[21] These algebras have degenerations generated by elements x, y satisfying $x^2 = y^2 = 0$. So they are tame by 2.6.

[22] In fact, the question of the tameness of some few algebras is left open.

where $e^1(x, y)$ denotes the number of arrows and $e^2(x, y)$ the number of morphisms ρ_s from x to y. In combination with Prop.1 of 2.8, this yields

$$0 \le q_\rho(D) := \sum_{x \in S} D(x)^2 - \sum_{x, y \in S} e^1(x, y) D(x) D(y) + \sum_{x, y \in S} e^2(x, y) D(x) D(y)$$

for all $\mathbb{N}$-valued functions D (defined) on (the points of) S.

We express these inequalities by saying that the quadratic form $q_\rho : \mathbb{Z}^S \to \mathbb{Z}$, whose restriction to $\mathbb{N}^S$ is considered above, is *weakly non-negative*.

In the particular case where S is *directed*, i.e. where $\mathbf{Q}_S$ admits no cycle, we can 'make' q_ρ independent of ρ. Indeed, in this case, the linear category of paths $k\mathbf{Q}_S$ is a spectroid, and ρ can be made *minimal*. For all $x, y \in S$, the morphisms ρ_s with domain x and range y then yield a basis of the vector space

$$\mathcal{K}(x, y)/(\mathcal{R}_{k\mathbf{Q}_S}\mathcal{K} + \mathcal{K}\mathcal{R}_{k\mathbf{Q}_S})(x, y) \xrightarrow{\sim} \mathrm{Ext}^2_S({}^S x, {}^S y)^\top .$$

Thus, *when S is directed and ρ minimal*, the numbers

$$e^1(x, y) = \dim \mathrm{Ext}^1_S({}^S x, {}^S y) \quad \text{and} \quad e^2(x, y) = \dim \mathrm{Ext}^2_S({}^S x, {}^S y)$$

depend only on S [9]. The function $q_S := q_\rho$ is then called the *Tits form* of S. It satisfies the statement:

Proposition [32]. *If the finite spectroid S is directed and tame, the Tits form q_S is weakly non-negative.*

The converse of this statement is true for some classes of spectroids, for instance if $S = kQ$, where Q is a finite directed quiver. We shall meet other examples later.

2.11 Bilinear Forms

When the finite spectroid S is *directed*, the Tits form q_S is close to the following bilinear form[23] on $\mathbb{Z}^S$ (, the *'Euler-Poincaré' form* of S):

$$\langle D, E \rangle_S = \sum_{x, y \in S} D(x) E(y) \sum_{p \in \mathbb{N}} (-1)^p \dim \mathrm{Ext}^p_S({}^S x, {}^S y) .$$

The formula implies that $q_S(D) = \langle D, D \rangle_S$ if $\mathrm{gldim}\, S \le 2$, hence if $S = k\mathbf{Q}_S$.

Whereas the Tits-form is related to the varieties $\mathbf{mod}^D_S$, the Euler-Poincaré-form is properly homological: The following formula reflects the long exact sequences of extension-groups:

$$\langle \mathrm{Dim}\, m, \mathrm{Dim}\, n \rangle_S = \sum_{p \in \mathbb{N}} (-1)^p \dim \mathrm{Ext}^p_S(m, n) , \quad \forall m, n \in \mathbf{mod}_S .$$

Here $\mathrm{Dim}\, m$ denotes the dimension-function $x \mapsto \dim m(x)$.

---

[23] If S is directed, its global dimension is finite; in particular, $\mathrm{Ext}^p_S({}^S x, {}^S y) = 0$ for large values of p. For details see [40].

The numbers $e^p(x, y) = \dim \operatorname{Ext}^p_S({}^S x, {}^S y)$ are the multiplicities occurring in the minimal projective and injective resolutions of the simple S-modules:

$$\cdots \longrightarrow \prod_{y \in S} {}^{\mathbf{P}}y^{\,e^1(x,y)} \longrightarrow \prod_{y \in S} {}^{\mathbf{P}}y^{\,e^0(x,y)} \longrightarrow {}^S x \longrightarrow 0,$$

$$0 \longrightarrow {}^S y \longrightarrow \prod_{x \in S} {}^{\mathbf{I}}x^{\,e^0(x,y)} \longrightarrow \prod_{x \in S} {}^{\mathbf{I}}x^{\,e^1(x,y)} \longrightarrow \cdots .$$

Setting $e(x, y) = \sum_{p \in \mathbb{N}} (-1)^p e^p(x, y)$, we infer that

$$\operatorname{Dim}{}^S x = \sum_{y \in S} e(x, y) \operatorname{Dim}{}^{\mathbf{P}}y = \sum_{y \in S} e(y, x) \operatorname{Dim}{}^{\mathbf{I}}y .$$

Accordingly, the matrix $[e(x, y)]$ (with *row*-index x) is the *inverse* of the matrix $\mathbf{C}_S^\top = [\dim S(x, y)]$, whose transpose $\mathbf{C}_S$ is usually called *Cartan matrix* of S.

2.12 Indecomposables and Roots

Let $S(D) = \{x \in S : D(x) \neq 0\}$ denote the *support* of a function $D \in \mathbb{N}^S$ on the finite spectroid S. If S is directed, $D \in \mathbb{N}^S$ is called a *root* of S provided $q_S(D)$ is ≤ 1 and $S(D)$ is the set of vertices of a connected subquiver of Q_S. The root D is *imaginary* if $q_S(D) \leq 0$, 'realistic' if $q_S(D) = 1$. The Tits form q_S is said to *control* mod_S if the following statement holds: For each $D \in \mathbb{N}^S$, the number of $m \in \mathrm{ind}_S$ with $\operatorname{Dim} m = D$ is 0 if D is not a root, 1 if D is a realistic root and ∞ if D is an imaginary root.

In the case of a finite directed quiver Q, q_{kQ} controls mod_{kQ} if the connected components of each proper subquiver of Q are of Dynkin or extended Dynkin type [24][24]. We shall meet other examples in the sequel.

3. Representation-Quivers

3.1 The Translation of ind_S

Let S be a finite spectroid and ind_S the spectroid formed by chosen representatives of the isoclasses of indecomposable S-modules (1.2). The *almost split sequences*

---

[24] According to V. Kač, $D \in \mathbb{N}^{kQ}$ is a *real* root if it belongs to the orbit of some $E_x = \operatorname{Dim}{}^S x$, $x \in Q$, under the action of the group generated by the 'reflections' $\sigma_x : \mathbb{Z}^{kQ} \to \mathbb{Z}^{kQ}$, $x \in Q$, where $\sigma_x(D) = D - (q_{kQ}(D + E_x) - q_{kQ}(D) - 1)E_x$. In the general case of a directed finite quiver Q, $D \in \mathbb{N}^{kQ}$ is the dimension-function of some $m \in \mathrm{ind}_{kQ}$ if and only if D is a real or imaginary root. In the real case, m is unique. In the imaginary case, $\deg^{\mathrm{i}}_{kQ}(D) = 1 - q_{kQ}(D)$ (2.4) [26].

of Auslander and Reiten provide precise information on the *representation-quiver* of S, i.e. the quiver $\Gamma_S = Q_{\mathrm{ind}_S}$ of ind_S (1.4).[25]

Let $\overset{r}{\underset{j=1}{\oplus}} P_{y_j} \overset{[P_{\mu_{ji}}]}{\longrightarrow} \overset{g}{\underset{i=1}{\oplus}} P_{x_i} \longrightarrow m \to 0$ be a minimal projective presentation of some $m \in \mathrm{mod}_S$. The *translate* $\tau_S m$ of m is by definition the kernel of the associated map $\overset{r}{\underset{j=1}{\oplus}} {}^{\mathrm{I}}y_j \overset{[{}^{\mathrm{I}}\mu_{ji}]}{\longrightarrow} \overset{g}{\underset{i=1}{\oplus}} {}^{\mathrm{I}}x_i$ between injective modules on S. The map thus defined is easily extended to an equivalence τ_S from the category $\underline{\mathrm{mod}}_S$ of S-modules *modulo projectives*[26] to the category $\overline{\mathrm{mod}}_S$ of S-modules *modulo injectives*. It induces a bijective *'translation'* τ_S from the set of *non-projective* vertices of Γ_S onto the set of *non-injective* vertices. The effect of τ_S on Γ_S is described in statement b) below:

a) Γ_S is *locally finite*: For each $m \in \Gamma_S$, the number of arrows $t\mu \overset{\mu}{\to} h\mu$ with head $h\mu = m$ or tail $t\mu = m$ is *finite*.

b) For each non-projective $m \in \Gamma_S$ and each $x \in \Gamma_S$, the number of arrows $x \to m$ of Γ_S equals the number of arrows $\tau_S m \to x$.

In the sequel, we call *translation-quiver* a *locally finite* quiver Γ together with two bijections τ and σ: The first, called *translation*, maps some set of vertices, called *non-projective*, onto a set of vertices called *non-injective*. The second maps each arrow $x \overset{\mu}{\to} m$ with tail x and non-projective head m to an arrow $\tau m \overset{\sigma\mu}{\to} x$ with head x and non-injective tail τm. In the case of Γ_S, the statements a) and b) above allow us to construct a (non-canonical) σ which makes (Γ_S, τ_S) a translation-quiver.[27]

---

[25] Almost split sequences belong to the most powerful tools of Representation Theory worked out in the past fifty years. Their theory, elaborated for general Artin algebras – hence in non-geometric style – by Auslander, Reiten and their pupils, is scattered through some miles of papers starting with [2] and the ensuing series in Commun. Algebra. Summaries can be found in surveys like [15], [40] or [18].

[26] Kill the morphisms of mod_S factorizing through projectives!

[27] The main point about *almost split sequences* is the following: Because of statement a), each display-functor $\Psi : k\Gamma_S \to \mathrm{ind}_S$ attaches maps

$$(\Psi\mu) : \underset{h\mu=m}{\oplus}\, t\mu \to m \quad \text{and} \quad (\Psi\nu) : m \to \underset{t\nu=m}{\oplus}\, h\nu$$

to each S-module m. Now, for each Ψ there is a display-functor Φ (and for each Φ there is a Ψ) such that the sequence

$$0 \to \tau_S m \overset{(\Phi\sigma\mu)}{\longrightarrow} \underset{h\mu=m}{\oplus}\, t\mu \overset{(\Psi\mu)}{\longrightarrow} m \to 0$$

is exact non-split for each non-projective $m \in \Gamma_S$. Moreover, the sequence is *almost split* in the sense that each non-invertible $n \to m$ with $n \in \mathrm{ind}_S$ factors through $(\Psi\mu)$, and each non-invertible $\tau_S m \to n$ factors through $(\Phi\sigma\mu)$.

3.2 The Hereditary Case

With each locally finite quiver Q we can associate a translation-quiver $\mathbb{Z}Q$. The vertices are the pairs (n, x), $n \in \mathbb{Z}$, $x \in Q$. The translation is defined everywhere and satisfies $\tau(n, x) = (n - 1, x)$. Finally, each arrow $x \xrightarrow{\alpha} y$ gives birth to two series of arrows $(n, x) \xrightarrow{(n, \alpha)} (n, y)$ and $(n - 1, y) \xrightarrow{\sigma(n, \alpha)} (n, x)$ of $\mathbb{Z}Q$ (Fig. 7) [37]. $\mathbb{Z}Q$ contains two sub-translation-quivers $\mathbb{Z}^- Q$ and $\mathbb{Z}^+ Q$ whose vertices are the pairs (n, x) such that $n \leq 0$ and $0 \leq n$ respectively. The pairs $(0, x)$ are projective in $\mathbb{Z}^+ Q$, which contains no injective vertex. They are injective in $\mathbb{Z}^- Q$, which has no projective.

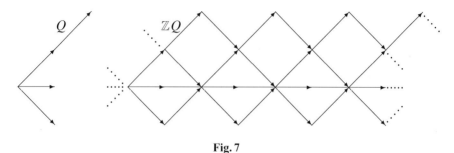

Fig. 7

Besides $\mathbb{Z}Q$ we also consider the translation quivers $\mathbb{Z}Q/p$, $0 \neq p \in \mathbb{N}$, which are obtained from $\mathbb{Z}Q$ by identifying each vertex $v = (n, x) \in \mathbb{Z}Q$ with its translates $\tau^{pr} v = (n - pr, x)$, $r \in \mathbb{Z}$, and each arrow α of $\mathbb{Z}Q$ with the translates $\sigma^{2pr} \alpha$. In particular, we are interested in the case where Q is the quiver

$$\mathbf{A}_\infty : \quad 1 \to 2 \to 3 \to 4 \to \cdots .$$

The translation-quiver $\mathbb{Z}\mathbf{A}_\infty/p$ is then called *tube of period p*.

Let now D be a connected finite quiver without cycle. Then kD is a spectroid of global (homological) dimension ≤ 1. If kD is *representation-finite*[28], Γ_{kD} is a finite sub-translation-quiver of $\mathbb{Z}^+ D^{op}$, whose projective vertices are $(0, x)$, $x \in D^{op}$ [18]. In the particular case where D^{op} is the quiver Q of Fig. 7, Γ_{kD} is the part of $\mathbb{Z}Q$ formed by the depicted full arrows (Fig. 7).

If D is an *extended Dynkin* quiver, Γ_{kD} has a (connected) component isomophic to $\mathbb{Z}^+ D^{op}$, which contains all the projective vertices of Γ_{kD}. It has a component isomorphic to $\mathbb{Z}^- D^{op}$, which embraces all the injectives. 'Between' these two components lies a family of periodic tubes indexed by $k \cup \infty$. The points of a reliable line L (2.5) belong to different tubes. Besides these tubes, there is one

---

[28] i.e. if the underlying graph of D is a Dynkin graph $\mathbf{A}_n$, $\mathbf{D}_n$, $\mathbf{E}_6$, $\mathbf{E}_7$ or $\mathbf{E}_8$.

more, which corresponds to the point of L at ∞^{29}. There are at most 3 tubes with period > 1.[30]

If D is *neither Dynkin nor extended Dynkin*, kD is wild. Then Γ_{kD} still has a component isomorphic to $\mathbb{Z}^+ D^{op}$ which includes the projectives and a component isomorphic to $\mathbb{Z}^- D^{op}$ with the injectives. All other components are isomorphic to $\mathbb{Z}\mathbf{A}_\infty$ [39].

3.3 Tubes and Tameness

Proposition [11]. *Let S be a tame finite spectroid. Then, for each dimension-function d, almost every[31] indecomposable $m \in \mathrm{ind}_S$ with dimension-function d belongs to a tube of period 1 of Γ_S.*

3.4 Regular Components

A translation-quiver Γ is *projective-free* if the translation is defined everywhere. It is *injective-free* if each vertex is a translate. If Γ is projective-free and injective-free, we call it *regular*. Our objective is to examine the regular (connected) components of the representation-quivers Γ_S.

Clearly, the translation-quivers $\mathbb{Z}Q$ of 3.2 are regular. We obtain other examples as follows: Consider the quiver

$$\mathbf{A}_\infty^\infty \; : \quad \cdots - 3 \to -2 \to -1 \to 0 \to 1 \to 2 \to 3 \cdots$$

and the associated translation-quiver $\mathbb{Z}\mathbf{A}_\infty^\infty$. Each function $f : \mathbb{Z} \to \mathbb{Z}$ then provides a *full*[32] sub-translation-quiver $\mathbb{Z}\mathbf{A}_\infty^f$ of $\mathbb{Z}\mathbf{A}_\infty^\infty$ which is formed by the vertices $(n, x) \in \mathbb{Z}\mathbf{A}_\infty^\infty$ with $x \geq f(n)$. In case $f(n) = 1, \forall n$, $\mathbb{Z}\mathbf{A}_\infty^f$ equals $\mathbb{Z}\mathbf{A}_\infty$. If f never increases, $\mathbb{Z}\mathbf{A}_\infty^f$ is injective-free.

Here we suppose that f never increases and that, for some $p \geq 1$,

$$f(n + p + 1) - f(n + p) = f(n + 1) - f(n), \quad \forall n \in \mathbb{Z}.$$

Then $q := f(n + p) - f(n)$ is independent of n, and $\mathbb{Z}\mathbf{A}_\infty^f$ is stable under the automorphism $(n, x) \mapsto (n + p, x + q)$ of $\mathbb{Z}\mathbf{A}_\infty^\infty$. Identifying vertices and arrows of

---

[29] In the first example of 2.5, Fig. 3, the tube at ∞ contains the indecomposable whose stalks are k, whose transition maps equal $\mathbb{1}$ with the exception of one 'vanishing clockwise arrow'; the period is 2 (=number of clockwise arrows) for the tube at ∞, 3 (=number of anticlockwise arrows) for the tube with index $t = 0$ and 1 for the remaining tubes. In the second example, the tube at ∞ is obtained by transposing 1 and t in $\left[\begin{smallmatrix}1\\t\end{smallmatrix}\right]$; the period is 2 for the tubes with index $t = 0, 1$ or ∞ and 1 in the remaining cases.

[30] For these results due to Donovan-Freislich and Nazarova see [40, 18].

[31] i.e. up to a finite number of exceptions.

[32] The arrows of $\mathbb{Z}\mathbf{A}_\infty^f$ are the arrows of $\mathbb{Z}\mathbf{A}_\infty^\infty$ whose heads and tails belong to $\mathbb{Z}\mathbf{A}_\infty^f$. The translate of a vertex x in $\mathbb{Z}\mathbf{A}_\infty^f$ is defined if the translate τx of x in $\mathbb{Z}\mathbf{A}_\infty^\infty$ lies in $\mathbb{Z}\mathbf{A}_\infty^f$. It then coincides with τx.

$\mathbb{Z}\mathbf{A}_\infty^f$ with their transforms, we obtain a *'radiant tube'* $\mathbb{Z}\mathbf{A}_\infty^f/p$, which is injective-free as $\mathbb{Z}\mathbf{A}_\infty^f$ is (Fig. 8).[33]

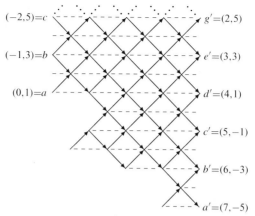

Fig. 8. $p = 7, q = -6$; to obtain $\mathbb{Z}\mathbf{A}_\infty^f/7$, identify $a, b, c, \ldots$ with $a', b', c', \ldots$

Theorem [22, 29]. *Let S be a finite spectroid and C an injective-free component of its representation-quiver. Then:*

a) If C contains a cycle and is projective-free, C is a periodic tube.

b) If C contains a cycle but no periodic vertex[34], C is a radiant tube.[35]

c) If C contains no cycle, C is isomorphic to a full sub-translation-quiver C' of $\mathbb{Z}Q$, where Q is a locally finite quiver without cycle and C' is closed in $\mathbb{Z}Q$ under predecessors.

3.5 Standard Components

If C is a translation-quiver, the *mesh-ideal* of $k\mathsf{C}$ is generated by the *mesh-sums* $\sum_{h\alpha=x} \sigma(\alpha)\circ\alpha$ attached to the non-projective vertices $x \in \mathsf{C}$. The associated residue-category $k(\mathsf{C}) = k\mathsf{C}/(\sum_{h\alpha=x} \sigma(\alpha)\circ\alpha)_x$ is the *mesh-category* of C [37].

We say that a component C of the representation-quiver Γ_S of a finite spectroid S is *standard* if there is a display-functor $\Phi : k\Gamma_S \to \mathrm{ind}_S$ which induces an isomorphism of the mesh-category $k(\mathsf{C})$ onto the full subcategory of ind_S supported by C. Standard components are important because they provide a direct information on the morphisms. They play a great role in the theory of representation-finite algebras. In general, we know that *postprojective*[36] and *preinjective* components of

---

[33] Acoording to our definition, periodic tubes (3.2) are radiant. In the sequel, we also consider *coradiant tubes*, i.e. translation-quivers whose duals are radiant tubes. For 'general' tubes, we refer to [13].

[34] i.e. no vertex x such that $x = \tau_S^s x$ for some $s > 0$.

[35] We are indebted to D. Vossieck for quite a few corrections, especially here and in 5.2 below.

[36] A component C of Γ_S is *postprojective* (resp. *preinjective*) if it contains no cycle and if each vertex $x \in \mathsf{C}$ has a projective translate $\tau_S^n x$, $n \in \mathbb{N}$ (resp. is the translate $\tau_S^n({}^1 y)$

Γ_S are standard. There are some other examples. For instance, if D is an extended Dynkin quiver, all the components of Γ_{kD} are standard[37].

Unfortunately, there is also a 'negative' result, since many components do not obey the following conclusions:

Proposition [43]. *Let S be a finite spectroid and C a component of Γ_S which is projective-free and injective-free. Let us further suppose that some display-functor induces surjections $kC(x, y) \to \mathrm{mod}_S(x, y)$ for all $x, y \in C$.*

Then C is a periodic tube or a sub-translation-quiver of some $\mathbb{Z}Q$, where Q is a quiver without cycle which has at most as many points as S.[38] ✓

4. Representations of Wild Algebras

4.1 Hereditary Wildness

Many of the algebras usually occurring in mathematics are wild, and representation men confronted with them usually stay bewildered. Some day, Algebraic Geometry will have to enter the contention seriously, in the third generation of 'hunters' maybe, when tame algebras will be tamed and computational Invariant Theory turned operative.

Nevertheless, some wild algebras have been examined so far, and the results deserve close examination. *Let us start with a connected finite quiver Q which has no cycle and is neither Dynkin nor extended Dynkin.* Let further C denote a *regular component* of Γ_{kQ}, which we may identify with $\mathbb{Z}\mathbf{A}_\infty$ (3.2). The vertices $(n, 1)$ at the *'border'* of C are the *quasi-simple* modules of [39], and each $X \in C$ has a filtration $X \supset X_1 \supset \cdots \supset X_s \supset X_{s+1} = 0$ whose terms 'lie' in C and yield quasi-simple quotients $X_i/X_{i+1} \xrightarrow{\sim} \tau_{kQ}^i(X/X_1) \in C$.

Indecomposables 'lying' in a regular component are called *'regular'*. The main result about their morphisms is:

Theorem. *Let the connected finite quiver Q have no cycle and kQ be wild. Then, for all regular $X, Y \in \mathrm{ind}_{kQ}$, there is an $N \in \mathbb{N}$ such that $m \geq N$ implies:*

---

of an injective vertex). Vertices of postprojective (resp. of preinjective) components are called *postprojective* (resp. *preinjective*).

[37] considered separately, since ind_{kD} is 'globally' not isomorphic to $k(\Gamma_{kD})$.

[38] The proof, obtained by Tilting (See below), is based on the formula $\mathrm{Ext}_S^1(x, y) = \overline{\mathrm{mod}_S(y, \tau_S x)}^\top$ [2, 18].

Indeed, if C is not a periodic tube, it is a sub-translation-quiver of some $\mathbb{Z}Q$, where Q has no cycle (3.4). Consider n points of C of the form $y_i = (r, x_i) \in \mathbb{Z}Q$. By the formula above, $\mathrm{Ext}_S^1(y_i, y_j) = 0$, $\forall i, j$. Moreover, if T is the quotient of S by the annihilator of $y = \oplus_i y_i$, each injective T-module m is a quotient of some y^s. We infer that

$$\mathrm{mod}_T(m, \tau_T y_i) \subset \mathrm{mod}_S(m, \tau_S y_i) \subset \mathrm{mod}_S(y, \tau_S y_i)^s = 0,$$

hence that $\{y_1 \ldots, y_n\}$ is part of a tilting set of T and that n is $\leq$ to the numbers of points of T and S [8]. ✓

a) $\mathrm{mod}_{kQ}(\tau_{kQ}^m X, Y) = 0\,[25]$,
b) $\mathrm{mod}_{kQ}(X, \tau_{kQ}^m Y) \neq 0\,[3]$.

Statement b) implies that there are non-zero morphisms from any regular component to any other, in definite contrast to the tame case. A proof will be sketched in 4.3.

4.2 Coxeter Matrices

Let Q be a connected finite quiver without cycle and $\mathbf{C}_{kQ} = [\dim kQ(y, x)]$ the Cartan matrix[39] of kQ. The *Coxeter matrix* of Q is then defined by $\Phi_Q = -\mathbf{C}_{kQ}^\top \mathbf{C}_{kQ}^{-1}$. Its raison d'être is the formula

$$\Phi_Q \mathrm{Dim}\, X = \mathrm{Dim}\, \tau_{kQ} X\,,$$

which holds for each non-projective $X \in \Gamma_{kQ}$ [40]. Via this formula, Φ_Q provides good information on mod_{kQ}.

The *spectral radius* ρ_Q of Φ_Q is defined as the maximum of the moduli of the complex eigenvalues. If Q is Dynkin, $\rho_Q = 1$ is not an eigenvalue of Φ_Q. If Q is extended Dynkin, $\rho_Q = 1$ is an eigenvalue.

Theorem [41]. *Let the quiver Q be connected, finite, without cycle and neither Dynkin nor extended Dynkin. Then:*

a) $\rho_Q > 1$, ρ_Q and ρ_Q^{-1} are eigenvalues of Φ_Q with multiplicity 1, and $\rho_Q^{-1} < |\lambda| < \rho_Q$ for any other eigenvalue λ.

b) There are eigenvectors $E^+ \gg 0$[40] and $E^- \gg 0$ with eigenvalue ρ_Q and ρ_Q^{-1} respectively.[41]

4.3 A Criterion of Regularity

Theorem [36]. *Let Q, E^-, E^+ be as in 4.2, $\langle\,,\,\rangle$ the extension of the Euler-Poincaré form of kQ (2.11) to $\mathbb{R}^{kQ}$ and $X \in \Gamma_{kQ}$. Then*[42]:

---

[39] $x \in Q$ is the row-index, y the column-index (2.11)

[40] These eigenvectors belong to $\mathbb{R}^{kQ}$, i.e. are $\mathbb{R}$-valued functions defined on the points of kQ. By $E^+ \gg 0$ we mean that all values of $E^+ \in \mathbb{R}^{kQ}$ are > 0.

[41] The proof is based on 'Perron-Frobenius' and uses a polyhedral cone $K \subset \mathbb{R}^{kQ}$ which is stable under Φ_Q and such that $\Phi_Q^m K \subset \overset{\circ}{K}$ (= interior of K) for some $m \in \mathbb{N}$.

[42] Let us show how this implies 4.1: If X, Y are indecomposable and regular,

$$\lim_{s \to \infty} \rho_Q^s \langle \mathrm{Dim}\, \tau_{kQ}^{-s} X, \mathrm{Dim}\, Y \rangle = \lambda_X^- \langle E^-, \mathrm{Dim}\, Y \rangle > 0\,.$$

Obviously, this implies b).

To prove a), we first remark with F. Lukas that the formula $\Phi_Q^m \mathrm{Dim}\, Z = \mathrm{Dim}\, \tau_{kQ}^m Z$ and statement c) imply the existence of some $N(X, Y) \in \mathbb{N}$ such that $\mathrm{Dim}\, \tau_{kQ}^{-m} Z > \mathrm{Dim}\, X$

a) X is postprojective (resp. preinjective) if and only if $\langle E^-, \text{Dim } X \rangle < 0$ (resp. $\langle \text{Dim } X, E^+ \rangle < 0$).

b) X is regular if and only if $\langle E^-, \text{Dim } X \rangle > 0$ and $\langle \text{Dim } X, E^+ \rangle > 0$.

c) If X is not preinjective (resp. not postprojective),

$$\lim_{s \to \infty} \rho_Q^{-s} \text{Dim } \tau_{kQ}^{-s} X = \lambda_X^- E^- \quad \text{for some } \lambda_X^- > 0,$$

$$(resp. \lim_{s \to \infty} \rho_Q^{-s} \text{Dim } \tau_{kQ}^{s} X = \lambda_X^+ E^+ \quad \text{for some } \lambda_X^+ > 0).$$

4.4 Tilting

Next to hereditary algebras, tilted algebras have unveiled some of their secrets and deserve presentation.

Lemma. *Let S be a spectroid with s points and $\mathcal{F}$ a full subspectroid of ind_S such that: a) $\text{Ext}_S^1(X, Y) = 0, \forall X, Y \in \mathcal{F}$; b) $\text{Ext}_S^2(Z, Y) = 0, \forall Z \in \Gamma_S, \forall Y \in \mathcal{F}$.*
Then the number of points of $\mathcal{F}$ is $\leq s$.

In the sequel, we suppose that $\mathcal{F}$ is a *cotilting*[43] subspectroid of ind_S, i.e. satisfies a) , b) and has cardinality s (Fig. 9). Then $\mathcal{F}$ has the following properties:

1) Set $^\perp\mathcal{F} = \{T \in \text{mod}_S : \text{mod}_S(T, X) = 0, \forall X \in \mathcal{F}\}$ and $^\perp\mathcal{F}^\perp = \{F \in \text{mod}_S : \text{mod}_S(T, F) = 0, \forall T \in {}^\perp\mathcal{F}\}$. It is equivalent to claim that a module $F \in \text{mod}_S$ belongs to $^\perp\mathcal{F}^\perp$ or that F is contained in a direct sum of modules from $\mathcal{F}$, or that $\text{Ext}_S^1(F, X) = 0, \forall X \in \mathcal{F}$. Under these conditions, we say that M is $\mathcal{F}$-torsionfree.

We say that a module $T \in \text{mod}_S$ is an $\mathcal{F}$-torsion-module if it belongs to $^\perp\mathcal{F}$. Each $M \in \text{mod}_S$ has a $\mathcal{F}$-torsion part $^\perp M \subset M$ in $^\perp\mathcal{F}$ and an $\mathcal{F}$-torsionfree part $M^\perp = M/^\perp M$ in $^\perp\mathcal{F}^\perp$.

2) Each $x \in S$ provides an $\mathcal{F}$-module $'x : Y \mapsto Y(x)$, and S is identified with a cotilting subspectroid $'S = \{'x : x \in S\}$ of $\text{ind}_{\mathcal{F}}$.

3) The functors

$$D_{\mathcal{F}}^0 : M \mapsto \text{mod}_S(M, ?)|\mathcal{F} \quad \text{and} \quad D_S^0 : N \mapsto \text{mod}_{\mathcal{F}}(N, ?)|S$$

map mod_S and $\text{mod}_{\mathcal{F}}$ to $^{\perp'}S^\perp$ and $^\perp\mathcal{F}^\perp$ respectively and induce quasi-inverse anti-equivalences

$$\text{mod}_S \supset {}^\perp\mathcal{F}^\perp \rightleftarrows {}^{\perp'}S^\perp \subset \text{mod}_{\mathcal{F}} .$$

---

for all $m \geq N(X, Y)$ and all indecomposables Z satisfying $\text{Dim } Z \leq \text{Dim } Y$.

Now consider a non-zero $f : \tau_{kQ}^m X \to Y$ and an indecomposable summand Z of $\text{Im } f$. Since Z is a quotient of $\tau_{kQ}^m X$, $\tau_{kQ}^{-m} Z$ is a quotient of X. Since $Z \subset Y$, we infer that $m < N(X, Y)$. $\checkmark$

[43] See [40]. For brevity, we stick to cotilting spectroids in this survey.

4) The functors

$$D^1_{\mathcal{F}} : M \mapsto \mathrm{Ext}^1_{\mathcal{S}}(M, ?)|\mathcal{F} \quad \text{and} \quad D^1_{\mathcal{S}} : N \mapsto \mathrm{Ext}^1_{\mathcal{F}}(N, ?)|\mathcal{S}$$

map $\mathrm{mod}_{\mathcal{S}}$ and $\mathrm{mod}_{\mathcal{F}}$ to $^\perp\!\mathcal{S}$ and $^\perp\!\mathcal{F}$ respectively and induce quasi-inverse anti-equivalences

$$\mathrm{mod}_{\mathcal{S}} \supset {}^\perp\!\mathcal{F} \rightleftarrows {}^\perp\!\mathcal{S} \subset \mathrm{mod}_{\mathcal{F}} .$$

5) Let x be a point of $\mathcal{S}$. Then $D^0_{\mathcal{F}} {}^\mathbf{P}\!x$ is injective if and only if $^\mathbf{I}\!x \in \mathcal{F}$. In case $^\mathbf{I}\!x \notin \mathcal{F}$, we have $\tau_{\mathcal{F}}^{-1} D^0_{\mathcal{F}} {}^\mathbf{P}\!x = D^1_{\mathcal{F}} {}^\mathbf{I}\!x$.

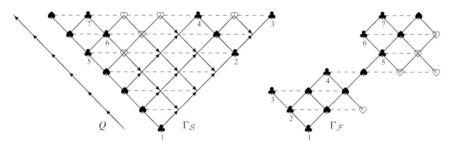

Fig. 9. $\mathcal{S} = kQ$. The points of $\mathcal{F}$ (and their images in $\Gamma_{\mathcal{F}}$) are marked with ♣, the remaining points of $^\perp\!\mathcal{F}^\perp$ with ♠, the points of $^\perp\!\mathcal{F}$ with ♡

4.5 Tilted Algebras

By definition, a *connected* finite spectroid $\mathcal{F}$ is called *tilted of type* Q if $\mathrm{ind}_{\mathcal{F}}$ contains a *slice*, i.e. a *hereditary*[44] cotilting subspectroid, $'\!\mathcal{S} \overset{\sim}{\to} kQ$. If this is so, we are in the situation examined in 4.4, and *each indecomposable $\mathcal{F}$-module is isomorphic to some* $D^0_{\mathcal{F}} F$ *or to some* $D^1_{\mathcal{F}} T$.[45] Thus $\mathcal{F}$ has 'less' indecomposables than $\mathcal{S}$, and we can transfer knowledge [40] from $\mathcal{S}$ to $\mathcal{F}$ (Fig. 9):

1) Each arrow $x \overset{\alpha}{\to} y$ of Q induces an arrow $'\!x \overset{'\!\alpha}{\to} '\!y$ of $\Gamma_{\mathcal{F}}$. The remaining arrows of $\Gamma_{\mathcal{F}}$ with tail $'\!x$ are associated with the arrows $t \to x$ of Q and stop at $D^1_{\mathcal{F}}('\!t)$. Accordingly, $'\!\mathcal{S}$ lies in *one* component of $\Gamma_{\mathcal{F}}$, the *connecting* component.

---

[44] *hereditary* = of global dimension ≤ 1; it is equivalent to require that the spectroid supported by $'\!\mathcal{S}$ is isomorphic to kQ for some finite quiver Q without cycle. In fact, a cotilting set $'\!\mathcal{S}$ gives rise to a hereditary spectroid if and only if, for all $x, z \in '\!\mathcal{S}$ and each arrow $y \to z$ of $\Gamma_{\mathcal{F}}$, $\mathrm{mod}_{\mathcal{F}}(x, y) \neq 0$ implies $y \in '\!\mathcal{S}$.

[45] By 4.4, it suffices to prove that $\mathrm{Ext}^1_{\mathcal{F}}(D^0_{\mathcal{F}} F, D^1_{\mathcal{F}} T) = 0, \forall T \in {}^\perp\!\mathcal{F}, \ F \in {}^\perp\!\mathcal{F}^\perp$. For this, we consider a projective resolution $0 \to R \to P \to T \to 0$ and the induced exact sequences

$$0 \to D^0_{\mathcal{F}} P \to D^0_{\mathcal{F}} R \to D^1_{\mathcal{F}} T \to 0,$$

$$\mathrm{Ext}^1_{\mathcal{F}}(D^0_{\mathcal{F}} F, D^0_{\mathcal{F}} R) \to \mathrm{Ext}^1_{\mathcal{F}}(D^0_{\mathcal{F}} F, D^1_{\mathcal{F}} T) \to \mathrm{Ext}^2_{\mathcal{F}}(D^0_{\mathcal{F}} F, D^0_{\mathcal{F}} P) .$$

The first term of the last sequence vanishes because $D^0_{\mathcal{F}}$ preserves extensions of $\mathcal{F}$-torsionfree modules. The last term is 0 because the indecomposable factors of $D^0_{\mathcal{F}} P$ lie in $'\!\mathcal{S}$, hence have injective dimension ≤ 1.

2) Set $\rangle'S = \{X \in \Gamma_{\mathcal{F}} \setminus 'S : \mathrm{mod}_{\mathcal{F}}(X, Y) \neq 0$ for some $Y \in 'S\}$ and $'S\langle = \{Z \in \Gamma_{\mathcal{F}} \setminus 'S : \mathrm{mod}_{\mathcal{F}}(Y, Z) \neq 0$ for some $Y \in 'S\}$. The sets $\rangle'S$, $'S$ and $'S\langle$ comprise all vertices of $\Gamma_{\mathcal{F}}$, and we have

$$0 = \mathrm{mod}_{\mathcal{F}}(Y, X) = \mathrm{mod}_{\mathcal{F}}(Z, Y) = \mathrm{mod}_{\mathcal{F}}(Z, X)$$

for all $X \in \rangle'S$, $Y \in 'S$, $Z \in 'S\langle$. Moreover, each morphism $X \to Z$ is a sum of morphisms factoring through points of $'S$.[46]

4.6 Wild Tilting

The general structure of the representation-quivers of tilted spectroids is fairly well known [25]. In the following short-cut, Q still denotes a connected finite quiver and $\mathcal{F}$ a cotilting subspectroid of ind_{kQ}. If $\mathcal{F}$ lies in the preinjective component of Γ_{kQ}, $\mathcal{F}$ and all isomorphic spectroids are called *concealed*.

Throughout 4.6 and 4.7, kQ is supposed to be wild.

The Reduced Case. The description of $\Gamma_{\mathcal{F}}$ can be reduced to the case – which we examine first – where $\mathcal{F}$ *contains no postprojective vertex* of Γ_{kQ}. In this case, $\Gamma_{\mathcal{F}}$ has *one* postprojective component, whose projectives form a wild concealed spectroid. This component contains all projectives of $\Gamma_{\mathcal{F}}$ provided $\mathcal{F}$ itself is concealed [47].

In the proposition below, we denote by $S\langle C|$ the full sub-translation-quiver of a component C of Γ_{kQ} which is formed by the termini of the paths of C with origin S. By $\Gamma_{\mathcal{F}}^{\mathrm{reg}}$ we denote the full sub-translation-quiver of $\Gamma_{\mathcal{F}}$ supported by the *regular*[47] vertices, i. e. the vertices X such that $\tau_{\mathcal{F}}^m X$ is defined for all $m \in \mathbb{Z}$.

Proposition [25]. *Let Q be a connected finite quiver which has no cycle and is neither Dynkin nor extended Dynkin. Set $kQ = S$, and let $\mathcal{F}$ be a cotilting subspectroid of ind_S which contains no postprojective vertex. Then:*

a) Each regular component C of Γ_S contains a quasi-simple vertex S such that $S\langle C| \subset {}^{\perp}\mathcal{F}^{\perp}$. It follows that $D_{\mathcal{F}}^0$ induces an anti-isomorphism of $S\langle C|$ onto a mesh-complete[48] sub-translation-quiver of $\Gamma_{\mathcal{F}}$. In particular, $D_{\mathcal{F}}^0(S\langle C|)$ lies in a connected component $C_{\mathcal{F}}^0$ of $\Gamma_{\mathcal{F}}^{\mathrm{reg}}$ which does not depend on S.

b) The map $C \mapsto C_{\mathcal{F}}^0$ is a bijection from the set of regular components of Γ_S onto the set formed by the connected components of $\Gamma_{\mathcal{F}}^{\mathrm{reg}}$ which intersect $\rangle'S$. All these connected components are isomorphic to $\mathbb{Z}\mathbf{A}_\infty$.

---

[46] As a consequence, $\mathrm{mod}_{\mathcal{F}}(\tau_{\mathcal{F}}^{-1}X, {}^{\mathbf{P}}p) = 0$, for each $p \in \mathcal{F}$ and each non-injective $X \in 'S$. It follows that $\mathrm{Ext}_{\mathcal{F}}^2(X, ?) = 0$ and that the modules $X^{\top} : p \to X(p)^{\top}$ form a slice of $\mathrm{ind}_{\mathcal{F}^{op}}$. Thus, $\mathcal{F}^{op}$ is a tilted spectroid if so is $\mathcal{F}$.

[47] Vertices of regular components (4.1) are regular, but the converse is not true.

[48] A full sub-translation-quiver Δ of a translation-quiver Γ is *mesh-complete* if $X \in \Delta$ and $\tau X \in \Delta$ imply that all arrows $\tau X \to Y$ of Γ and all $Y \to X$ belong to Δ.

The General Case. If $\mathcal{F}$ is a connected finite spectroid, we say that a slice $'S$ of $\mathrm{ind}_{\mathcal{F}}$ is *reduced* if we are in the reduced case considered above, i.e. if no $D_S^0{}^{\mathbf{P}}x$, $x \in \mathcal{F}$, is postprojective in Γ_S.

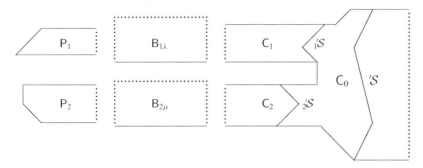

Fig. 10. In our picture – where the right hand side of $\Gamma_{\mathcal{F}}$ has to be completed in a dual manner –, s is equal to 2. Each $_i'S$ is wild, has one component C_i in C_0, one postprojective component P_i and infinitely many injective-free components $B_{i\nu}$. Among them, a finite number may contain projective vertices; the remaining ones are isomorphic to $\mathbb{Z}\mathbf{A}_\infty$. In general, s may be 0, or some $_i'S$ may be tame; in this case, the structure of $)_i'S$ is still to be elucidated in Sect. 5

Theorem [25]. *Suppose that $'S$ is a wild slice of a connected finite spectroid $\mathcal{F}$. Let C_0 be the connecting component of $\Gamma_{\mathcal{F}}$ (, which contains $'S$), $_\infty\mathcal{F}$ the full sub-spectroid of $\mathcal{F}$ formed by the points x such that $^{\mathbf{P}}x \notin C_0$ and $_1\mathcal{F}$, $_2\mathcal{F}$, $\dots$, $_s\mathcal{F}$ the connected components of $_\infty\mathcal{F}$. Then, the connecting component C_0 of $\mathcal{F}$ contains[49] reduced slices $_1'S$, $_2'S$, $\dots$, $_s'S$ of $\mathrm{ind}_{1\mathcal{F}}$, $\mathrm{ind}_{2\mathcal{F}}$, $\dots$, $\mathrm{ind}_{s\mathcal{F}}$ with the following properties:*

For each i, the full sub-translation-quiver $)_i'S \cup {}_i'S$ of $\Gamma_{i\mathcal{F}}$ is identified with a mesh-complete sub-translation-quiver of $)'S \cup 'S$ (4.5).

The complement of $\bigcup_{1 \le i \le s} ( )_i'S \cup {}_i'S)$ in $)'S \cup 'S$ is a finite subset of the connecting component C_0.

The theorem implies that each component $\neq C_0$ of $\Gamma_{\mathcal{F}}$ formed by $'S$-torsionfree modules coincides, for some i, with a component of $\Gamma_{i\mathcal{F}}$ formed by $_i'S$-torsionfree modules. The desciption of these components follows from the reduced case, wild (See above and Fig. 10) or tame (See Sect. 5).

Dual statements hold for the components of $\Gamma_{\mathcal{F}}$ 'at the right hand' of C_0.

---

[49] We here identify a module over $_i\mathcal{F}$ with its zero-extension to $\mathcal{F}$.

4.7 Tame Offsprings of Wild Hereditary Algebras

In 4.6 $\mathcal{F}$ may be tame even though $'S$ is wild (Fig. 11). To be precise, we introduce the dual of s, i.e. the number r of connected components of the full subspectroid $\mathcal{F}_\infty$ of $\mathcal{F}$ formed by the points y such that $^1y \notin C_0$.

Theorem [25, 34]. *Under the assumptions of 4.6, $\mathcal{F}$ is tame if and only if the Tits form $q_{\mathcal{F}}$ is weakly non-negative. If this is so, $q_{\mathcal{F}}$ controls $\mathrm{mod}_{\mathcal{F}}$, and $\mathcal{F}$ is domestic with $s + r$ reliable punched lines.*

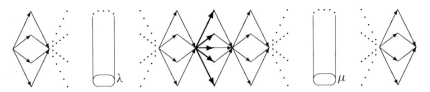

Fig. 11. $Q_{\mathcal{F}}$ is isomorphic to the quiver formed by the first six vertices at the left and is subjected to 'all' commutativity relations. The figure depicts $\Gamma_{\mathcal{F}}$. The projective vertices are the first five vertices and the center of the picture. The indices λ, μ range over $k \cup \infty$. The associated tubes have period 2 for λ, $\mu = 0, 1, \infty$ and 1 otherwise. The slice $'S$ is given by the bold face vectors

The rôle of tilted spectroids goes beyond providing examples. Their importance has a concealed cause, unveiled in the statement below, where a module $M \in \mathrm{ind}_{\mathcal{F}}$ is called *omnipresent* if $M(x) \neq 0$ for all $x \in \mathcal{F}$, and *directing* if $\mathrm{ind}_{\mathcal{F}}$ admits no 'cycle'

$$M = M_0 \overset{\mu_1}{\to} M_1 \to \mu_2 \ldots \to \mu_{n-1} M_{n-1} \to \mu_n M_n = M, \ n > 0,$$

of non-invertible non-zero morphisms μ_i.

Proposition [40]. *A finite spectroid which has an omnipresent directing module is tilted.*

Tame spectroids $\mathcal{F}$ with an omnipresent directing module are well understood. Under the conditions of the preceding theorem,

a) if $s = r = 0$, $\mathcal{F}$ is representation-finite and belongs to a celebrated list established by Bongartz, completed by Dräxler [14];

b) if $s = 1$ and $r = 0$, $_\infty\mathcal{F}$ is a tilted spectroid of extended Dynkin type and supports all the indecomposables of $\mathcal{F}$ with the exception of some preinjectives;

c) if $s + r \geq 2$, then $s = 1 = r$. In this case, $\Gamma_{\mathcal{F}}$ has only finitely many omnipresent vertices, all of them in the connecting component.[50]

---

[50] In case $s = 1 = r$, all spectroids $\mathcal{F}$ with an omnipresent directing module and 20 points at least have been listed [34].

5. Representations of Tame Algebras

In search of some classification of tame algebras, the experts roughly seem to follow the tracks of the representation-finite case, even though there are quite a few new phenomena. But so far, the booty is promising, contours of a theory are discernible, and the end still justifies the means.

5.1 Tame Concealed Algebras

Let Q be a connected finite quiver without cycle and $\mathcal{F}$ a cotilting subset of the preinjective component of Γ_{kQ}.[51] If Q is not a Dynkin quiver, all postprojective and regular vertices of Γ_{kQ} are $\mathcal{F}$-torsionfree. Therefore, $\mathcal{F}$ is representation-finite, tame or wild if and only if so is kQ.

Let us now assume that the underlying graph of Q is an extended Dynkin graph Δ. The representation-quiver of $\mathcal{F}$ then behaves as in the hereditary case (3.2):

a) There is a postprojective component $\mathsf{P}_{\mathcal{F}}$ formed by the images $D_{\mathcal{F}}^0 Y$ of the $\mathcal{F}$-torsionfree preinjective vertices $Y \in \Gamma_{kQ}$. In particular, all projective vertices $D_{\mathcal{F}}^0 F$, $F \in \mathcal{F}$, of $\Gamma_{\mathcal{F}}$ lie in $\mathsf{P}_{\mathcal{F}}$.

b) There is a preinjective component $\mathsf{I}_{\mathcal{F}}$ formed by the images $D_{\mathcal{F}}^0 X$ of the postprojective vertices $X \in \Gamma_{kQ}$ and by the images $D_{\mathcal{F}}^1 T$ of the (preinjective) $\mathcal{F}$-torsion-modules T. All the injective vertices of $\Gamma_{\mathcal{F}}$ are in $\mathsf{I}_{\mathcal{F}}$.

c) The regular vertices of $\Gamma_{\mathcal{F}}$ form a family of tubes $\mathsf{T}_{\mathcal{F}\lambda}$ of period, say, $p(\lambda)$, where $\lambda \in k \cup \infty$. Each $\mathsf{T}_{\mathcal{F}\lambda}$ is the anti-isomorphic image of a tube of Γ_{kQ} under the map induced by $D_{\mathcal{F}}^0$. The cardinalities of the fibres of p can be memorized as follows: To each $\lambda \in k \cup \infty$ we attach a linear graph with $p(\lambda)$ vertices, among which one starting vertex is common to all λ. We thus obtain a 'star'[52] $\mathbf{T}_p$ which, in the considered case, is a Dynkin graph admitting Δ as an extension.

d) All the components of $\Gamma_{\mathcal{F}}$ are standard. Morphisms between vertices of different tubes vanish. The assumptions $P \in \mathsf{P}_{\mathcal{F}}$, $T \in \mathsf{T}_{\mathcal{F}\lambda}$ and $I \in \mathsf{I}_{\mathcal{F}}$ imply

$$0 = \mathrm{mod}_{\mathcal{F}}(T, P) = \mathrm{mod}_{\mathcal{F}}(I, P) = \mathrm{mod}_{\mathcal{F}}(I, T).$$

Moreover, for each λ, each morphism $P \to I$ is a sum of morphisms factoring through points of $\mathsf{T}_{\mathcal{F}\lambda}$.

The Tits form $q_{\mathcal{F}}$ controls $\mathrm{mod}_{\mathcal{F}}$ and is non-negative. The isotropic functions of $\mathbb{Z}^{\mathcal{F}}$ form a subgroup of rank 1 generated by a function $Z_{\mathcal{F}}$ whose values $Z_{\mathcal{F}}(x)$ are > 0, $\forall x \in \mathcal{F}$. In case $X \in \Gamma_{\mathcal{F}}$, the inequality $\langle Z_{\mathcal{F}}, \mathrm{Dim}\, X \rangle_{\mathcal{F}} < 0$ (resp. $= 0$, resp. > 0) is equivalent to $X \in \mathsf{P}_{\mathcal{F}}$ (resp. to $X \in \mathsf{T}_{\mathcal{F}\lambda}$ for some $\lambda \in k \cup \infty$, resp. to $X \in \mathsf{I}_{\mathcal{F}}$).[53]

---

[51] In the terminology of [40], each spectroid isomorphic to $\mathcal{F}$ is then called *concealed of type Q*

[52] Up to isomorphism, the graph $\mathbf{T}_p$ is uniquely determined by the family of tubes $(\mathsf{T}_{\mathcal{F}\lambda})_{\lambda \in k \cup \infty}$. We therefore call it the *star* of the family.

[53] For the classification of the tame concealed spectroids, due to Happel and Vossieck, see [23]. Their importance proceeds from their coincidence with the 'critical' spectroids, discovered by Bongartz and involved in his Finiteness Criterion.

5.2 Tubular Algebras

A *branch* is by definition a finite connected full subspectroid of $\mathcal{C}$ (Fig. 12) which contains the *branching vertex* b.

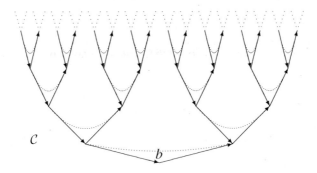

$\mathcal{C}$

b

Fig. 12. $\mathcal{C}$ is defined by the *infinite* quiver depicted above and by the relations $\alpha\beta = 0$, where α and β denote any two composable arrows connected by a dotted curve

Let now M be a finite set of vertices located at the mouths[54] of some tubes $\mathsf{T}_{\mathcal{F}\lambda}$ of a tame concealed spectroid $\mathcal{F}$ (5.1). Let further $\mathcal{B} = (\mathcal{B}_m)_{m \in M}$ be a family of 'copies' of branches with branching vertices b_m. By $\mathcal{BF}$ we then denote a spectroid whose set of points is the disjoint union of the sets of points of $\mathcal{F}$ and of all $\mathcal{B}_m$. The morphisms are such that $\mathcal{BF}(x, y) = \mathcal{F}(x, y)$ if $x, y \in \mathcal{F}$, $\mathcal{BF}(x, y) = \mathcal{B}_m(x, y)$ if $x, y \in \mathcal{B}_m$, $\mathcal{BF}(x, y) = m(y) \otimes_k \mathcal{B}_m(x, b_m)$ if $x \in \mathcal{B}_m$, $y \in \mathcal{F}$, and $\mathcal{BF}(x, y) = 0$ in the remaining cases. The composition is the obvious one.

The spectroid $\mathcal{BF}$ is called a *'tubular source-extension'* of $\mathcal{F}$. All tame spectroids isomorphic to some $\mathcal{BF}$ are called *tubular*.

In the following theorem, we set $q(\lambda) = p(\lambda) + \sum_{m \in M \cap T_\lambda} |\mathcal{B}_m|$, where p is the period of T_λ and $|\mathcal{B}_m|$ the number of points of $\mathcal{B}_m$. As in 5.1c) above, the function q gives birth to a graph $\mathbf{T}_q$.

Fig. 13. As an example, the picture exhibits up to duality some non-domestic tubular spectroids with star $\tilde{\mathbf{E}}_6 = $ ●→●—●—● whose quiver is a tree. As usually, a dotted line means that the path along it is zero. The three edges may be given any orientation

---

[54] The *mouth* of a tube $T \overset{\sim}{\to} \mathbb{Z}\mathbf{A}_\infty/p$ is formed by the classes modulo p of the vertices $(n, 1)$ (3.2).

Theorem [40]. *Let $\mathcal{F}$ be the tame concealed spectroid and $\mathcal{BF}$ the tubular source-extension considered above (5.1–5.2).*

a) $\mathcal{BF}$ is domestic if and only if $\mathbf{T}_q$ is a Dynkin graph. Up to isomorphism, the domestic tubular spectroids coincide with the 'postprojective-free' cotilting subspectroids of ind_{kQ}, where Q ranges over the extended Dynkin quivers. In particular, if $\mathcal{BF}$ is domestic, the Tits form $q_{\mathcal{BF}}$ is non-negative of corank 1 and controls $\mathrm{mod}_{\mathcal{BF}}$.

b) $\mathcal{BF}$ is wild if and only if $\mathbf{T}_q$ is neither Dynkin nor extended Dynkin. If this is so, there is a dimension-function $D \in \mathbb{N}^{\mathcal{BF}}$ with $q_{\mathcal{BF}}(D) < 0$.

c) $\mathcal{BF}$ is non-domestic of polynomial growth if and only if $\mathbf{T}_q$ is an extended Dynkin graph. In this case, we have the following three statements:

$\Gamma_{\mathcal{BF}}$ consists of a postprojective component $\mathsf{P}_{\mathcal{BF}}$, a preinjective component $\mathsf{I}_{\mathcal{BF}}$ and families of components $(\mathsf{T}_\lambda^\gamma)_{\lambda \in k \cup \infty}$, where $\gamma \in \mathbb{Q}_0^\infty = \{\kappa \in \mathbb{Q} \cup \infty : 0 \leq \kappa \leq \infty\}$. For a fixed value of $\gamma \notin \{0, \infty\}$, $(\mathsf{T}_\lambda^\gamma)_{\lambda \in k \cup \infty}$ is a family of periodic tubes with star $\mathbf{T}_q$. The components T_λ^0 are radiant tubes, the $\mathsf{T}_\lambda^\infty$ coradiant tubes (Fig. 13).

The assumptions

$$T \in \mathsf{T}_\lambda^\gamma, \quad X \in \mathsf{P}_{\mathcal{BF}} \cup \bigcup_{\beta < \gamma} \bigcup_\lambda \mathsf{T}_\lambda^\beta, \quad Z \in \mathsf{I}_{\mathcal{BF}} \cup \bigcup_{\gamma < \delta} \bigcup_\lambda \mathsf{T}_\lambda^\delta$$

imply

$$0 = \mathrm{mod}_{\mathcal{BF}}(T, X) = \mathrm{mod}_{\mathcal{BF}}(Z, T) = \mathrm{mod}_{\mathcal{BF}}(Z, X),$$

and each morphism $X \to Z$ is a sum of morphisms factoring through $\mathsf{T}_\lambda^\gamma$. $q_{\mathcal{BF}}$ is non-negative of corank 2 and controls $\mathrm{mod}_{\mathcal{BF}}$.[55]

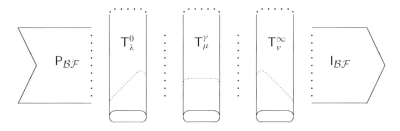

Fig. 14. The representation-quiver of a non-domestic tubular algebra

---

[55] If $\mathbf{T}_q$ is an extended Dynkin graph, there are functions D_0 and D_∞ in $\mathbb{N}^{\mathcal{BF}}$ such that $\mathrm{Dim}\, m = D_0$ and $\mathrm{Dim}\, n = D_\infty$ for almost all vertices m and n at the 'mouths' of the tubes T_μ^0 and T_ν^∞ respectively. Since $q_{\mathcal{BF}}$ controls $\mathrm{mod}_{\mathcal{BF}}$, these functions satisfy $\mathbb{Z} = \sum_{x \in \mathcal{BF}} \mathbb{Z} D_0(x) = \sum_{x \in \mathcal{BF}} \mathbb{Z} D_\infty(x)$. If $a, b \in \mathbb{N}$ are > 0, almost all vertices at the mouths of the tubes $\mathsf{T}_\mu^{a/b}$ have the dimension-function $\frac{1}{g}(bD_0 + aD_\infty)$, where g is the greatest common divisor of the numbers $bD_0(x) + aD_\infty(x)$, $x \in \mathcal{BF}$.

5.3 Tame Selfinjective Algebras

With each finite spectroid S we associate a new spectroid, the *repetition* $S^{\mathbb{Z}}$ of S. Its objects are the pairs $x^m = (x, m)$ where $x \in S$ and $m \in \mathbb{Z}$. The morphisms $x^m \to y^n$ are the pairs $f^n = (f, n)$ with $f \in S(x, y)$ if $m = n$, the pairs $\varphi^n = (\varphi, n)$ with $\varphi \in S(y, x)^{\top}$ if $m = n - 1$. They are zero if $m \notin \{n, n - 1\}$. The composition is the obvious one (Fig. 15).

Fig. 15. Here, S is the first tubular spectroid of Fig. 13. The 3 zero-relations binding the arrows of S give rise to 3 zero-relations in each copy S^n of S. The remaining relations are relations of commutativity, three for each n, which bind the arrows connecting S^{n-1} to S^n

The repetition $S^{\mathbb{Z}}$ is infinite, yet *locally bounded*[56]. It is *self-injective* in the sense that the projective modules coincide with the injective ones.[57]

The *shifting* $x^n \mapsto x^{n+1}$ provides a natural automorphism of $S^{\mathbb{Z}}$, and the associated action of $\mathbb{Z}$ on $S^{\mathbb{Z}}$ is *free* (on the vertices, hence on the morphisms). As Fig. 15 shows, other free group actions may exist: For instance, if S has an automorphism ρ, we obtain a free action of $\mathbb{Z}$ on $S^{\mathbb{Z}}$ by mapping x^n to $\rho(x^{n+1})$.

As a short excursus, we now consider a general group G acting on a locally bounded spectroid T. Then we can construct a new spectroid T/G: The points of T/G are the orbits Gx of a points x of T. The morphisms $Gx \to Gy$ are the elements of $\prod_{u \in Gx, v \in Gy} T(u, v)$ which stay fixed under the induced action of G. They are composed like matrices. The *residue-spectroid* T/G thus obtained is locally bounded.

Theorem [44]. *Let the finite spectroid S be tubular. Then each group $G \neq 1$ acting freely on $S^{\mathbb{Z}}$ is isomorphic to $\mathbb{Z}$, and the residue-spectroid $S^{\mathbb{Z}}/G$ is finite, selfinjective and of polynomial growth; it is domestic if and only if so is S.*

The statement can be made more precise, in the sense that a spectroid T which is not *locally representation-finite*[58] is isomorphic to $S^{\mathbb{Z}}$ for some tubular S if and only if it is locally bounded, selfinjective, *support-finite, locally of polynomial*

---

[56] For each $u \in S^{\mathbb{Z}}$, there are only finitely many v such that $S^{\mathbb{Z}}(u, v) \neq 0$ or $S^{\mathbb{Z}}(v, u) \neq 0$.

[57] In the present case, we have ${}^{\mathrm{P}}x^n = {}^{\mathrm{I}}x^{n+1}$.

[58] A spectroid T is *support-finite* (resp. *locally representation-finite*) if each $u \in T$ is contained in a finite full (resp. full representation-finite) subspectroid $T_u \subset T$ such that all indecomposable T-modules M with $M(u) \neq 0$ are supported by T_u.

growth[59] and *simply connected*[60] [44]. The theorem provides a weak generalization of Riedtmann's classification of the representation-finite selfinjective algebras, which was – in the late seventies – the first big step towards a 'general settlement' of the representation-finite question. Riedtmann's result might suggest that any selfinjective, finite, connected spectroid of polynomial growth is a deformation of $S^{\mathbb{Z}}/G$ for some tubular S and some G.

6. Aus Frischen Waben

Skowronski's classification and the preceding statements belong to the well-established results on tame algebras. To season our picture, we shall add some snapshots on events still permeated by the smell of powder.

6.1 The Componental Quiver

Let S be a finite spectroid and c, d vertices of two different components C, D of Γ_S. The definition of Γ_S then implies[61]

$$\mathrm{ind}_S(c, d) = \bigcap_{n \in \mathbb{N}} \mathcal{R}^n_{\mathrm{ind}_S}(c, d) =: \mathcal{R}^\infty_{\mathrm{ind}_S}(c, d) .$$

As it seems, there is a correlation between the 'representation type' of S and the location of the 'big jumps', the non-zero morphisms in $\mathcal{R}^\infty_{\mathrm{ind}_S}$. These big jumps are roughly depicted in the *componental quiver* Υ_S, whose vertices are the connected components of Γ_S, components C and D (equal or not) being connected by an arrow C $\to$ D (*one only!*) if $\mathcal{R}^\infty_{\mathrm{ind}_S}(c, d) \neq 0$ for some $c \in$ C, $d \in$ D. Here we examine the case where 'big jumps admit no return'.

Theorem [45]. *A finite spectroid without intercomponental*[62] *cycle which has an omnipresent module in a periodic tube of its representation-quiver is either tame concealed or non-domestic tubular.*

Corollary 1 [45]. *A finite spectroid without intercomponental cycle is of polynomial growth.*[63]

---

[59] i.e. each finite full subspectroid is of polynomial growth.

[60] A spectroid T is *simply connected* if it is directed (2.10), connected and not isomorphic to some $\mathcal{U}/H$, where $\mathcal{U}$ is connected and $H \neq \{1\}$ acts freely on $\mathcal{U}$.

[61] In fact, this follows from the basic properties of the almost split sequences (See footnote 30). We may recall here that $\mathcal{R}_{\mathrm{ind}_S}$ denotes the radical of ind_S, i.e. the ideal formed by the non-invertible morphisms (1.4).

[62] i.e. without cycle in Υ_S

[63] Skowronski has announced the converse under the additional assumption that S is simply connected throughout (6.2). The proof involves so-called *multicoil* algebras, which are of polynomial growth [1].

Corollary 2 [45]. *A non-domestic finite spectroid without intercomponental cycle contains a non-domestic tubular convex[64] subspectroid.*

6.2 Simply Connected Algebras

A spectroid is called *simply connected throughout* if each convex subspectroid is simply connected (5.3).[65]

Proposition [31]. *Let the finite spectroid S be simply connected throughout. Then the Tits form q_S is weakly non-negative if and only if no convex subspectroid of S is concealed (5.1) of type Δ, where the underlying graph of Δ belongs to Fig. 16.*

The concealed spectroids of type Δ are classified in [48]. As it seems, they still look for some 'regisseur', hoping to find on tame stages a rôle comparable to the part played by tame concealed spectroids in the representation-finite scenery.

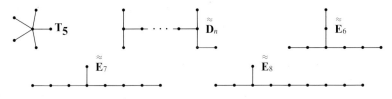

Fig. 16. $\mathbf{D}_n$ has $n + 2$ vertices, where $4 \leq n \leq 8$

6.3 Tame Simply Connected Algebras and Tits Forms

If T is a full subspectroid of a finite spectroid S, we identify a function on T with its extension to S by zero. Thus we obtain an embedding $\mathbb{Z}^T \subset \mathbb{Z}^S$.

Suppose now that S is directed and *tame* and that $D \in \mathbb{N}^S$ is *isotropic* with respect to the Tits form : $q_S(D) = 0$. Since q_S is weakly non-negative (2.10), its restriction to the *support* $S(D)$ (2.12) of D must be non-negative, and the isotropic functions of $\mathbb{N}^{S(D)}$ form a 'convex polyhedral' cone: the intersection of $\mathbb{N}^{S(D)}$ with the kernel of the quadratic form $q_S|\mathbb{Z}^{S(D)}$. Accordingly, the isotropic part of $\mathbb{N}^S$ is a finite union of convex polyhedral cones.

Since the vectors of salient convex polyhedral cones of $\mathbb{Z}^S$ are positive *rational* linear combinations of 'extremal' vectors, the isotropic dimension-functions admit the following description: Call an isotropic $E \in \mathbb{N}^S$ *extremal* if $q_S|S(E)$ has corank

---

[64] A subspectroid T of S is called *convex* if it is full and connected and if each path of Q_S with origin and terminus in T has all its vertices in T.

[65] It is equivalent to require that, for each convex $T \subset S$, the selfextensions of the T^{op}-T-bimodule $(x, y) \mapsto T(x, y)$ are trivial (nullity of the first Hochschild cohomology group) [46].

1 and if $\sum_{x \in S} \mathbb{Z}E(x) = \mathbb{Z}.$[66] Then the isotropic dimension-functions are just the positive rational linear combinations of extremal isotropic functions.

Proposition. *Suppose that the finite spectroid* S *is simply connected throughout.* *Then*[67]:

   a) S *is domestic if and only if the Tits form* q_S *is weakly non-negative and all isotropic dimension-functions with connected support are multiples of extremal isotropic functions.*
   b) If S *has no intercomponental cycle, but has an omnipresent indecomposable and three pairwise orthogonal extremal isotropic dimension-functions* $E_1, E_2, E_3,$ *then the full subquiver of* Q_S *supported by* $\bigcup_{1 \leq i \leq 3} S(E_i)$ *is not connected.*

6.4 Tame Algebras of Non-polynomial Growth

Theorem [35]. *Suppose that the finite spectroid* S *is tame of non-polynomial growth, is simply connected throughout and has an omnipresent indecomposable. Then* S *contains no convex subspectroid which is non-domestic tubular or tame concealed of extended Dynkin type* $\tilde{\mathbf{E}}_p,$ $p \in \{6, 7, 8\}.$

6.5 'Canonical' Algebras

A *'canonical'* spectroid S of *'type'*[68] $(n_1, n_2, \ldots, n_s)$ is defined by the quiver of Fig. 17 and relations

$$\alpha_{in_i} \cdots \alpha_{i1}\alpha_{i0} = \alpha_{1n_1} \cdots \alpha_{11}\alpha_{10} + \lambda_i \, \alpha_{2n_2} \cdots \alpha_{21}\alpha_{20}, \quad 3 \leq i,$$

where $\lambda_i \in k \setminus \{0, 1\}$ and $\prod_{i<j}(\lambda_i - \lambda_j) \neq 0.$
   The projective module $^Pa \in \mathrm{ind}_S$ has a socle of dimension 2 which is isomorphic to $^Sz^2$ and is generated by the morphisms $\alpha_1, \alpha_2 \in S(a, z)$ associated with the paths $\alpha_{1n_1} \cdots \alpha_{11}\alpha_{10},$ $\alpha_{2n_2} \cdots \alpha_{21}\alpha_{20}.$ The quotients $^Pa_\lambda = {}^Pa/(\alpha_1 + \lambda\alpha_2),$ $\lambda \in k,$ and $^Pa_\infty = {}^Pa/\alpha_2$ have the same dimension-function: the constant function I with value 1. They belong to different periodic tubes T_λ of $\Gamma_S,$ whose period is 1 if $\lambda \notin \Lambda = \{\lambda_1 = 0, \lambda_2 = \infty, \lambda_3, \ldots, \lambda_s\};$ under this condition, $^Pa_\lambda$ is the vertex at the mouth of $T_\lambda.$ In case $\lambda = \lambda_i,$ T_λ has period $n_i,$ and the mouth is formed by the vertex $^Sb_{i1}$ and by the sequence of its transforms $^Sb_{i2}, {}^Sb_{i3}, \ldots, {}^Sb_{in_i}, {}^Pa/{}^Pb_{i1}$ under $\tau_S.$
   The union Γ_S^0 of the components T_λ of Γ_S has the following properties:

   a) The full subcategory of mod_S formed by the modules isomorphic to direct sums of vertices of Γ_S^0 is closed under kernels, cokernels and extensions.

---

[66] It follows from a result of Ovsienko that the supports $S(E)$ of the extremal isotropic dimension-functions E are precisely the tame concealed full subspectroids of $S.$
[67] This unpublished result is taken from an article in preparation of J.A. de la Peña.
[68] Face à tous ces 'canons' aux titres inquiétants, on serait tenté d'écrire 'calibre'.

208   José A. de la Peña and Peter Gabriel

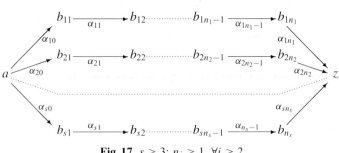

Fig. 17. $s \geq 3$; $n_i \geq 1$, $\forall i \geq 2$

b) Γ_S is the union of Γ_S^0, of $\Gamma_S^+ = \{x \in \Gamma_S : \mathrm{mod}_S(y, x) = 0, \forall y \in \Gamma_S^0\}$ and of $\Gamma_S^- = \{z \in \Gamma_S : \mathrm{mod}_S(z, y) = 0, \forall y \in \Gamma_S^0\}$. Moreover, each projective vertex of Γ_S lies in Γ_S^+, each injective in Γ_S^-.

c) The assumptions $x \in \Gamma_S^+$, $z \in \Gamma_S^-$ imply $\mathrm{mod}_S(z, x) = 0$, and each morphism $x \to z$ is a sum of morphisms factorized through Γ_S^0.

It follows from these conditions that Γ_S^+ and Γ_S^- are unions of components of Γ_S, as Γ_S^0 is. In fact, Γ_S^+, Γ_S^0 and Γ_S^- are formed by the vertices v such that $\dim v(a) < \dim v(z)$, $\dim v(a) = \dim v(z)$ or $\dim v(a) > \dim v(z)$ respectively. Γ_S^+ contains *the* postprojective component of Γ_S, which includes ^Pz and all $^Pb_{ij}$, but not ^Pa in general. Dually, Γ_S^- contains the preinjective component ...

Set $p(\lambda_i) = n_i$ and $p(\lambda) = 1$ if $\lambda \notin \Lambda$. The star $\mathbf{T}_p$ (5.1c) is then identified with the underlying graph of the quiver Q formed by the vertices z and b_{ij} of Fig. 17. If $\mathbf{T}_p$ is Dynkin or extended Dynkin, S is tame concealed or non-domestic tubular (respectively). Then Γ_S is already familiar to us. In all other cases, S is wild. Then all regular components of Γ_S are isomorphic to $\mathbb{Z}\mathbf{A}_\infty$.

Theorem [27]. *Let S be a wild canonical spectroid defined as above, Q the quiver formed by the vertices z and b_{ij} of Fig. 17, ρ_Q the spectral radius of the Coxeter matrix Φ_Q, $E^- \gg 0$ an eigenvector of Φ_Q with eigenvalue ρ_Q^{-1} (4.2) and $I \in \mathbb{R}^S$ the constant function with value 1. Then, for some $\lambda_X^-, \mu_X^+ > 0$, each regular vertex $X \in \Gamma_S^+$ satisfies*

$$\lim_{m \to \infty} \rho_Q^{-m} \operatorname{Dim} \tau_S^{-m} X = \lambda_X^- E^- \quad and \quad \lim_{m \to \infty} \frac{1}{m} \operatorname{Dim} \tau_S^m X = \mu_X^+ I.$$

6.6 Trisections

Tame concealed, tubular and canonical spectroids S share in the following features: The representation-quiver Γ_S contains a union Γ_S^0 of periodic tubes T_v, $v \in N$, which satisfies the conditions a), b) and c) of 6.5. Moreover, there is a dimension-function I such that all $x \in \Gamma_S^0$ fixed by τ_S satisfy $\operatorname{Dim} x \in \mathbb{N}I$. We express these properties by saying that $(T_v)_{v \in N}$ is a *trisecting* family of tubes.

Theorem [28]. *Let S be a finite spectroid with n points and $(\mathsf{T}_\nu)_{\nu \in N}$ a trisecting family of tubes T_ν with periods $p(\nu)$. Then:*

a) $\sum_{\nu \in N}(p(\nu) - 1) = n - 2$.

b) S is isomorphic to a cotilting subspectroid of ind_T, *where T is a canonical spectroid.*

References

1. Assem, I., Skowronski, A.: Indecomposable modules over multicoil algebras. Math. Scand. **71** (1992) 31–61
2. Auslander M., Reiten I.: Representation Theory of Artin algebras III. Commun. Algebra **3** (1975) 239–294
3. Baer, D.: Wild hereditary Artin algebras and linear methods. Manuscr. Math. **55** (1986) 69–82
4. Bangming, D.: On a problem of Nazarova and Roiter. Diss. Uni. Zürich, 1993
5. Bautista, R.: On algebras of strongly unbounded representation type. Comment. Math. Helv. **60** (1985) 392–399
6. Bondarjenko, V.M.: Svjaski poluzepnich mnojestv i ich prjedstavljenija. Inst. Mat. Ukrainian Ac. Sc., 1988, 32 pp. (preprint 88.60)
7. Bondarjenko, V.M., Drozd, Ju.A.: The representation type of finite groups (russian). Zapiski nauchn. seminar. LOMI **71** (1977) 24–42
8. Bongartz, K.: Tilted algebras. Lecture Notes in Mathematics, vol. 903 (1981) 26–38
9. Bongartz, K.: Algebras and quadratic forms. J. Lond. Math. Soc., II Ser **28** (1983) 461–469
10. Bongartz, K.: Indecomposables are standard. Comment. Math. Helv. **60** (1985) 400–410
11. Crawley-Boevey, W.W.: On tame algebras and bocses. Proc. Lond. Math. Soc. III. **56** (1988) 451–483
12. Crawley-Boevey, W.W.: Functorial filtrations II: clans and the Gelfand problem. J. London Math. Soc. **40** (1989) 9–30
13. d'Este, G., Ringel, C.M.: Coherent tubes. J. Algebra **87** (1984) 150–201
14. Dräxler, P.: Aufrichtige gerichtete Ausnahmealgebren. Bayreuther Math. Schr. **29** (1989). Sur les algèbres exceptionnelles de Bongartz. Comptes-Rendus Acad. Sc. Paris, Sér. I **311** (1990) 495–498
15. Gabriel, P.: Auslander-Reiten sequences and representation-finite algebras. Lecture Notes in Mathematics, vol. 831 (1980) 1–70
16. Gabriel, P.: Darstellungen endlichdimensionaler Algebren. Proc. I.C.M., Berkeley 1984
17. Gabriel, P., Nazarova, L.A., Roiter, A.V., Sergeichuk, V.V., Vossieck, D.: Tame and wild subspace problems. Ukrainian Math. J. (1993) 313–352
18. Gabriel, P., Keller, B., Roiter, A.V.: Representations of finite-dimensional algebras. Encycl. Math. Sc. **73** (1992) 1–176
19. Geiss, Chr., On degenerations of tame and wild algebras. Preprint 1992, 11 pp.
20. Geiss, Chr.: Tame distributive 2-points algebras. Proceedings ICRA VI, Carleton University (Ottawa) (to appear)
21. Geiss, Chr., de la Peña, J.A.: An interesting family of algebras. Arch. Math. **60** (1993) 25–35
22. Happel, D., Preiser, U., Ringel, C.M.: Vinberg's characterisation of Dynkin diagrams using subadditive functions with an application to DTr-periodic modules. Lecture Notes in Mathematics, vol. 832 (1980) 280–284
23. Happel, D., Vossieck, D.: Minimal algebras of infinite representation type with preprojective component. Manuscripta Math. **42** (1983) 221–243
24. Kač, V.: Infinite root systems, representations of graphs and invariant theory. Inv. Math. **56** (1980) 57–92

25. Kerner, O.: Tilting wild algebras. J. London Math. Soc. **39** (1989) 29–37
26. Kraft, H., Riedtmann, Chr.: Geometry of representations of quivers. In: Representations of Algebras. London Math. Soc. Lect. Notes Ser. **116** (1986) 109–145
27. Lenzing, H., de la Peña, J.A.: Wild canonical algebras. Preprint 1991
28. Lenzing, H., de la Peña, J.A.: Separating exact subcategories. In preparation
29. Liu, Shiping: Semi-stable Components of an Auslander-Reiten Quiver. J. London Math. Soc. (to appear)
30. Nazarova, L.A., Zavadski A.G.: Partially ordered sets of finite growth. Funkts. Anal. Prilozh. **16** (1982) 72–73. English transl.: Funct. Anal. Appl. **16** (1982) 135–137
31. de la Peña, J.A.: Algebras with hypercritical Tits form. Topics in Algebra, Banach Center Publications **26**, I (1990) 353–369
32. de la Peña, J.A.: On the dimension of the module-varieties of tame and wild algebras. Commun. Algebra **19** (1991) 1795–1807
33. de la Peña, J.A.: Sur les degrés de liberté des indécomposables. C. R. Ac. Sci. Paris **312**, I (1991) 545–548
34. de la Peña, J.A.: The families of 2-parametric domestic algebras with a sincere directing module. Proceedings ICRA VI, Carleton University (Ottawa) (to appear)
35. de la Peña, J.A., Skowronski, A.: Forbidden subalgebras of algebras of non-polynomial growth. In preparation
36. de la Peña, J.A., Takane, M.: Spectral properties of Coxeter transformations and applications. Archiv Math. **55** (1990) 120–134
37. Riedtmann, Chr.: Algebren, Darstellungsköcher, Überlagerungen und zurück. Comment. Math. Helv. **55** (1980) 199–224
38. Ringel, C.M.: The representation type of local algebras. Lecture Notes in Mathematics **488** (1975) 282–305
39. Ringel, C.M.: Finite-dimensional hereditary algebras of wild representation type. Math. Z. **161** (1978) 235–255
40. Ringel, C.M.: Tame algebras and integral quadratic forms. Lecture Notes in Mathematics **1099** (1984)
41. Ringel, C.M.: The spectral radius of the Coxeter transformation for a generalized Cartan matrix. Preprint 1992
42. Simson, D.: Linear representations of partially ordered sets and vector space categories. Gordon and Breach, 1991, 498 pp.
43. Skowronski, A.: Generalized standard Auslander-Reiten components. J. Math. Soc. Japan (to appear)
44. Skowronski, A.: Selfinjective algebras of polynomial growth. Math. Ann. **285** (1988) 177–199
45. Skowronski, A.: Cycle-finite algebras. Preprint 1993
46. Skowronski, A.: Simply connected algebras and Hochschild cohomology. Proceedings of ICRA VI, Ottawa (Carleton University) (to appear)
47. Strauss, H.: The perpendicular category of a partial tilting module. J. Algebra **144** (1991) 43–66
48. Unger, L.: The concealed algebras of the minimal wild hereditary algebras. Bayreuther Math. Schriften **32** (1990) 145–154
49. Zavadski, A.G.: Differentiation algorithm and classification of representations. Izv. Akad. Nauk SSSR, Ser. Matem. **55** (1991) 975–1012

Oberwolfach zum Dank. Unser Bericht aus subjektiver Sicht ist eine Momentaufnahme, eine Aufzeichnung des Wachstums eines Kleinzweiges der Oberwolfacher Eiche. Doch auch wer an Zweigen wirkt, glaubt an sein Werk und hofft, daß einst der Honig fließen wird aus seinen Blüten. Denn das Leben quillt zuerst im unermeßlich kleinen. Ob es strömen wird, entscheidet der Gott Zukunft, in fünfzig Jahren vielleicht, beim nächsten Jubiläum. Einstweilen schuldet der Zweiggebundene Dank dem Gönner für sein Vertrauen und seine Zuversicht.

Number Theory and Cryptography

Neal Koblitz

In several branches of number theory – algebraic, analytic, and computational – certain questions have acquired great practical importance in the science of cryptography. Broadly speaking, the term *cryptography* refers to a wide range of security issues in the transmission and safeguarding of information. Most of the applications of number theory have arisen since 1976 as a result of the development of *public-key* cryptography.

Except for a brief discussion of the historical role of number theory in *private-key* cryptography (pre-1976), we shall devote most of this survey to the (generally more interesting) questions that arise in the study of public-key cryptosystems. After discussing the idea of public-key cryptography and its importance, we next describe certain prototypical public-key constructions that use number theory. Then we survey the progress that has been made in the three number-theoretic problems which are at the heart of these constructions – primality testing, factorization of integers, and discrete logarithms in a finite field. Finally, we describe elliptic curve cryptosystems and the corresponding discrete logarithm problem. We conclude with a list of open questions.

§ 1. Early History

A cryptosystem for message transmission means a map from units of ordinary text called *plaintext message units* (each consisting of a letter or block of letters) to units of coded text called *ciphertext message units*. The idea of using arithmetic operations to construct such a map goes back at least to the Romans. In modern terminology, they used the operation of addition modulo N, where N is the number of letters in the alphabet, which we suppose has been put in one-to-one correspondence with $\mathbf{Z}/N\mathbf{Z}$. For example, if $N = 26$ (i.e., messages are in the usual Latin alphabet, with no additional symbols for punctuation, numerals, capital letters, etc.), the Romans might encipher single letter message units according to the formula $C \equiv P + 3 \pmod{26}$. This means that we replace each plaintext letter by the letter three positions farther down the alphabet (with the convention that $X \mapsto A$, $Y \mapsto B$, $Z \mapsto C$). It is not hard to see that the Roman system – or in fact any cryptosystem based on a permutation of single letter message units – is easy to break.

In the 16th Century, the French cryptographer Vigenère invented a variant on the Roman system that is not quite so easy to break. He took a message unit to be a block of k letters – in modern terminology, a k-vector over $\mathbf{Z}/N\mathbf{Z}$. He then shifted each block by a "code word" of length k; in other words, his map from plaintext to ciphertext message units was translation of $(\mathbf{Z}/N\mathbf{Z})^k$ by a fixed vector.

Much later, Hill [1931] noted that the map from $(\mathbf{Z}/N\mathbf{Z})^k$ to $(\mathbf{Z}/N\mathbf{Z})^k$ given by an invertible matrix over the ring $\mathbf{Z}/N\mathbf{Z}$ would be more likely to be secure than Vigenère's simple translation map. Here "secure" means that one cannot easily figure out the map knowing only the ciphertext. (The Vigenère cipher, on the other hand, can easily be broken if one has a long string of ciphertext, by analyzing the frequency of occurrence of the letters in each arithmetic progression with difference k. It should be noted that, even though the Hill system cannot be easily broken by frequency analysis, it is easy to break using linear algebra modulo N if you know or can guess a few plaintext/ciphertext pairs.)

For the most part, until about 20 years ago only rather elementary number theory was used in cryptography. A possible exception was the use of shift register sequences (see [Golomb 1982] and Chapter 6 and 9.2 of [Lidl and Niederreiter 1986]).

Perhaps the most sophisticated mathematical result in cryptography before the 1970s was the famous theorem of information theory that said, roughly speaking, that the only way to obtain perfect secrecy is to use a *one-time pad*. (A "one-time pad" is a Vigenère cipher with period $k = \infty$.) See [Shannon 1949].

The first harbinger of a new type of cryptography seems to have been a passage in a book about time-sharing systems that was published in 1968 [Wilkes 1968, p. 91-92]. In it, the author describes a new *one-way cipher* used by R. M. Needham in order to make it possible for a computer to verify passwords without storing information that could be used by an intruder to impersonate a legitimate user.

In Needham's system, when the user first sets his password, or whenever he changes it, it is immediately subjected to the enciphering process, and it is the enciphered form that is stored in the computer. Whenever the password is typed in response to a demand from the supervisor for the user's identity to be established, it is again enciphered and the result compared with the stored version. It would be of no immediate use to a would-be malefactor to obtain a copy of the list of enciphered passwords, since he would have to decipher them before he could use them. For this purpose, he would need access to a computer and even if full details of the enciphering algorithm were available, the deciphering process would take a long time.

In [Purdy 1974] the first detailed description of such a *one-way function* was published. The original passwords and their enciphered forms are regarded as integers modulo a large prime p, and the "one-way" map $\mathbf{F}_p \longrightarrow \mathbf{F}_p$ is given by a polynomial $f(x)$ which is not hard to evaluate by computer but which takes an unreasonably long time to invert. Purdy used $p = 2^{64} - 59$, $f(x) = x^{2^{24}+17} + a_1 x^{2^{24}+3} + a_2 x^3 + a_3 x^2 + a_4 x + a_5$, where the coefficients a_i were arbitrary 19-digit integers.

§ 2. The Idea of Public-Key Cryptography

Until 1976, all cryptographic message transmission was by what can be called *private key*. This means that someone who has enough information to encrypt messages automatically has enough information to decipher messages as well. As a result, any two users of the system who want to communicate secretly must have exchanged keys in a safe way, e.g., using a trusted courier.

The face of cryptography was radically altered when Diffie and Hellman invented an entirely new type of cryptography, called *public-key* [Diffie and Hellman 1976]. At the heart of this concept is the idea of using a one-way function for encryption. Speaking informally, we can define a public-key encryption function (also called a "trapdoor" function) as a map from plaintext message units to ciphertext message units that can be feasibly computed by anyone having the so-called "public" key but whose inverse function (which deciphers the ciphertext message units) cannot be computed in a reasonable amount of time without some additional information (the "private" key). This means that everyone can send a message to a given user using the same enciphering key, which they simply look up in a public directory. There is no need for the sender to have made any prior secret arrangement with the recipient; indeed, the recipient need never have had any prior contact with the sender at all.

It was the invention of public-key cryptography that led to a dramatic expansion of the role of number theory in cryptography. The reason is that number theory seems to be the best source of one-way functions. Later we shall discuss the most important examples.

A curious historical question is why public-key cryptography had to wait until 1976 to be invented. Nothing involved in the idea of public-key cryptography or the early public-key cryptosystems required the use of 20th century mathematics. The first public-key cryptosystem to be used in the real world – the RSA system (see below) – uses number theory that was well understood by Euler. Why had it not occurred to Euler to invent RSA and offer it to the military advisers of Catherine the Great in gratitude for her generous support for the Russian Imperial Academy of Sciences, of which he was a beneficiary?

One possible reason for the late development of the concept of public key is that until the 1970s cryptography was used mainly for military and diplomatic purposes, for which private-key cryptography was well-suited. However, with the increased computerization of economic life, new needs for cryptography arose. To cite just one obvious example, when large sums of money are transferred electronically, one must be able to prevent white-collar thieves from stealing funds and nosy computer hackers (or business competitors) from monitoring what others are doing with their money. Another example of a relatively new use for cryptography is to protect the privacy of data (medical records, credit ratings, etc.). Unlike in the military or diplomatic situation – with rigid hierarchies, long-term lists of authorized users, and systems of couriers – in the applications to business transactions and data privacy one encounters a much larger and more fluid structure of cryptography users. Thus, perhaps public-key cryptography was not invented earlier simply because there was no real need for it until quite recently.

Another reason why RSA was not likely to have been discovered in Euler's time is that in those days all computations had to be done by hand. To achieve an acceptable level of security using RSA, it would have been necessary to work with rather large integers, for which computations would have been cumbersome. Thus, Euler would have had difficulty in selling the merits of RSA to a committee of skeptical tsarist generals.

In practice, the great value of public-key cryptography today is intimately connected with the proliferation of powerful computer technology.

Tasks for Public-Key Cryptography. The most common purposes for which public-key cryptography has been applied are:

(1) *message transmission*;

(2) *authentication* (verification that the message was sent by the person claimed); *digital signatures*; and *password* systems (proving authorization to have access to a facility or some data);

(3) *key exchange*, i.e., two people using the open airwaves want to agree upon a secret key for use in some private-key cryptosystem;

(4) *coin flip*, for example, two chess players in different cities want to determine by telephone (or e-mail) who plays white;

(5) *secret sharing*, e.g., you want some secret information (such as the password to launch a missile) to be available to k subordinates working together but not to $k - 1$ of them;

(6) *zero knowledge proof*, i.e., you want to convince someone that you have successfully solved a number-theoretic or combinatorial problem (e.g., finding the square root of an integer modulo a large unfactored integer, or 3-coloring a map) without conveying any knowledge whatsoever of what the solution is;

(7) *pseudorandom number generator*, i.e., you want to generate a sequence of 0's and 1's that for all practical purposes is random.

Probabilistic Encryption. Most of the number theory based cryptosystems for message transmission are *deterministic*, in the sense that a given plaintext will always be encrypted into the same ciphertext by anyone. However, deterministic encryption has two disadvantages: (1) if an eavesdropper knows that the plaintext message belongs to a small set (for example, the message is either "yes" or "no"), then she can simply encrypt all possibilities in order to determine which is the supposedly secret message; and (2) it seems to be very difficult to *prove* anything about the security of a system if the encryption is deterministic. For these reasons, *probabilistic encryption* was introduced. We will not discuss this further or give examples in the present paper. For more information, see the fundamental papers on the subject by Goldwasser and Micali [1982, 1984].

We will now discuss two examples of public-key cryptosystems – RSA and Diffie-Hellman – that have been at the center of research in public key cryptography. Both are connected with fundamental questions in number theory – factoring integers and discrete logarithms, respectively. Although the systems can be modified to perform most or all of the 7 tasks listed above, we will describe a protocol for only one of these tasks in each case (message transmission in the case of RSA, and key exchange in the case of discrete log).

§3. The RSA Cryptosystem

Suppose we have a large number of users of our system, each of whom might want to send a secret message to any one of the other users. We shall assume that the message units m have been identified with integers in the range $0 \leq m < N$. For example, a message might be a block of k letters in the Latin alphabet, regarded as an integer to the base 26 with the letters of the alphabet as digits; in that case $N = 26^k$. In practice, in the RSA system N is a number of between 200 and 300 digits.

Each user A (traditionally named Alice) selects two extremely large primes p and q whose product n is greater than N. Alice keeps the individual primes secret, but she publishes the value of n in a directory under her name. She also chooses at random an exponent e which must have no common factor with $p - 1$ or $q - 1$ (and presumably has the same order of magnitude as n), and publishes that value along with n in the directory. Thus, her *public key* is the pair (n, e).

Suppose that another user B (Bob) wants to send Alice a message m. He looks up her public key in the directory, computes the least nonnegative residue of m^e modulo n, and sends Alice this value (let c denote this *ciphertext* value). Bob can perform the modular exponentiation $c = m^e$ (mod n) very rapidly (in $O(\log^3 n)$ bit operations), using the so-called "repeated squaring method" (see 4.6.3 of [Knuth 1981]).

To decipher the message, Alice uses her secret *deciphering key* d, which is any integer with the property that $de \equiv 1$ (mod $p - 1$) and $de \equiv 1$ (mod $q - 1$). She can find such a d easily by applying the extended Euclidean algorithm to the two numbers e and l.c.m.$(p - 1, q - 1)$. It is easy to check (using Fermat's Little Theorem) that if Alice computes the least nonnegative residue of c^d modulo n, the result will be the original message m.

What would prevent an unauthorized person C (Cynthia) from using the public key (n, e) to decipher the message? The problem for Cynthia is that without knowing the factors p and q of n there is apparently no way to find a deciphering exponent d that inverts the operation $m \mapsto m^e$ (mod n). Nor does there seem to be any way of inverting the encryption other than through a deciphering exponent. Here I use the words "apparently" and "seem" because these assertions have not been proved. Thus, one can only say that *apparently* breaking the RSA cryptosystem is as hard as factoring n.

Remark. There is another version of RSA encryption [Hugh Williams 1980] that *provably* cannot be broken unless one can factor n. At first glance, its security is on more solid ground, because there is no possibility that breaking it is easier than factoring. However, it is extremely vulnerable to "chosen ciphertext" attack. This means that if one is able to see the decryption of a ciphertext of one's own choosing, then one can break the system (in this case, by factoring n). For this reason, the Williams version of RSA is not used in practice.

We will not give the details of Williams' construction. The idea, roughly speaking, is to allow the exponent e in RSA to be an even number (and hence obviously not prime to l.c.m.$(p-1, q-1)$). The equivalence between breaking the system and factoring n, and also the system's vulnerability to chosen ciphertext attack, are based on the following observation.

There's a simple and clever argument that shows that anyone who has an algorithm (any algorithm whatsoever) to find square roots modulo n can use the same algorithm to factor $n = pq$. Namely, suppose that Cynthia has such a square root algorithm. Here is how she can use it to factor n. She chooses a random number x prime to n, and applies the algorithm to $x^2 \bmod n$. The algorithm gives one of the 4 square roots of x^2 modulo n, say x'. There is a 50% chance that x' is <u>not</u> congruent to either $\pm x$ modulo n. In that case Cynthia can quickly find the prime factors by computing g.c.d.$(n, x + x')= p$ or q. If x' happens to be $\equiv \pm x$ (mod n), then she simply repeats the procedure. The odds that she will fail to factor n after k repetitions of the procedure are only 1 out of 2^k. Thus, with 99.9999% certainty she will be able to factor n after at most 20 different choices of x. In other words, any square root algorithm immediately gives a probabilistic factoring algorithm.

§ 4. Diffie-Hellman Key Exchange

The second landmark example of a public-key cryptographic system is based on the discrete logarithm problem. First we define this problem.

Let $\mathbf{F}_q^*$ denote the multiplicative group of the finite field of q elements. Let $g \in \mathbf{F}_q^*$ be a fixed element ("base"). The *discrete log problem* in $\mathbf{F}_q^*$ to the base g is the problem, given $x \in \mathbf{F}_q^*$, of determining an integer y such that $x = g^y$ (if such y exists; otherwise, one must receive an output to the effect that x is not in the group generated by g).

The Diffie-Hellman key exchange works as follows. Suppose that Alice and Bob want to agree upon a large integer to serve as a key for some private-key cryptosystem. This must be done using open communication channels, i.e., any eavesdropper (Cynthia) knows everything Alice sends to Bob and everything Bob sends to Alice. Alice and Bob first agree on the field $\mathbf{F}_q$ and a base element g. They also agree on a map from $\mathbf{F}_q^*$ to the positive integers; for example, in the case $q = p$ it could be the obvious map $\mathbf{F}_p^* \longrightarrow \{1, 2, \ldots, p-1\}$. All of this is agreed upon publicly, so that Cynthia also has this information at her disposal.

Next, Alice and Bob choose random secret integers k_A and k_B, respectively. Alice sends Bob the group element g^{k_A} (but not her secret integer k_A), and Bob sends Alice the element g^{k_B} (but not the integer k_B). The common key is then

$g^{k_A k_B} \in \mathbf{F}_q$ (or rather, the integer associated to $g^{k_A k_B}$ under the agreed upon correspondence). Alice determines this key by taking the element g^{k_B} received from Bob and raising it to the k_A-th power; Bob determines the key by taking the element g^{k_A} that he received from Alice and raising it to the k_B-th power. But Cynthia is in the unenviable position of having to determine the element $g^{k_A k_B}$ knowing only g^{k_A} and g^{k_B} but not k_A or k_B.

The problem facing Cynthia is the so-called *Diffie-Hellman problem*: Given g^{k_A}, $g^{k_B} \in \mathbf{F}_q^*$, find $g^{k_A k_B}$. It is easy to see that anyone who can solve the discrete log problem in $\mathbf{F}_q^*$ can then immediately solve the Diffie-Hellman problem as well. The converse is not known. That is, it is conceivable (though generally thought to be unlikely) that someone could invent a way to solve the Diffie-Hellman problem without being able to find discrete logarithms. In other words, breaking the Diffie-Hellman key exchange has not been *proved* to be equivalent to solving the discrete log problem. But for practical purposes it is probably safe to assume that the Diffie-Hellman key exchange in a group G is secure provided that the discrete logarithm problem in G is intractable.

Several related systems based on the discrete logarithm problem have also been proposed. A particularly important variant is due to C. P. Schnorr (see [Schnorr 1990]); his signature scheme is essentially the one that has been adopted by the U.S. National Institute of Standards and Technology as the new Digital Signature Standard.

§5. Knapsack

One of the earliest and most efficient types of public-key cryptosystem was based on the *knapsack* or *subset-sum* problem [Merkle and Hellman 1978]. Unfortunately, these systems turned out to be insecure. But because of the historical importance of the idea – and also because of the possibility that a cryptosystem will be found that is based on similar ideas and is secure as well as efficient – we will briefly discuss knapsack systems. For a more detailed survey see [Odlyzko 1990].

The knapsack problem is as follows: Given an n-tuple of positive integers $a_1, \ldots, a_n$ and an integer c, find a subset of the a_j whose sum is c (or else determine that no such subset exists). The only algorithm known for the general knapsack problem takes approximately $n2^{n/2}$ arithmetic operations. On the other hand, the knapsack problem is trivial if the sequence a_j happens to be *superincreasing*; this means that each a_j is greater than the sum of all of the a_i for $i < j$. Merkle and Hellman constructed a cryptosystem in which the authorized recipient of the message needs only solve a superincreasing knapsack problem, whereas an eavesdropper must solve a non-superincreasing knapsack problem.

Namely, suppose that Alice wants to be able to receive messages. She selects a secret superincreasing sequence $b_1, \ldots, b_n$, a large secret modulus $M > 2b_n$, and a secret multiplier w prime to M. Her public key is the n-tuple $a_1, \ldots, a_n$ consisting of the least positive residues of wb_j modulo M (more precisely, a secret permutation of these integers). Thus, the operation of multiplication by w modulo M disguises the superincreasing sequence, transforming it into what is at first glance a random–looking sequence.

In the Merkle-Hellman system we take a message unit to be an n-tuple of bits $m_1, \ldots, m_n$. When Bob wants to transmit such a message to Alice, he computes the corresponding subset sum of the a_j: $c = \sum m_j a_j$. To decipher c, Alice simply multiplies it by the inverse of w modulo M, thereby obtaining $\sum m_j b_j$, from which she can quickly obtain each $m_j = 0$ or 1.

At first, it seemed that there was a good chance that the Merkle-Hellman system is secure. Since the general knapsack problem is NP-complete (see [Garey and Johnson 1979]), it is unlikely that an efficient algorithm will be found that solves the knapsack problem in all cases. However, a problem can be NP-complete even if most of its instances are easy to solve [Wilf 1984], provided only that the most difficult instances are hard to solve. Thus, NP-completeness is not the most relevant concept for cryptographic security, and the above argument for confidence in knapsack systems is not a good one.

It turned out that the knapsack problems that arise in the Merkle-Hellman system are of a very special type, and in fact are solvable in polynomial time. Shamir showed this in 1984, thereby breaking the Merkle-Hellman cryptosystem. Soon after, a modified version, involving iterations of the disguising operation (with a sequence of different multipliers w and moduli M), was also broken [Brickell 1985]. Later work on breaking various special types of knapsacks (see [Brickell and Odlyzko 1988] for a survey of these results) has led to widespread skepticism about whether the knapsack problem can ever be used to construct a secure public-key cryptosystem.

An As Yet Unbroken Knapsack. We now describe a method of message transmission based on a knapsack-type one-way function that uses polynomials over a finite field. The cryptosystem is due to Chor and Rivest [1988] (see also [Odlyzko 1990]); we shall describe a slightly simplified (and less efficient) version of their construction.

Again suppose that Alice wants to be able to receive messages that are n-tuples of bits $m_1, \ldots, m_n$. (The number n is selected by Alice, as described below.) Her public key, as before, is a sequence of positive integers $a_1, \ldots, a_n$, constructed in the way described below. This time Bob must send her not only the integer $c = \sum m_j a_j$ but also the sum of the bits $c' = \sum m_j$.

Alice constructs the sequence a_j as follows. All of the choices described in this paragraph can be kept secret, since it is only the final n-tuple $a_1, \ldots, a_n$ that Bob needs to know in order to send a message. First, Alice chooses a prime power $q = p^f$ such that $q - 1$ has no large prime factors (in which case discrete logs can feasibly be computed in $\mathbf{F}_q^*$, see 8 below) and such that both p and f are of intermediate size (e.g., 2 or 3 digits). In [Chor and Rivest 1988] the value $q = 197^{24}$ is suggested. Next, Alice chooses a monic irreducible polynomial $F(X) \in \mathbf{F}_p[X]$ of degree f, so that $\mathbf{F}_q$ may be regarded as $\mathbf{F}_p[X]/F(X)$. She also chooses a generator g of $\mathbf{F}_q^*$, and an integer z. Alice makes these choices of F, g, and z in some random way.

Let $t \in \mathbf{F}_q = \mathbf{F}_p[X]/F(X)$ denote the residue class of X. Alice chooses n to be any integer less than both p and f. For $j = 1, \ldots, n$, she computes the nonnegative integer $b_j < q - 1$ such that $g^{b_j} = t + j - 1$. (By assumption, Alice can easily find discrete logarithms in $\mathbf{F}_q^*$.) Finally, Alice chooses at random a permutation π of $\{1, \ldots, n\}$, and sets a_j equal to the least nonnegative residue of $b_{\pi(j)} + z$ modulo $q - 1$. She publishes the n-tuple $(a_1, \ldots, a_n)$ as her public key.

Deciphering works as follows. After receiving c and c' from Bob, she first computes $g^{c - zc'}$, which is represented as a unique polynomial $G(X) \in \mathbf{F}_p[X]$ of degree $< f$. But she knows that this element must also be equal to $\prod g^{m_j b_{\pi(j)}} = \prod (t + \pi(j) - 1)^{m_j}$, which is represented by the polynomial $\prod (X + \pi(j) - 1)^{m_j}$. Since both $G(X)$ and $\prod (X + \pi(j) - 1)^{m_j}$ have degree $< f$ and represent the same element modulo $F(X)$, she must have

$$G(X) = \prod (X + \pi(j) - 1)^{m_j},$$

from which she can determine the m_j by factoring $G(X)$ (for which efficient algorithms are available [Knuth 1981]).

§6. Primality Testing

In the implementation of several of the most popular cryptosystems (RSA, Diffie-Hellman in the multiplicative group of a prime field, etc.) one has to find large prime numbers, usually of 100 to 200 digits. Suppose we generate a random 100-digit odd number n. By the Prime Number Theorem, n has approximately a 1 out of 115 chance of being prime. We then apply a *test of primality* to n. If n turns out to be composite, we choose another random odd number (or perhaps simply replace n by $n + 2$). If n passes the primality test, then we are happy, and choose $p = n$ to be our prime.

Depending on how demanding we are, the term "primality test" means one of three things:

(1) a sequence of easily verified tests of compositeness, each of which is a congruence that is likely (but not certain) to fail if n is composite; thus, if n satisfies all of these congruences, it is a *probable prime*;

(2) an efficient algorithm that tells us with certainty if n is prime;

(3) an efficient algorithm that not only guarantees the primality of n but also gives us a *certificate of primality*, i.e., a list of number theoretic relations that can all be true only if n is prime and that can be verified very rapidly.

We will discuss each type of primality test in turn.

Compositeness Tests. The simplest such test is *trial division*. Suppose, for example, that we divide a large random odd number n by all odd primes less than 100. If n is composite, then it is easy to see that there is less than a 25% probability that this trial division will fail to reveal a factor of n. However, trial division is practical only as a preliminary step to eliminate trivial composites.

A more promising compositeness test is based on Fermat's Little Theorem, according to which a^{n-1} must be $\equiv 1 \pmod{n}$ for any a prime to n, whenever n is prime. For any a, recall that a^{n-1} can be computed modulo n in $O(\log^3 n)$ bit operations, i.e., very rapidly. If a is chosen at random, then most composite numbers n will fail the Fermat test. However, this test has a serious drawback in that there exist composite numbers n – called *Carmichael numbers* – for which $a^{n-1} \equiv 1 \pmod{n}$ for *all* a prime to n. Carmichael numbers tend to be rather rare (the smallest is $561 = 3 \cdot 11 \cdot 17$); however, it was recently proved that there are infinitely many Carmichael numbers (see [Alford, Granville, and Pomerance 1992] and [Granville 1992]).

There is a refined Fermat test [Miller 1976] that does not suffer from this defect and can be performed as rapidly as the a^{n-1} test. Namely, the Miller test has the feature that *no* composite odd number n passes it for more than 25% of the choices of residue class a prime to n. (For most composite n, the chance of passing the Miller test is actually much less than 25%.) The Miller test proceeds as follows. Write $n - 1 = 2^s t$, where t is odd. For any given a prime to n, first compute a^t modulo n. If this is equal to ± 1, then n passes the test. If not, then repeatedly take the square modulo n, obtaining the residue of $a^{2^j t}$, $j = 1, 2, \ldots, s - 1$. If any of these residues is -1 modulo n, then n passes the test. If not, then n fails the test. (Notice that a prime number n would pass the test, for two reasons: (i) $a^{2^s t} = a^{n-1} \equiv 1 \pmod{n}$, if n is prime; and (ii) -1 is the only square root of 1 modulo n other than 1 itself, if n is prime.)

Applying the Miller test with 10 different random values of a gives – for all practical purposes – an infallible test that n is prime. In fact, n is almost certain to be prime if it passes the Miller test for just the four values $a = 2, 3, 5, 7$; it turns out that there is only one odd composite number less than $25 \cdot 10^9$ that passes the Miller test for all four of these values of a [Pomerance, Selfridge, and Wagstaff 1980].

Proofs of Primality. If the generalized Riemann hypothesis is true, then any composite number n must fail the Miller test for some a in the interval $1 < a < 2 \log^2 n$ ([Miller 1976] and [Bach 1985]). Thus, applying the Miller test for all of these values of a – a procedure which takes only polynomial time, namely $O(\log^5 n)$ bit operations – will (under the generalized Riemann hypothesis) give a proof either that n is prime or that it is composite.

In the early 1980s, an elegant and efficient deterministic algorithm was developed that gives a proof of primality using Gauss sums [Adleman, Pomerance, and Rumely 1983]. This algorithm was then improved upon, by using Jacobi sums in place of Gauss sums, in [Cohen and Lenstra 1984]. The idea was to perform a series of Fermat–like tests with Jacobi sums, working in the ring of integers of the cyclotomic field generated by the n-th roots of unity. The Jacobi sum test requires $(\log n)^{O(\log \log \log n)}$ bit operations. Strictly speaking, the algorithm is not quite polynomial time; and the existence of an unconditional polynomial time deterministic proof of primality is still an open question. However, in practice, the Jacobi sum algorithm is very fast in the range of primes of fewer than 200 digits. Even though it is asymptotically slower than the polynomial time Miller test algorithm described in the last paragraph (that depends upon the Riemann hypothesis), the latter test does not become faster than the Jacobi sum algorithm until one gets beyond the range where either algorithm is feasible, i.e., in the range of primes having tens of thousands of digits. This shows that whether or not a given algorithm is polynomial time, while of great theoretical importance, is not always what determines its practical utility. For a detailed description of the algorithm and a discussion of implementation, see the two articles cited above.

Certificates of Primality. Given a probable prime n, we now want not only to prove primality, but also to do it in such a way that the validity of our proof can be checked in far less time than it took to produce the proof. At present, the most efficient way to do this uses elliptic curves [Goldwasser and Kilian 1986]. (We shall need some basic facts about elliptic curves over finite fields; for an introduction to this subject, see Chapter VI of [Koblitz 1987b], [Koblitz 1993], [Silverman 1986], or [Husemöller 1987].)

The point of departure is an analogue of the following special case of a theorem of Pocklington [1914–1916]: *If $q > \sqrt{n}$ is a prime divisor of $n - 1$ and if there exists an integer a satisfying $a^{n-1} \equiv 1$ (mod n) and g.c.d.$(a^{(n-1)/q} - 1, n) = 1$, then n is prime.* This result is easy to prove, by observing that if p is any prime divisor of n, then the assumptions imply that the element $a^{(n-1)/q}$ has exact order q modulo p; but then $q|p - 1$, and so $p > q > \sqrt{n}$.

Pocklington's theorem can be of practical use in proving primality of certain special primes n, but it cannot be applied to prove primality of an arbitrary prime n (because $n - 1$ must be factored, and it must be divisible by a prime $> \sqrt{n}$). On the other hand, the elliptic curve analogue of Pocklington's theorem does in fact lead to a primality proof that is generally applicable. This analogue is as follows: *Let E be an elliptic curve defined over the ring $\mathbf{Z}/n\mathbf{Z}$, and let m be an integer having a prime divisor $q > (n^{1/4} + 1)^2$. If there exists a point P on E such that $m \cdot P = O$ and $(m/q) \cdot P$ is defined and different from O (where O denotes the identity point of E), then n is prime.* We make some remarks about how this statement is converted to a primality algorithm.

1. The integer chosen as m has the property that, if n is prime, then m is the order of the group E. Both the group operation on E and the computation of m are performed under the supposition that n is prime; if the corresponding operations break down because of compositeness of n, then at the same time a nontrivial factor of n will be revealed. In practice, of course, this will never happen, since one goes to the trouble of proving primality only for a probable prime n (e.g., n has already passed the Miller test for $a = 2, 3, 5, 7$).

2. If n is prime, then Hasse's theorem says that the order of the group E is between $n+1-2\sqrt{n}$ and $n+1+2\sqrt{n}$. If n is not prime and if $p|n$ is a prime $< \sqrt{n}$, then the order of E mod p (the group of $\mathbf{F}_p$-points of E) is $\leq p+1+2\sqrt{p} < (n^{1/4}+1)^2$.

3. Multiplying a point on E mod n by a large integer such as m or m/q can be done very efficiently, using the "repeated doubling" method that is analogous to the "repeated squaring" method for exponentiation modulo n.

4. The reason why the elliptic curve analogue of Pocklington – and not the original Pocklington theorem – can be used to test any n is that the number m that is analogous to $n-1$ varies as we vary the elliptic curve E. Instead of being stuck with $n-1$, we thus have flexibility in arriving at a number m that is divisible by $q > (n^{1/4}+1)^2$.

5. Because of the difficulty of factoring m, in practice one looks for an m which is a small multiple of a prime q, i.e., q will actually be much greater than $(n^{1/4}+1)^2$.

6. The most time-consuming part of the algorithm is finding a curve E with suitable m. Once we find E and m (and also a point P with the desired property), then it takes much less time to check that with this choice the conditions of the theorem are satisfied, and so n is prime.

7. The point P is selected at random. Once E is found with the desired property (namely, its order m is divisible by a large prime q), almost any point P on E has order dividing m but not m/q.

8. In practice, q, like n itself, is known only to be a *probable prime*. After completing the algorithm for n, one has a proof of primality contingent upon q being prime. One then applies the procedure recursively to prove primality of q.

9. To find a suitable E, two methods have been proposed, respectively by Goldwasser-Kilian and by A. O. L. Atkin: (1) random selection of the coefficients of the defining equation, and then computation of $m = \#E$ using [Schoof 1985] (or one of the recent improvements – see the discussion below); (2) selection of a complex multiplication field first, and then m, followed by the construction of E. See [Lenstra 1990] for more details, and [Atkin and Morain 1993] for further information on implementation.

Whichever method is used, the final *certificate of primality* of n will be a sequence of 4-tuples (E_i, m_i, q_i, P_i), the first of which satisfies the conditions of the elliptic Pocklington theorem for the probable prime n, the second of which satisfies the conditions for the probable prime q_1 dividing m_1, the third of which satisfies the conditions for the probable prime q_2 dividing m_2, and so on.

We conclude this section by remarking that in cryptographic applications one does not usually need a certificate of primality. Thus, one normally requires only that n pass the Miller test for several values of a, or (if one wants to be completely certain of primality) that it pass the Jacobi sum test.

§7. Factoring

More work has been devoted to the problem of factoring large integers than to any other problem in computational number theory. Research on this question has been particularly intense since the invention of the RSA cryptosystem in 1978, because the security of the system depends upon the assumed difficulty of factoring. Despite some major advances in factoring, RSA seems still to be secure if the modulus n is the product of two primes of about 100 digits.

All of the subexponential algorithms to factor general integers (i.e., integers not of any special form) are probabilistic rather than deterministic. Most of them (but not all) are based upon the following simple observation. Suppose that n is an odd number having r distinct prime factors. Then every square in $(\mathbf{Z}/n\mathbf{Z})^*$ has 2^r square roots modulo n. Thus, if we can produce a congruence of the form $x^2 \equiv y^2 \pmod{n}$, where x and y are obtained independently of one another, then there is a $1 - 2^{-(r-1)}$ probability that $x \not\equiv \pm y \pmod{n}$. It is easy to see that as soon as that happens, we need only compute g.c.d.$(x + y, n)$ to obtain a nontrivial factor of n. Even in the "worst" case $r = 2$ (as in RSA), we have a 50% chance of factoring n any time we obtain a congruence $x^2 \equiv y^2 \pmod{n}$. If this congruence does not lead to a factorization of n, then we simply repeat the procedure with different random choices, and we again have a 50% chance of success. The chance is only about 1 in 1000 that we will fail to factor n within 10 repetitions of the algorithm.

We shall describe one of the basic factoring algorithms of this type, the *quadratic sieve*. Consider the quadratic polynomial $f(X) = (X + [\sqrt{n}])^2 - n \in \mathbf{Z}[X]$. This polynomial has two obvious properties: (1) $f(x) \equiv (x+[\sqrt{n}])^2 \pmod{n}$ for any $x \in \mathbf{Z}$; and (2) $f(x)$ is significantly smaller than n in absolute value (of order roughly $\sqrt{n}$) if x is small in absolute value. Suppose that we evaluate $f(x_i)$ at a sequence of fairly small integers $x_1, x_2, \ldots$, and find that for some subsequence $x_{i_1}, x_{i_2}, \ldots$ the product of the $f(x_{i_l})$ happens to be a perfect square. Then all we have to do is set x equal to the least positive residue modulo n of $\prod_l (x_{i_l} + [\sqrt{n}])$, and set y equal to the least positive residue modulo n of $\sqrt{\prod_l f(x_{i_l})}$; then we have $x^2 \equiv y^2 \pmod{n}$. Note that x and y were formed in quite different ways, so there is no greater-than-random likelihood that $x \equiv \pm y \pmod{n}$. That is, we obtain a congruence of the desired form.

The hard part is to select the x_{i_l} so that $\prod_l f(x_{i_l})$ is a perfect square. To do this we select a set of primes – called a *factor base* – consisting of $p_1 = 2$ and the first $r - 1$ odd primes $p_2, p_3, \ldots, p_r$ for which $\left(\frac{n}{p_j}\right) = 1$ (i.e., n is a quadratic residue modulo p_j). The choice of r will be made later in a certain optimal way. If x_i is small, then there is a good chance that $f(x_i)$ is not divisible by a prime greater than p_r, in which case it is a product of the primes in the factor base. (Note that if $p|f(x_i)$, then n is a square modulo p; so $f(x_i)$ cannot be divisible by any prime p for which $\left(\frac{n}{p}\right) = -1$. This is why we took only primes for which n is a quadratic residue.) Thus, suppose we have a sequence x_i of s different values of x such that

$$f(x_i) = \prod_{j=1}^{r} p_j^{a_{ij}}, \qquad i = 1, \ldots, s.$$

If s is large enough, then it should be possible to find a subset $\{x_{i_l}\}_l \subset \{x_i\}_i$ such that

$$\prod_l f(x_{i_l}) = \prod_{l,j} p_j^{a_{i_l j}}$$

is a perfect square, i.e., $\sum_l a_{i_l j}$ is an even number for each $j = 1, \ldots, r$. This method of factoring seems to have first appeared in print in [Kraitchik 1926].

A systematic way to find the required subset $\{x_{i_l}\}$ of $x_1, \ldots, x_s$ is due to Brillhart and Morrison [1975]. One simply regards the last condition – that the sum of the exponents $a_{i_l j}$ be an even number – as an equation over the field $\mathbf{F}_2$, and use linear algebra. More precisely, let $u_1, \ldots, u_s$ be unknowns taking the value 0 (if the corresponding x_i is not in the subsequence $\{x_{i_l}\}$) or 1 (if the corresponding x_i is in the subsequence). Then the condition we want is that

$$\sum_{i=1}^{s} a_{ij} u_i \equiv 0 \pmod{2} \qquad \text{for} \quad j = 1, \ldots, r.$$

If s is somewhat greater than r, then this system has a nontrivial solution – in fact, several solutions. Each solution gives a different subset $\{x_{i_l}\}$, and hence a different congruence of the form $x^2 \equiv y^2 \pmod{n}$. As explained above, at least one of these congruences is almost certain to give us a nontrivial factor of n.

It is important to choose r wisely. If r is too small, then it will take us a very long time to find even a single x_i for which $f(x_i)$ can be written as a product of primes in the factor base. On the other hand, if r is too large, then our $r \times s$ system of linear congruences modulo 2 will take us too long to solve (and we might also have difficulty storing such a large system). One can use a heuristic argument to show that r should be of order of magnitude roughly equal to $\exp(\frac{1}{2}\sqrt{\log n \log \log n})$ (where log denotes natural log). In the case of the largest numbers factored by this method (for which $\log n \approx 200$), r was chosen to be several tens of thousands.

In 1981, Pomerance developed an important technique – called *sieving* – for speeding up the process of finding the large sequence $\{x_i\}_{i=1,\ldots,s}$ that is needed in the above algorithm. We briefly describe one version of this technique. The idea is to find many of the x_i at once. Consider a long interval of T consecutive values of x. In each of T positions corresponding to the x's, we store an approximation to $\log f(x)$. Next, for each odd prime p in our factor base, we find the two solutions mod p of the congruence $f(x) \equiv 0 \pmod{p}$. (This amounts to finding the two square roots of n modulo p.) We now go to the interval of T consecutive values of x, and run through the two arithmetic progressions with difference p corresponding to the two solutions of $f(x) \equiv 0 \pmod{p}$. For each such value of x, we subtract an approximation to $\log p$ from the stored value. (The stored value is equal to the approximation to $\log f(x)$ minus the sum of the approximations to $\log p_j$ for each of the earlier p_j for which x was in one of the two corresponding arithmetic progressions, i.e., for which $p_j | f(x)$.) After completing this procedure for all of the p_j in the factor base, we look once more through the interval of T consecutive values of x. The locations where the stored number is close to zero will be the ones for which $f(x)$ is a product of the p_j.

It turns out that the expected running time for the quadratic sieve factoring method is of the form $\exp\big((1+o(1))\sqrt{\log n \log \log n}\big)$. For a discussion of running time estimates for factoring algorithms – some of which have been rigorously proved, and others of which are based only on heuristic arguments – see [Pomerance 1982] and [Lenstra and Pomerance 1992]. For a more precise discussion of sieving in the above setting and also in the so-called *multiple polynomial* version, see [Pomerance 1990].

Elliptic Curve Factorization. A watershed in the application to cryptography of supposedly "pure" algebraic number theory was the discovery by H. W. Lenstra, Jr. of an ingenious method of using elliptic curves to factor integers. We now outline Lenstra's algorithm.

Let n be the large odd integer that we want to factor, and let p be an as yet unknown prime factor of n. Let $E_{a,b}$ denote the elliptic curve $y^2 = x^3 + ax + b$ considered modulo n. Strictly speaking, $E_{a,b}$ modulo n is not a group; however, it is a group when considered modulo any prime $p|n$. As in the case of elliptic curve primality testing, we use the usual formulas for the group operation on an elliptic curve, working always modulo n, i.e., we "pretend" that n is prime. In the steps described below, if we ever find ourselves unable to apply those formulas (because of a denominator that is not prime to n and hence not invertible modulo n), we are almost certain to obtain a nontrivial factor of n, and we are done. (Namely, we take the g.c.d. of n with the troublesome denominator.)

Let r be a positive integer chosen in a certain optimal way, let $p_1, \dots, p_r$ be the first r primes, and let $m = m(r)$ denote $\prod_{j=1}^{r} p_j^{\alpha_j}$, where $p_j^{\alpha_j}$ is the highest power of p_j that is less than $(n^{1/4} + 1)^2$.

We start by choosing a random pair consisting of the coefficient a of our elliptic curve and a point $P = (x_0, y_0)$ in the xy-plane modulo n. We want P to lie on the curve $E_{a,b}$: $y^2 = x^3 + ax + b$, so the coefficient b is determined by P and a: $b = y_0^2 - x_0^3 - ax_0$. (We should check that $4a^3 + 27b^2$ is prime to n, so that our elliptic curve is not degenerate.) We next compute (or rather, try to compute) the point mP on $E_{a,b}$. If we are able to compute the multiple mP (i.e., if all of the denominators we encounter in the addition and doubling formulas are prime to n), then we have *not* succeeded in factoring n, and we must make another random choice of P and a.

Eventually we are lucky, in that our curve $E_{a,b}$ has the following property: for some prime $p|n$, the order of $E_{a,b}$ modulo p (i.e., the order of the group of $\mathbf{F}_p$-points) is p_r-smooth (this means that $\#E_{a,b}(\mathbf{F}_p)$ has no prime factors greater than p_r). If we suppose that $p < \sqrt{n}$, this means that $\#E_{a,b}(\mathbf{F}_p)$ divides m, because by Hasse's theorem this order is $< (n^{1/4} + 1)^2$. In that case the point mP considered modulo p is the identity, i.e., the point at infinity. This situation will show up in our computations in the form of a denominator divisible by p. At the same time, it is unlikely that the same denominator is divisible by n. Thus, in computing mP we will encounter a denominator whose g.c.d. with n gives a nontrivial factor (most likely the prime factor p).

It remains for us to discuss what the probability is that the randomly chosen $E_{a,b}$ has the desired property. According to Hasse's theorem, the order of $E_{a,b}(\mathbf{F}_p)$ is in the interval $[p + 1 - 2\sqrt{p}, p + 1 + 2\sqrt{p}]$. Moreover, it follows from work of Deuring, Waterhouse, and Schoof that as a and b vary, most of these orders fall in the interval $(p - \sqrt{p}, p + \sqrt{p})$, where they are roughly uniformly distributed (see [Lenstra 1987] for details). Thus, the chance that $\#E_{a,b}(\mathbf{F}_p)$ is p_r-smooth is about the same as the chance that a random integer in the interval $(p - \sqrt{p}, p + \sqrt{p})$ is p_r-smooth.

Note that r must be chosen in an optimal way. If it is too small, then we might never find an elliptic curve whose order modulo p is p_r-smooth. On the other hand, if r is too big, then it will take us too long to compute mP, since $m = m(r)$ grows rather rapidly with r.

At this point Lenstra needs an unproved conjecture to the effect that the standard results on the distribution of p_r-smooth integers in the interval $(0, x)$ as $x \longrightarrow \infty$ [Canfield, Erdös, and Pomerance 1983] also apply to the smaller interval $(x - \sqrt{x}, x + \sqrt{x})$ as $x \longrightarrow \infty$. Assuming this plausible conjecture, one can find an optimal strategy (i.e., an optimal choice of r) so as to find the desired elliptic curve and factor n as rapidly as possible, and one can rigorously analyze the running time. As in the case of the quadratic sieve factoring algorithm, the expected running time turns out to have the form $\exp\big((1 + o(1))\sqrt{\log n \log \log n}\big)$.

In the case of the hardest numbers to factor, in practice the elliptic curve factorization algorithm is not quite as efficient as its main competitor, the quadratic sieve method. For example, in factoring a number n which is the product of two primes of size roughly $\sqrt{n}$, as in RSA, the best versions of the quadratic sieve are somewhat faster [Pomerance 1990]. However, the elliptic curve method is the only subexponential general purpose factoring algorithm whose expected running time actually depends on the size of the smallest prime factor p of n. If $p \ll \sqrt{n}$, then the fastest way to factor n is probably by elliptic curves.

Until recently, all of the contenders for the best general purpose factoring algorithm had running time of the form $\exp\big(O(\sqrt{\log n \log \log n})\big)$. Some people even thought that this function of n might be a natural lower bound on the running time. However, during the last few years a new method – called the *number field sieve* – has been developed that has a heuristic running time that is much better (asymptotically), namely:

$$\exp\big(O((\log n)^{1/3}(\log \log n)^{2/3})\big) \, .$$

The number field sieve is similar to the earlier algorithms that attempt to combine congruences so as to obtain a relation of the form $x^2 \equiv y^2 \pmod{n}$. However, one uses a "factor base" in the ring of integers of a suitably chosen algebraic number field. The details are quite complicated, so we shall not describe the algorithm here. See the articles in [Lenstra and Lenstra 1993].

§8. Discrete Logarithm

Recall that the discrete logarithm problem in a finite field $\mathbf{F}_q$ is the problem, given $x \in \mathbf{F}_q^*$, of solving the relation $g^y = x$ for the integer y. There are two types of algorithms for solving this problem: (1) algorithms that are inefficient unless $q - 1$ happens to have no large prime factors; and (2) index calculus algorithms.

An example of the first type is D. Shanks' "giant step – baby step" method (see [Knuth 1973, pp. 9, 575-576]) combined with the Silver-Pohlig-Hellman algorithm [Pohlig and Hellman 1978]. We give a sketch of this algorithm. First, a routine argument based on the Chinese Remainder Theorem reduces the problem to that of solving an equation of the form $g^y = x$ when g and x are p-th roots of 1 in $\mathbf{F}_q$, where p is a prime divisor of $q - 1$. Now compute the two sets $S_1 = \{xg^i\}_{0 \le i < \sqrt{p}}$ and $S_2 = \{g^{j([\sqrt{p}]+1)}\}_{0 \le j < \sqrt{p}}$, and compare them. When you have a match, i.e., $xg^i = g^{j([\sqrt{p}]+1)}$, you are done – just take $y = j([\sqrt{p}] + 1) - i$. This method is practical if $q - 1$ is a product of small primes, but not if $q - 1$ is divisible by a prime p of 40 digits or more [Odlyzko 1985].

For a description of another important algorithm of the first type, due to Pollard, see [Pollard 1978] or [McCurley 1990a].

In the general case, when $q - 1$ has a large prime divisor, the most efficient algorithms are all of index calculus type. There are two variants of these algorithms: those that apply when $q = p^f$ is a power of a small prime, and those that apply when $q = p$ is prime itself or a very small power of p. It was only very recently that an index calculus algorithm was found that is subexponential in $q = p^f$ as both p and f increase [Adleman and DeMarrais 1993].

Index calculus algorithms are probabilistic rather than deterministic, and in spirit they resemble the "combination of congruences" procedure used to produce a relationship of the form $x^2 \equiv y^2 \pmod{n}$ in such factoring algorithms as the quadratic sieve (see above).

We shall outline the index calculus method in the case when q is prime. For simplicity we shall suppose that g is a generator of $\mathbf{F}_q^*$. The first part of the algorithm is called *precomputation*. This means that it needs to be performed only once for a given field $\mathbf{F}_q$ and base g. The second part of the algorithm must be performed for each individual logarithm, i.e., it depends upon the particular x whose discrete logarithm we want.

In the precomputation, let $p_1, \ldots, p_r$ be the first r primes, where r is chosen in a certain optimal way. Then for random positive integers $l < q - 1$ consider the least positive residue of g^l. If this integer is divisible by a prime $> p_r$, we move on to another l. But any time that the least positive residue of g^l factors into a product of the $p_1, \ldots, p_r$ we obtain a relation of the form

$$\prod_{j=1}^{r} p_j^{a_{ij}} = g^{l_i} \qquad \text{in} \quad \mathbf{F}_q^*.$$

For $a \in \mathbf{F}_q^*$, let $\text{ind}_g a$ denote the discrete logarithm (also called the "index") of a to the base g. Then the last formula implies the following relation of exponents:

$$\sum_{j=1}^{r} a_{ij}\,\text{ind}_g\,p_j \equiv l_i \pmod{q-1}.$$

This is a system of linear congruences in the unknowns $\text{ind}_g\,p_j$, $j = 1,\ldots,r$. Once we have enough independent congruences of this form, we can solve for the discrete logs of the p_j. It should be noted that, although solving a system of congruences is quite routine in principle, there are subtle questions that arise when one tries to decrease the time spent on this part of the algorithm (see [Pomerance 1987]). (For a recent discussion of index calculus implementation in the case of the fields $\mathbf{F}_{2^k}$, see [Gordon and McCurley 1993].)

Once the precomputation is complete, it is relatively easy to find $y = \text{ind}_g x$. Namely, we choose random l' and compute $x g^{l'}$ until we find a least positive residue of $x g^{l'}$ that has no prime divisor $> p_r$. Writing $x g^{l'}$ as a product of the p_j and using the values of $\text{ind}_g\,p_j$ from the precomputation, we find the discrete log of $x g^{l'}$. Then we simply subtract l' to get the desired $y = \text{ind}_g x$.

It remains to explain how r should be chosen. If r is too small, it will take too long to find g^{l_i} whose least positive residue has no prime factor $> p_r$. On the other hand, if r is too large, the linear system in the r unknowns $\text{ind}_g\,p_j$ will be too hard to solve. An optimal value of r can be chosen using the asymptotic behavior of the function $\psi(X, s) = \#\{s\text{-smooth positive integers} \le X\}$. (Recall that a positive integer is said to be s-smooth if it has no prime factors $> s$.) The result of this analysis is that, very roughly speaking, $r \sim \exp(\sqrt{\log q \log\log q})$.

Using such an optimal r, one can show that the running time of the algorithm is of the form $\exp\big(O(\sqrt{\log q \log\log q})\big)$. Most of this running time is devoted to (1) finding p_r-smooth values of g^{l_i}, and (2) solving the linear system of congruences for $\text{ind}_g\,p_j$.

Recently, D. Gordon has modified the number field sieve factoring algorithm so as to obtain a discrete logarithm algorithm for $\mathbf{F}_q$. The result is also an index calculus algorithm, but it is much more complicated, because one works in the ring of integers of an algebraic number field. Although much work, both theoretical and practical, remains to be done, it seems that the number field sieve has a much better asymptotic running time, namely $\exp\big(O(\log^{1/3} q \log\log^{2/3} q)\big)$.

We conclude our discussion of algorithms for factoring and discrete log by observing that there is thought to be a close relationship between the complexity of factoring an integer n and the complexity of solving the discrete log problem in $\mathbf{F}_q$, where q and n have the same order of magnitude. See [Odlyzko 1985] and [van Oorschot 1992] for discussions of this question. Until recently, all of the best algorithms for both problems had time estimates of the form $L(n, 1/2)$, where for $0 \le \gamma \le 1$ one defines $L(n, \gamma) = \exp\big(O(\log^{\gamma} n \log\log^{1-\gamma} n)\big)$. Then with the invention of the number field sieve for factoring [Lenstra and Lenstra 1993], the time estimate for factoring was brought down to $L(n, 1/3)$. Soon after, the number field sieve was also applied to the discrete log problem [Gordon 1993, and Gordon, to appear], bringing the time estimate for discrete log down to $L(q, 1/3)$ as well.

However, at present there seems to be one basic difference between the complexity of the two problems. If n has a relatively small prime factor p, then we have a factoring algorithm – namely, Lenstra's elliptic curve method – that works much faster than in the general case. For example, if $p < \log^k n$, where the parameter k is fixed, then n can be factored in polynomial time (with degree of the polynomial independent of k), i.e., factoring is "fixed-parameter tractable" [Fellows and Koblitz 1993]. On the other hand, in the case of the problem of solving $g^y = x$ in $\mathbf{F}_q^*$, we know of no subexponential algorithm that can take advantage of either the circumstance that $y < \log^k q$ or the circumstance that y has Hamming weight $\leq k$. (At present, one cannot do better than roughly $\log^{k/2} q$.) The latter question is of some practical interest, because some proposed implementations of Diffie-Hellman use exponents of fairly small Hamming weight [Agnew, Mullin, Onyszchuk, and Vanstone 1991].

§9. Elliptic Curve Cryptosystems

Influenced by Lenstra's use of elliptic curves, Koblitz [1987a] and Miller [1986] independently proposed using the abelian group of an elliptic curve as the basis for cryptosystems such as the Diffie-Hellman key exchange. We now describe how this works, and discuss the corresponding discrete logarithm problem.

We first define the discrete log problem in an arbitrary abelian group G. If the group law in G is written additively, and if $P \in G$ denotes a fixed base element, then the *discrete log problem in G* consists, given $X \in G$, of determining an integer y such that $X = yP$ in G (or else determining that X is not in the subgroup generated by P).

The Diffie-Hellman key exchange was first proposed for the group $\mathbf{F}_q^*$, but it can just as well be stated in any finite abelian group G. As before, Alice and Bob want to agree upon a large integer to serve as a key for some private-key cryptosystem. Alice and Bob first select a group G, a map from elements of G to the integers, and a base element $P \in G$. All of this is discussed publicly.

Next, Alice and Bob choose random secret integers k_A and k_B, respectively. Alice sends Bob the group element $k_A P$ (but not her secret integer k_A), and Bob sends Alice the element $k_B P$ (but not the integer k_B). The common key is then $k_A k_B P \in G$ (or rather, the integer associated to $k_A k_B P$ under the agreed upon correspondence). Alice determines this key by multiplying her secret integer k_A times the element $k_B P$ received from Bob; Bob determines the key by multiplying k_B times the element $k_A P$ that he received from Alice. An eavesdropper is in the difficult position of having to determine the element $k_A k_B P$ knowing only $k_A P$ and $k_B P$ but not k_A or k_B.

Now take $G = E$ to be an elliptic curve defined over a finite field $\mathbf{F}_q$. Let $P \in E$ be a fixed base point. The multiples $k_A P$, $k_B P$, and $k_A k_B P$ are computed using the addition law on the curve and the "repeated doubling" method. The secret key (known by Alice and Bob but no one else) is the point $k_A k_B P$ – or, more precisely, a certain positive integer corresponding to this point, for example, its x-coordinate in the case when $\mathbf{F}_q$ is a prime field.

The main potential advantage of an elliptic curve cryptosystem is that the discrete logarithm problem seems to be harder to solve in the group E than in the group $\mathbf{F}_q^*$. In the latter group, the generally applicable discrete log algorithms that do not require exponential time – all of which can be categorized as "index calculus" algorithms – rely on the fact that one can find a fairly large set of "small" group elements in terms of which any other element can be expressed with small exponents. For example, if $q = p$ is a large prime, then this set (called a "factor basis") might consist of all prime numbers less than a certain bound. On the other hand, if $q = p^f$, where p is small and f is large, and if $\mathbf{F}_q$ is regarded as $\mathbf{F}_p[X]/F(X)$ for a fixed irreducible polynomial $F(X)$ of degree f, then the factor basis might consist of all irreducible polynomials of degree less than a certain bound, regarded modulo $F(X)$. In the case of the group E, however, it does not seem to be possible to find such a factor basis. As argued in [Miller 1986], the most natural notion of "small" (the reduction modulo p of a point in characteristic zero having small height) does not work.

Until 1990, the only discrete log algorithms known for an elliptic curve were the ones that work in any group, irrespective of any particular structure. These are exponential time algorithms, provided that the order of the group is divisible by a large prime factor. But in [Menezes, Okamoto, and Vanstone 1993] the authors found a new approach to the discrete log problem on an elliptic curve E defined over $\mathbf{F}_q$. Namely, they used the Weil pairing (see III.8 of [Silverman 1986]) to imbed the group E into the multiplicative group of some extension field $\mathbf{F}_{q^k}$. This imbedding reduces the discrete log problem on E to the discrete log problem in $\mathbf{F}_{q^k}^*$.

However, in order for the Weil pairing reduction to help, it is essential for the extension degree k to be small. Essentially the only elliptic curves for which k is small are the so-called "supersingular" elliptic curves, the most familiar examples of which are curves of the form $y^2 = x^3 + ax$ when the characteristic p of $\mathbf{F}_q$ is $\equiv -1 \pmod 4$, and curves of the form $y^2 = x^3 + b$ when $p \equiv -1 \pmod 3$. The vast majority of elliptic curves, however, are nonsupersingular. For them, the reduction almost never leads to a subexponential algorithm (see [Koblitz 1991b]).

Thus, the first advantage of elliptic curve cryptosystems is that no subexponential algorithm is known that breaks the system, provided that we avoid supersingular curves and also curves whose order has no large prime factor. A second selling point is that, unlike in the case of the groups $\mathbf{F}_q^*$, there is a tremendous variety of elliptic curves from which to choose. To take advantage of this, we must have an efficient way to determine the order of an elliptic curve E defined over $\mathbf{F}_q$.

The first polynomial time algorithm to compute $\#E$ was discovered by René Schoof. Schoof's algorithm is even deterministic. It is based on the idea of finding the value of $\#E$ modulo l for all primes l less than a certain bound. This is done by examining the action of the "Frobenius" (the p-th power map) on points of order l.

In the original paper [Schoof 1985] the bound for running time was $O(\log^9 q)$, which is polynomial but quite unpleasant. At first it looked like the algorithm was not practical. However, since then many people have worked on speeding up Schoof's algorithm (V. Miller, N. Elkies, J. Buchmann, V. Müller, A. Menezes, L. Charlap, R. Coley, and D. Robbins). In addition, A. O. L. Atkins has developed a somewhat different method that, while not guaranteed to work in polynomial time, functions extremely well in practice. As a result of all of these efforts it has become feasible to compute the order of an arbitrary elliptic curve over $\mathbf{F}_q$ if q is, say, a 50-digit prime power. If this order is found to have a large prime factor, then the discrete log problem on it is intractable at our present level of knowledge and technology, and so the curve can be used for a secure cryptosystem.

Several methods of choosing elliptic curves have been suggested. One possibility is to stick with a simple equation, such as $y^2 = x^3 + ax$ or $y^2 = x^3 + b$. If the curve is supersingular, then it is trivial to compute $\#E$. For example, if $q \equiv -1 \pmod 4$ for the first equation or $q \equiv -1 \pmod 3$ for the second equation, then $\#E = q + 1$. But in this case the discrete logarithm problem reduces to discrete logs in $\mathbf{F}_{q^2}^*$ [Menezes, Okamoto, and Vanstone 1993], so cryptosystems using E have no strong advantage over cryptosystems using the multiplicative group of a finite field. On the other hand, even if the curve is nonsupersingular, because of the special form of the equation there are algorithms to find $\#E$ that are simpler and faster than Schoof's algorithm and its variants. For example, computing the order of the curve $y^2 = x^3 + ax$ over $\mathbf{F}_p$, where $p \equiv 1 \pmod 4$, amounts to expressing p as a sum of two squares (see, for example, II.2 of [Koblitz 1993]), and this can easily be done [Brillhart 1972].

A second possibility is to start with a fixed elliptic curve E_1 over a small finite field $\mathbf{F}_q$, and then consider the sequence E_j obtained by regarding the same curve over the extension field $\mathbf{F}_{q^j}$. The sequence $N_j = \#E_j$ is determined in a simple way from $N_1 = \#E_1$, using the congruence zeta-function (see, e.g., II.1 of [Koblitz 1993]).

Another possibility is to generate elliptic curves in some random manner over a fixed finite field $\mathbf{F}_q$. One could take $q = p$ to be an arbitrary large prime whose magnitude is the same as that of the group E you need. (By Hasse's theorem, $|\#E - p - 1|$ is bounded by $2\sqrt{p}$.) Alternately, one could take $q = 2^f$ to be a fixed large power of 2. The choice of an elliptic curve in characteristic 2 has some advantages in practice, because arithmetic can be performed somewhat more efficiently in such fields. It has been suggested that a random elliptic curve over $\mathbf{F}_{2^f}$ would be particularly suitable for a cryptosystem that could be implemented on a "smart card," where the space limitation is severe [Menezes and Vanstone 1990].

In each case, one must search for an elliptic curve whose order has a large prime factor; otherwise, discrete logs can be found quickly by the giant step – baby step algorithm.

In conclusion, it should be mentioned that, besides elliptic curves and the multiplicative groups of finite fields, other groups have also been proposed for Diffie-Hellman type cryptosystems: the jacobians of hyperelliptic curves [Koblitz 1989], the class groups of imaginary quadratic number fields [Buchmann and Williams 1988], and others.

§ 10. Some Open Problems

1. *Determine whether breaking RSA is equivalent to factoring.* That is, prove or disprove: If one has any algorithm that, given (n, e) and c (where n is a product of two primes p and q, and g.c.d.$(e, p-1)$=g.c.d.$(e, q-1) = 1$), produces m such that $m^e \equiv c \pmod{n}$, then one can use this algorithm with at most a polynomial (in $\log n$) number of additional steps to obtain a probabilistic algorithm that factors n.

2. *Find a deterministic polynomial time algorithm to test primality.* The algorithm must be provably polynomial time (without depending upon any heuristic argument or unproved conjecture).

3. *Determine whether breaking the Diffie-Hellman key exchange over a finite field is equivalent to finding discrete logarithms.* That is, prove or disprove: Any algorithm that can find g^{ab} when given $g, g^a, g^b \in \mathbf{F}_q^*$, can be used to obtain an algorithm that finds y when given $g, g^y \in \mathbf{F}_q^*$.

4. *Find an algorithm for the discrete logarithm problem in a finite field $\mathbf{F}_q$, $q = p^f$, that, as p and f both vary, has the same type of time estimate as in the number field sieve, i.e.,* $\exp\left(C \log^{1/3} q \, \log\log^{2/3} q\right)$. *Alternately, from a practical point of view it would be nice to find an algorithm that, for variable p and f, can in practice compute discrete logs in $\mathbf{F}_q$ for $q = p^f$ of roughly the same size as $q = p$ in the case of the fastest algorithm for prime fields (currently about 100 digits).*

5. *Is the bounded Hamming weight discrete log problem in a finite field a fixed-parameter tractable problem?* In other words, look for an algorithm which, given $g, g^y \in \mathbf{F}_q$, where the sum of the bits in y is k, finds y in time bounded by a function of the form $f(k) \log^c q$, where $f(k)$ is arbitrary and the constant c is independent of k.

6. *Find a subexponential time algorithm for finding discrete logarithms on a non-supersingular elliptic curve.* For example, let E be the group of $\mathbf{F}_p$-points on the curve $y^2 = x^3 + 1$, where $p \equiv 1 \pmod 3$ is an n-bit prime. Find an algorithm taking time bounded by $e^{o(n)}$ which can find the integer y if it is given $P, yP \in E$.

7. *Try to break the Chor–Rivest knapsack.*

References

L. M. Adleman: A subexponential algorithm for the discrete logarithm problem with applications to cryptography. Proc. 20th IEEE Symp. Foundations of Computer Science 1979, pp. 55–60

L. M. Adleman and J. DeMarrais: A subexponential algorithm for discrete logarithms over all finite fields. Math. Computation **61** (1993) 1–15

L. M. Adleman, C. Pomerance, R. S. Rumely: On distinguishing prime numbers from composite numbers. Annals Math. **117** (1983), 173–206

G. B. Agnew, R. C. Mullin, I. M. Onyszchuk, S. A. Vanstone: An implementation for a fast public-key cryptosystem. J. Cryptology **3** (1991), 63–79

G. B. Agnew, R. C. Mullin, S. A. Vanstone: An implementation of elliptic curve cryptosystems over $F_{2^{155}}$. IEEE Journal on Selected Areas in Communications **11** (1993) 804–813

R. Alford, A. Granville, C. Pomerance: There are infinitely many Carmichael numbers. Preprint 1992

A. O. L. Atkin: The number of points on an elliptic curve modulo a prime. Preprint 1991

A. O. L. Atkin, F. Morain: Elliptic curves and primality proving. Math. Computation **61** (1993) 29–68

E. Bach: Analytic Methods in the Analysis and Design of Number-Theoretic Algorithms. MIT Press, 1985

E. Bach: Number-theoretic algorithms. Annual Reviews in Computer Science **4** (1990) 112–172

I. F. Blake, R. Fuji-Hara, R. C. Mullin, S. A. Vanstone: Computing logarithms in fields of characteristic two. SIAM J. Algebraic and Discrete Methods **5** (1984) 276–285

W. Bosma, M.-P. van der Hulst: Primality proving with cyclotomy. Univ. of Amsterdam, 1990

E. F. Brickell: Breaking iterated knapsacks. Advances in Cryptology – Crypto '84. Spinger, 1985, pp. 342–358

E. F. Brickell, A. M. Odlyzko: Cryptanalysis: A survey of recent results. Proc. IEEE **76** (1988) 578–593

J. Brillhart: Note on representing a prime as a sum of two squares. Math. Computation **26** (1972) 1011–1013

J. Brillhart, D. H. Lehmer, J. L. Selfridge, B. Tuckerman, S. S. Wagstaff, Jr.: Factorizations of $b^n \pm 1$, $b = 2, 3, 5, 6, 7, 10, 11, 12$ up to high powers. Amer. Math. Soc., 1988

J. Brillhart, M. Morrison: A method of factoring and the factorization of F_7. Math. Computation **29** (1975) 183–205

J. Buchmann, V. Müller: Computing the number of points of elliptic curves over finite fields. Presented at International Symposium on Symbolic and Algebraic Computation, Bonn, July 1991

J. Buchmann, H. C. Williams: A key exchange system based on imaginary quadratic fields. J. Cryptology **1** (1988) 107–118

J. Buchmann, H. C. Williams: A key exchange system based on real quadratic fields. Advances in Cryptology – Crypto '89. Springer, 1990, pp. 335–343

E. R. Canfield, Paul Erdös, C. Pomerance: On a problem of Oppenheim concerning "Factorisatio Numerorum". J. Number Theory **17** (1983) 1–28

T. R. Caron, R. S. Silverman: Parallel implementation of the quadratic sieve. J. Supercomputing **1** (1988) 273–290

L. Charlap, R. Coley, D. Robbins: Enumeration of rational points on elliptic curves over finite fields. Preprint 1991

D. Chaum: Achieving electronic privacy. Scientific American **267** (1992) 96–101

B. Chor, R. Rivest: A knapsack-type public key cryptosystem based on arithmetic in finite fields. Advances in Cryptology – Crypto '84. Springer, 1985, pp. 54–65; revised version in IEEE Trans. Information Theory **34** (1988) 901–909

H. Cohen, A. K. Lenstra: Implementation of a new primality test. Math. Computation **48** (1987) 103–121

H. Cohen, H. W. Lenstra, Jr.: Primality testing and Jacobi sums. Math. Computation **42** (1984) 297–330

D. Coppersmith: Fast evaluation of logarithms in fields of characteristic two. IEEE Trans. Information Theory **30** (1984) 587–594

D. Coppersmith, A. M. Odlyzko, R. Schroeppel: Discrete logarithms in $GF(p)$. Algorithmica **1** (1986) 1–15

W. Diffie, M. Hellman: New directions in cryptography. IEEE Trans. Information Theory **22** (1976) 644–654

J. D. Dixon: Factorization and primality testing. Amer. Math. Monthly **91** (1984) 333–352

T. ElGamal: A public key cryptosystem and a signature scheme based on discrete logarithms. IEEE Trans. Information Theory **31** (1985) 469–472

T. ElGamal: A subexponential-time algorithm for computing discrete logarithms over $GF(p^2)$. IEEE Trans. Information Theory **31** (1985) 473–481

M. R. Fellows, N. Koblitz: Fixed-parameter complexity and cryptography. Proc. Tenth Intern. Symp. Appl. Algebra, Algebraic Algorithms and Error Correcting Codes, San Juan, Puerto Rico, 1993

M. R. Garcy, D. S. Johnson: Computers and Intractability: A Guide to the Theory of NP-Completeness. W. H. Freeman & Co., 1979

J. Gerver: Factoring large numbers with a quadratic sieve. Math. Computation **41** (1983) 287–294

S. Goldwasser, J. Kilian: Almost all primes can be quickly certified. Proc. 18th ACM Symp. Theory of Computing 1986, pp. 316–329

S. Goldwasser, J. Kilian, S. Micali: Probabilistic encryption and how to play mental poker keeping secret all partial information. Proc. 14th ACM Symp. Theory of Computing, 1982, pp. 365–377

S. Goldwasser, J. Kilian, S. Micali: Probabilistic encryption. J. Comput. System Sci. **28** (1984) 270–299

S. Golomb: Shift Register Sequences (2nd ed.). Aegean Park Press, 1982

D. Gordon: Discrete logarithms in $GF(p)$ using the number field sieve. SIAM J. Discrete Math. **6** (1993) 124–138

D. Gordon: Discrete logarithms in $GF(p^n)$ using the number field sieve. Preprint

D. Gordon, K. McCurley: Massively parallel computation of discrete logarithms. Advances in Cryptology – Crypto '92. Springer, 1993

A. Granville: Primality testing and Carmichael numbers. Notices of the Amer. Math. Soc. **39** (1992) 696–700

M. E. Hellman, J. M. Reyneri: Fast computation of discrete logarithms in $GF(q)$. Advances in Cryptology – Crypto '82. Plenum Press, 1983, pp. 3–13

L. S. Hill: Concerning certain linear transformation apparatus of cryptography. Amer. Math. Monthly **38** (1931) 135–154

D. Husemöller: Elliptic Curves. Springer, 1987

B. Kaliski: A pseudorandom bit generator based on elliptic logarithms. Advances in Cryptology – Crypto '86. Springer, 1987, pp. 84–103

D. E. Knuth: The Art of Computer Programming (vol. 2, 2nd ed.). Addison-Wesley, 1981

D. E. Knuth: The Art of Computer Programming (vol. 3). Addison-Wesley, 1973

N. Koblitz: Elliptic curve cryptosystems. Math. Computation **48** (1987) 203–209

N. Koblitz: A Course in Number Theory and Cryptography. Springer, 1987

N. Koblitz: Hyperelliptic cryptosystems. J. Cryptology **1** (1989) 139–150

N. Koblitz: Constructing elliptic curve cryptosystems in characteristic 2. Advances in Cryptology – Crypto '90. Springer, 1991, pp. 156–167

N. Koblitz: Elliptic curve implementation of zero-knowledge blobs. J. Cryptology **4** (1991) 207–213

N. Koblitz: CM-curves with good cryptographic properties. Advances in Cryptology – Crypto '91. Springer, 1992, pp. 279–287

N. Koblitz: Introduction to Elliptic Curves and Modular Forms (2nd ed.). Springer, 1993

M. Kraitchik: Théorie des nombres (vol. 1). Gauthier-Villars, 1922

J. C. Lagarias: Pseudorandom number generators in cryptography and number theory. Cryptology and Computational Number Theory, Proc. Symp. Appl. Math. **42**, 1990, pp. 115–143

B. LaMacchia, A. M. Odlyzko: Computation of discrete logarithms in prime fields. Designs, Codes and Cryptography **1** (1991) 47–62

A. K. Lenstra: Primality testing. Cryptology and Computational Number Theory, Proc. Symp. Appl. Math. **42**, 1990, pp. 13–25

A. K. Lenstra, H. W. Lenstra, Jr.: Algorithms in number theory. Handbook of Theoretical Computer Science. Elsevier, 1990, pp. 673–715

A. K. Lenstra, H. W. Lenstra, Jr. (eds.): The Development of the Number Field Sieve. Springer, 1993

H. W. Lenstra, Jr.: Factoring integers with elliptic curves. Annals Math. **126** (1987) 649–673

H. W. Lenstra, Jr., C. Pomerance: A rigorous time bound for factoring integers. J. Amer. Math. Soc. **5** (1992) 483–516

R. Lidl, H. Niederreiter: Introduction to Finite Fields and Their Applications. Cambridge Univ. Press, 1986

K. McCurley: Cryptographic key distribution and computation in class groups. In: Number Theory and Applications, R. Mollin (ed.). Kluwer, Boston 1989

K. McCurley: The discrete logarithm problem. Cryptology and Computational Number Theory, Proc. Symp. Appl. Math. **42**, 1990, pp. 49–74

K. McCurley: Odds and ends from cryptology and computational number theory. Cryptology and Computational Number Theory, Proc. Symp. Appl. Math. **42**, 1990, pp. 49–74

A. Menezes: Elliptic Curve Public Key Cryptosystems. Kluwer Acad. Pub., 1993

A. Menezes, S. A. Vanstone: The implementation of elliptic curve cryptosystems. Advances in Cryptology – Auscrypt '90. Springer, 1990, pp. 2–13

A. Menezes, T. Okamoto, S. A. Vanstone: Reducing elliptic curve logarithms to logarithms in a finite field. IEEE Trans. Information Theory **39** (1993) 1639–1646

R. C. Merkle, M. E. Hellman: Hiding information and signatures in trapdoor knapsacks. IEEE Trans. Information Theory **24** (1978) 525–530

G. L. Miller: Riemann's hypothesis and tests for primality. J. Comput. System Sci. **13** (1976) 300–317

V. Miller: Uses of elliptic curves in cryptography. Advances in Cryptology – Crypto '85. Springer, 1986, pp. 417–426

P. L. Montgomery: Speeding the Pollard and elliptic curve methods of factorization. Math. Computation **48** (1987) 243–264

F. Morain: Implementation of the Goldwasser-Kilian-Atkin primality testing algorithm. INRIA report 911, 1988

R. Mullin, I. Onyszchuk, S. A. Vanstone, R. Wilson: Optimal normal bases in $GF(p^n)$. Discrete Appl. Math. **22** (1988/89) 149–161

A. M. Odlyzko: Discrete logarithms and their cryptographic significance. Advances in Cryptology – Eurocrypt '84. Springer, 1985, pp. 224–314

A. M. Odlyzko: The rise and fall of knapsack cryptosystems. Cryptology and Computational Number Theory, Proc. Symp. Appl. Math. **42**, 1990, 75–88

H. C. Pocklington: The determination of the prime and composite nature of large numbers by Fermat's theorem. Proc. Cambridge Philos. Soc. **18**, 1914–1916, pp. 29–30

S. Pohlig, M. Hellman: An improved algorithm for computing logarithms over $GF(p)$ and its cryptographic significance. IEEE Trans. Information Theory **24** (1978) 106–110

J. Pollard: Monte Carlo methods for index computation mod p. Math. Computation **32** (1978) 918–924

C. Pomerance: Recent developments in primality testing. Math. Intelligencer **3** (1981) 97–105

C. Pomerance: Analysis and comparison of some integer factoring algorithms. In: Computational Methods in Number Theory, H. W. Lenstra, Jr., and R. Tijdeman (eds.). Math. Centrum, Amsterdam **154** (1982) 89–139

C. Pomerance: The quadratic sieve factoring algorithm. Advances in Cryptology – Eurocrypt '84. Springer, 1985, pp. 169–182

C. Pomerance: Fast, rigorous factorization and discrete logarithms algorithms. Discrete Algorithms and Complexity (1987) 119–143

C. Pomerance: Factoring. Cryptology and Computational Number Theory, Proc. Symp. Appl. Math. **42**, 1990, pp. 27–47

C. Pomerance, J. L. Selfridge, S. S. Wagstaff, Jr.: The pseudoprimes to $25 \cdot 10^9$. Math. Computation **35** (1980) 1003–1026

G. Purdy: A high-security log-in procedure. Communications of the ACM **17** (1974) 442–445

M. O. Rabin: Probabilistic algorithms for testing primality. J. Number Theory **12** (1980) 128–138

H. Riesel: Prime Numbers and Computer Methods for Factorization. Birkhäuser, 1985

R. Rivest: Cryptography. In: Handbook of Theoretical Computer Science, vol. A. Elsevier, 1990, pp. 717–755

R. Rivest, A. Shamir, L. Adleman: A method for obtaining digital signatures and public–key cryptosystems. Communications of the ACM **21** (1978) 120–126

R. S. Rumely: Recent advances in primality testing. Notices Amer. Math. Soc. **30** (1983) 475–477

C. P. Schnorr: Efficient identification and signatures for smart cards. Advances in Cryptology – Crypto '89. Springer, 1990, pp. 239–251

R. Schoof: Elliptic curves over finite fields and the computation of square roots mod p. Math. Computation **44** (1985) 483–494

A. Shamir: A polynomial time algorithm for breaking the basic Merkle-Hellman cryptosystem. IEEE Trans. Information Theory **30** (1984) 699–704

A. Shamir: An efficient identification scheme based on permuted kernels. Advances in Cryptology – Crypto '89. Springer, 1990, pp. 606–609

C. E. Shannon: Communication theory of secrecy systems. Bell Syst. Tech. J. **28** (1949) 656–715

J. Silverman: The Arithmetic of Elliptic Curves. Springer, 1986

R. D. Silverman, S. S. Wagstaff, Jr.: A practical analysis of the elliptic curve factoring algorithm. Math. Computation **61** (1993) 445–462

A. Sinkov: Elementary Cryptanalysis: A Mathematical Approach. Math. Assoc. Amer., 1966

P. van Oorschot: A comparison of practical public-key cryptosystems based on integer factorization and discrete logarithms. In: Contemporary Cryptology: The Science of Information Integrity, G. Simmons (ed.). IEEE Press, 1992, pp. 289–322

H. S. Wilf: Backtrack: An $O(1)$ expected time graph coloring algorithm. Inform. Process Lett. **18** (1984) 119–122

M. V. Wilkes: Time-Sharing Computer Systems. Elsevier, 1968

H. C. Williams: A modification of the RSA public-key encryption procedure. IEEE Trans. Information Theory **26** (1980) 726–729

"Ist 4 denn noch normal?"

Matthias Kreck

Yang- Mills, Do- nald- son,

welch ei- ne Wahl, welch ei- ne Qual!

Ist 4 denn noch nor - mal?

This song in the style of a moritat was composed by Wulf-Dieter Geyer for the Geyer-Harder Arbeitsgemeinschaft in Oberwolfach in the fall of 1984. The Arbeitsgemeinschaft is one of the most attractive activities at Mathematisches Forschungsinstitut Oberwolfach (see the article by J. Neukirch in this book). It meets twice a year and brings together mathematicians from different areas to learn a recent particularly interesting development in mathematics. At the end of each such meeting the participants decide in a long democratic procedure (with a complicated system of voting, only very experienced people like Geyer and Harder can conduct) what should be the topic of the next meeting. Besides being of general interest the topic has to be accessible to a wider range of mathematicians. The participants of the spring meeting 1984 decided that it would be time to learn the fascinating new results on 4-manifolds. But would it be accessible to normal mathematicians? My personal opinion at this time was that the delicate shrinking arguments of Freedman's proof were almost inaccessible (my standard phrase was: "Only accessible to those selected by god"). But, his statements were easily understandable and being a tool for studying the homeomorphism classification of 4-manifolds one could along familiar lines apply it to solve very interesting problems like the topological Poincaré conjecture.

On the other hand, Donaldson's results were more unexpected. The reason for this is, that we had a great experience with manifolds of dimension ≥ 5. They led to the general picture that smooth manifolds in these dimensions can completely be analyzed by homotopy theoretic methods plus secondary invariants which typically sit in algebraic K- or L-groups. Examples are the Whitehead torsion distinguishing simple homotopy equivalence from ordinary homotopy equivalence or Wall's surgery obstructions or invariants distinguishing pseudoisotopies from isotopies sitting in Waldhausen's K-theory. This picture should not leave the impression that we really understand manifolds of dimension ≥ 5: just think about open questions in the classification of exotic spheres and the understanding of their group of diffeomorphisms. But we had settled down in this comfortable world of higher-dimensional manifolds and tried to improve our knowledge by playing with the established tools. Similarly, although less understood, there was a certain understanding, in part conjectural, of 3-dimensional manifolds, based on the work of mathematicians like Waldhausen in the sixties and Thurston in the seventies.

Donaldson showed us that in dimension 4 there are completely new obstructions and invariants which had no counterpart in other dimensions. Learning these new methods and results the question in our song sounds quite understandable:

Ist Dimension 4 denn noch normal?

In the following article I want to discuss the questions:

1. What should we have expected?
2. What were the surprises?
3. Do we have a good conception of 4-manifolds by now?
4. What would we still like to know?

1. What Should We Have Expected?

What we expect depends on time and social background. Thus we ask more precisely: What was considered among topologists as well-behaved according to our expectations in dimension 4 around 1980? To answer this question we have to describe the fundamental known results and methods. The first result I would like to mention is Whitehead's homotopy classification of simply-connected 4-manifolds [Wh] (he actually classified simply-connected polyhedra but we will concentrate on manifolds). Let M be a smooth closed oriented 4-manifold. Consider differentiable maps $f_i : F_i \to M$, where F_i are oriented closed 2-manifolds, i.e. surfaces. One can approximate f_i by immersions, i.e. embeddings with self-intersections. Even more one can assume that $f_0(F_0)$ and $f_1(F_1)$ intersect in M transversally away from the self-intersections. If p is an intersection point we can assign to p the sign $+1$, if the orientation of $f_0(F_0)$ and $f_1(F_1)$ form the orientation of M and -1 otherwise. The following picture (see Fig. 1) indicates this for immersions of circles into the torus:

The **intersection number** is the sum of these signs over all intersection points. It turns out that this number depends only on the homology classes α and β represented by $f_0(F_1)$ and $f_1(F_1)$ and is denoted as $\alpha \cdot \beta$. Since every 2-dimensional

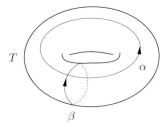

Fig. 1

homology class can be represented by an immersed surface we obtain the **inter-section form**

$$H_2(M; \mathbb{Z}) \otimes H_2(M; \mathbb{Z}) \to \mathbb{Z}$$
$$\alpha \otimes \beta \longmapsto \quad x \cdot y$$

By construction this is a symmetric bilinear form and by Poincaré duality it is unimodular after dividing out the torsion in $H_2(M; \mathbb{Z})$. Using the cup product on cohomolgy one can define the intersection form also for oriented topological manifolds. In a reformulation by Milnor the classification result of Whitehead is the following:

Theorem 1 [Wh, Mi]. *Two closed oriented topological simply-connected 4-manifolds are orientation preserving homotopy equivalent if and only if they have isomorphic intersection forms.*

This theorem leads naturally to the question which symmetric unimodular bilinear forms over the integers occur as intersection form of 4-manifolds, in particular of simply-connected ones. For a long time, namely until Freedman's and Donaldson's breakthroughs in 1982 the following result was the only partial answer by Rohlin.

Theorem 2 [Ro, 1952]. *The signature of a **smooth** closed oriented 4-dimensional spin manifold is divisible by 16.*

Here the signature is the signature of the intersection form, i.e. the difference of the number of positive and negative eigenvalues. A smooth manifold admits a spin structure if the structure group of the tangent bundle can be reduced to the spinor group. For simply-connected 4-manifolds this is equivalent to having an **even** intersection form, i.e. $x \cdot x$ is an even number for all x. It is known that the signature of an even symmetric unimodular bilinear form is always divisible by 8 and that 8 or better -8 is realized by the famous E_8 matrix (whose associated graph is the Dynkin diagram of the corresponding Lie group):

$$E_8 = \begin{pmatrix} -2 & 1 & 0 & 0 & 0 & 0 & 0 & 0 \\ 1 & -2 & 1 & 0 & 0 & 0 & 0 & 0 \\ 0 & 1 & -2 & 1 & 0 & 0 & 0 & 0 \\ 0 & 0 & 1 & -2 & 1 & 0 & 0 & 0 \\ 0 & 0 & 0 & 1 & -2 & 1 & 0 & 1 \\ 0 & 0 & 0 & 0 & 1 & -2 & 1 & 0 \\ 0 & 0 & 0 & 0 & 0 & 1 & -2 & 1 \\ 0 & 0 & 0 & 0 & 1 & 0 & 1 & -2 \end{pmatrix}$$

As a consequence of Rohlin's theorem this form cannot be the intersection form of a closed simply-connected smooth 4-manifold.

In addition to these two and a few other results like Wall's stable (i.e. up to connected sum with $S^2 \times S^2$'s) classification of simply-connected 4-manifolds [Wa1] the expectations, how 4-manifolds should be classified, came from analogies to other dimensions. In dimension ≥ 5 surgery theory [Wa2] gives on one hand a method for classifying a few but very interesting manifolds explicitly like Kervaire and Milnor's classification of exotic spheres bounding a parallelizable manifold. On the other hand it gives a frame for a certain understanding of arbitrary manifolds. Roughly speaking surgery theory reduces the decision if two closed manifolds M_0 and M_1 of dimension ≥ 5 are diffeomorphic to three steps:

1. Are M_0 and M_1 simply homotopy equivalent?
2. Are they normally bordant, i.e. does there exist a compact manifold W with $\partial W = M_0 + M_1$ and a continous map $f : W \longrightarrow M_0$ such that $f|M_i$ is a simple homotopy equivalence and the normal bundle of M_0 pulls back to the normal bundle of W?
3. An invariant $\Theta(W) \in L^S_{n+1}(\pi_1(W), w_1(W))$ decides if W is normally bordant rel. boundary to an s-cobordism. Here $L^S_{n+1}(\pi_1(W), w_1(W))$ is a sort of an algebraic K-group where the objects are equipped with quadratic form. If $n+1$ is even $\Theta(W)$ comes from the intersection form on W and if $n + 1$ is odd it comes from the **linking form**.

If these three steps can be carried out successfully, the s-cobordism theorem of Smale (in the simply-connected case), Barden-Mazur-Stallings (compare [Ke]) implies that M_0 and M_1 are diffeomorphic.

Let's look at dimensions 2 and 3. In dimension 2 one has to distinguish between orientable and non-orientable manifolds. Then the Euler characteristic classifies a surface. In dimension 3 we don't have a general answer. But Thurston has developed a big program whose basic idea is to decompose 3-manifolds into pieces with a geometric structure. For most of the pieces the geometric structure implies that the piece is rigid and thus determined by a simple invariant like the fundamental group. If this program will be carried out successfully the classification of closed 3-manifolds is the following: Except for a few special cases like lens spaces where additional but simple invariants have to be controlled the fundamental group should be the only invariant.

Given these results and programs in dimension $\neq 4$ and the fact that the known results in dimension 4 could easily be integrated into the picture of manifolds of arbitrary dimension we can explain what was considered as expected around 1980:

Closed 4-manifolds were supposed to be classified by their homotopy types, the normal bordism classes plus a generalized linking form invariant.

Around that time I worked on a modification of surgery theory aiming for a more effective classification process. The following result, formulated in a slightly sketchy way, gives a program for reducing the classification of a large class of 4-manifolds to the 4-dimensional s-corbordism problem:

Theorem 3. *Two closed smooth oriented odd 4-manifolds M_0 and M_1 are s-cobordant if and only if they have same signature, Euler characteristic, image of fundamental class in $H_4(\pi; \mathbb{Z})$ and if a secondary linking obstruction vanishes.*

Here a manifold is called odd if the second Stiefel-Whitney class of the universal covering is non-trivial (in the simply-connected case this means that the intersection form is odd). The fundamental group of M is denoted by π. The image of the fundamental class in $H_4(\pi; \mathbb{Z})$ means the image under a classifying map of the universal covering. The secondary linking obstruction is a refinement of Wall's surgery obstruction.

This result helps to focus the question: Ist Dimension 4 denn noch normal? It would be as expected if the 5-dimensional s-cobordism theorem were true.

2. What Were the Surprises?

A research institute like Oberwolfach intends to reflect important developments in mathematics (without taking up short-lived trends and fashions). It is impressive to see how well the institute functioned in connection with the development in 4-manifolds in the eighties. Four conferences were devoted to this topic in the years 1984, 1986, 1988, 1992. In addition the area was represented in most of the annual topology meetings.

If one wants to find out what was new one only has to look at the conference reports of these meetings. Let's do this. The meeting in 1984 was already mentioned in the beginning of this article. Let me quote from the report (in an English translation): "After presenting some "classical" results in dimension 4, Casson handles, the basic tool for Freedman's work, were introduced. Then a sketch of the technically very complicated proof of Freedman's main result was given, saying that a Casson handle is homeomorphic to a standard handle. This result allows to extend methods from higher-dimensional manifolds to the topological classification of 4-manifolds."

Let me explain this a bit more. Two of the very powerful higher-dimensional theories, the proof of the h- or s-cobordism theorem and surgery make use of the so-called Whitney trick. Consider for example a continuous map $f : S^n \longrightarrow M^{2n}$, where M is simply-connected. We want to replace f by an embedding which is homotopic to f. By general position one can first replace it by an immersion whose self-intersection consists of transversal double points. Assume that the normal bundle of the immersed sphere is trivial. Then the double points occur in pairs

242   Matthias Kreck

with opposite local orientations which can be joined by two embedded arcs on the sphere. Since M is simply-connected one can extend the union of these arcs to a map from D^2 to M. The situation is indicated in the following picture:

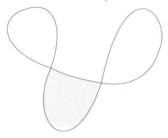

Fig. 2

Now, if $n \geq 3$, one can choose $D^2 \to M$ as an embedding and use it to remove the two double points by "pulling" the finger back:

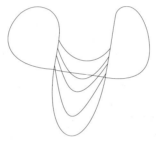

Fig. 3

If n is 2 this is a priori not possible. In this situation, A. Casson [Ca] in the beginning of the seventies had the idea to consider instead of an embedding an immersed disk D^2 bounding the two arcs, and to proceed with its double points as before hoping that the resulting infinite process would converge in such a way that the finger trick would finally work. He proved that the process could be carried out with enough control to obtain a so-called Casson handle (more precisely a Casson handle is a thickened version of this infinite construction).

Here is a schematic picture of a Casson handle (see Fig. 4).

Here 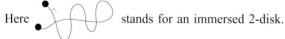 stands for an immersed 2-disk.

A good language for describing Casson handles or more generally 4-manifolds is the so-called **Kirby calculus**. This is not only a way for describing 4-manifolds in terms of 3-dimensional pictures (links), one can in principle use it to decide if two manifolds are diffeomorphic. Unfortunately this is very complicated and far from effective.

In this situation, Freedman [Fr1] in an enormous tour de force managed to show that a Casson handle is **homeomorphic** to a standard handle. Note, that Casson handles are smooth objects but Freedman's proof works only in the topological

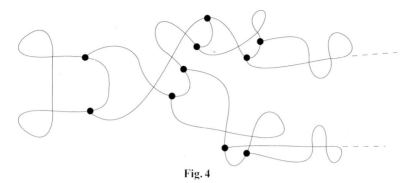

Fig. 4

category (as we will see, Donaldson's results imply that in general a Casson handle is not diffeomorphic to the standard handle). Originally Freedman proved this, if the 4-dimensional manifold M is simply-connected, a little bit later he extended it to manifolds with poly- (finite or cyclic) fundamental group (i.e. there is a finite ascending sequence of subgroups, each normal in the next, whose successive quotients are finite or cyclic) [Fr2].

Freedman's proof is so complicated that it took even an expert like L. Siebenmann a whole morning at this 1984 meeting to give a sketch. An article like this is not the place for an attempt to do this, we refer instead to [FQ]. But the article is a good place for formulating some consequences of Freedman's theorem which were proved in the last 10 years. I will give some examples in the next section. Here I only want to mention that as a consequence Freedman proves that every symmetric unimodular form occurs as the intersection form of a simply-connected **topological** 4-manifold and that the intersection form together with a $\mathbb{Z}/2$-valued invariant, the Kirby-Siebenmann obstruction, determines the homeomorphism type. The most spectacular application is:

Theorem 4 [Fr1]. *The 4-dimensional topological Poincaré conjecture is true.*

Let me return to the 1984 meeting and come to the second main topic. Again I quote from the report. "The second part of the conference dealt with Donaldson's theorem saying that if the intersection form of a **smooth** 4-manifold is positive definite, it is the standard Euclidean form. This theorem should be seen in the context of Freedman's result mentioned above. We began with a talk about the necessary differential geometric background (for Donaldson's proof). Then we reported on the moduli space of the Yang-Mills equations, in particular on its local structure. The moduli space has two types of singularities, on one hand cones over $\mathbb{C}P^2$ and on the other hand singularities which disappear after generic perturbations. This and the orientability of the moduli space were reported. The technically most difficult part of Donaldson's proof is the compactification of the moduli space by the given 4-manifold."

Again, I would like to explain this a bit more. Let M be a closed connected oriented smooth 4-manifold. Consider a compact Lie group G and a smooth principal G bundle $\pi : P \rightarrow M$. Typically $G = SU(2)$ and we restrict ourselves for a moment to this case. Such bundles are classified by their second Chern class $k = c_2(P) \in H^4(M; \mathbb{Z}) = \mathbb{Z}$ and we denote P as P_k. A **connection** A on P_k is a splitting of the tangent bundle into the bundle along the fibres and a horizontal complement which can best be be given by a 1-form on P_k with values in the Lie algebra of G. We denote the space of all connections by $\mathcal{A}_k$. The **gauge group** $\mathcal{G} = \mathrm{Aut}(P_k)$ acts on $\mathcal{A}_k$ by pulling connections back. The curvature F_A of $A \in \mathcal{A}_k$ is the covariant derivative applied to A or equivalently the horizontal component of the ordinary differential of A. This is an invariant form and thus a 2-form on M with values in the Lie algebra: $F_A \in \Omega^2(M; su(2))$.

Now, choose a Riemannian metric g on M and the corresponding Hodge-operator

$$* : \Omega^k(M; g) \longrightarrow \Omega^{4-k}(M; g).$$

The **Yang-Mills** equation is

$$*F_A = -F_A,$$

i.e. the solutions are **anti-self-dual connections** or instantons. The **moduli space of asd-connections** is

$$\mathcal{M}(M, k, g) = \{A \in \mathcal{A}_k/\mathcal{G} \mid *F_A = -F_A\}.$$

This space is the central tool for Donaldson's work. One can show that under certain assumptions on the intersection form the space is not empty and, for generic metrics, is a manifold with rather mild singularities (often without singularities) of dimension $8k - 3(1 + b_2^+(M))$, if M is simply-connected, and where $b_2^+(M)$ is the number of positive eigenvalues of the intersection form of M. The space admits a natural compactification in terms of the original manifold. For example if the intersection form of M is negative definite (if we would consider the positive definite case we would have to take the moduli space of self-dual connections) and $k = 1$, the moduli space has dimension 5. All singularities are in the generic situation cones over $\mathbb{C}P^2$ and the number of them is equal to the second Betti number of M. The compactification is roughly obtained as follows. Consider sequences of anti self-dual connection which are flat away from the neighborhood of a point in M. Of course near this point the curvature gets larger and larger the smaller we make the neighborhood. Thus we get the picture of a flat manifold with a bubble near some point.

It turns out that such sequences exist for each point in M and this leads to a compactification of $\mathcal{M}(M, k, g)$ by M itself. After removing open neighborhoods of the singularlities one obtains an oriented compact 5-manifold W with boundary M and $b_2(M)$ copies of $\mathbb{C}P^2$ or $\overline{\mathbb{C}P}^2$. Then a simple consideration in linear algebra combined with the fact that the signature of M is equal to the signature of the rest of the boundary of W implies that the form on M is the negative of the standard Euclidean form. A little bit later Donaldson extended this result to general closed oriented 4-manifolds.

Theorem 5 [Do1, Do2]. *If the intersection form of a smooth closed 4-manifold is negative definite then it is the negative definite Euclidean form.*

Freedman's very powerful result was in so far a surprise as almost nobody hoped that one could actually prove that the higher dimensional methods would hold in dimension 4. But in short the message was that **topological** 4-manifolds behave as expected (at least for a large class of fundamental groups). In contrast Donaldson showed us that **smooth** 4-manifolds are different from manifolds in all other dimensions. This is most dramatically demonstrated by a consequence of the existence of Casson handles, Freeedman's topological h-cobordism theorem (for open manifolds) and Donaldson's result that there is no smooth simply-connected 4-manifold with intersection from $E_8 \oplus E_8$. Using these tools one can prove that $\mathbb{R}^4$ has at least two smooth structures. Now, we even know that there is a continuum of smooth structures.

Theorem 6 [Ta]. *The number of smooth structures on $\mathbb{R}^4$ is not countable.*

Comparing with other dimensions we know that $\mathbb{R}^n$ for $n \neq 4$ has a unique smooth structure and that a manifold homotopy equivalent to a finite CW-complex has at most finitely many smooth structures if the dimension is bigger than 4 [KS] and a unique smooth structure in dimension 3 [Moi].

3. Do We Have a Good Conception of 4-Manifolds by Now?

Given the development described in the last section the topologists had to reorient themselves in the world of 4-manifolds. The situation is like that of a person who stayed for a rather long time in a dark room in which suddenly a strong light is switched on. One looks for things which are familiar like the formulation of Freedman's result and tries to check how dominant the really new phenomena are. Here the main question is: How large is the difference between smooth and topological 4-manifolds? In particular, is the existence of completely unexpected "exotic" structures restricted to non-compact manifolds like $\mathbb{R}^4$? If no, the smooth h- (or s- in the non simply-connected case) theorem is wrong in dimension 5.

The latter question was answered by Donaldson [Do3] around 1985. He constructed a new diffeomorphism invariant based on the moduli space of asd-connections and proved that it distinguishes $\mathbb{C}P^2 \# 9\overline{\mathbb{C}P^2}$ from another simply-connected algebraic surface, one of the so-called Dolgatchev surfaces, obtained from it by a certain logarithmic transformation. Since invariants like Euler characteristic and signature are unchanged under logarithmic transformations and determine the intersect ion form in this situation, the manifolds are homeomorphic by Freedman.

This was the beginning of a fast development. If we look at the report of the second meeting in Oberwolfach devoted to 4-manifolds, the 1986 Geyer-Harder activity, we can find the following results.

Theorem 7 (Donaldson [Do3], Friedman-Morgan [FM1], Okonek-Van de Ven [OV]). *There are infinitely many smooth structures on $\mathbb{C}P^2 \# 9\overline{\mathbb{C}P^2}$.*

This shows a dramatic failure of the 5-dimensional h-cobordism theorem. If this theorem were true there would be a unique smooth structure on $\mathbb{C}P^2\#9\overline{\mathbb{C}P}^2$. In all other dimensions the number of smooth structures is finite.

The other result is the indecomposibility result for 1-connected algebraic surfaces which was recently extended by Morgan and Mrowka to arbitrary surfaces.

Theorem 8 (Donaldson [Do4], Morgan-Mrowka [MM]). *Let X be a non-singular complex algebraic surface. Then X is not diffeomorphic to a connected sum $Y_1\#Y_2$ unless perhaps if Y_1 or Y_2 have negative definite intersection form.*

The main ingredient of these results is the definition and computation of Donaldson's new invariants. Given the construction and compactification of the moduli space the definition of these invariants is in principle (although in detail complicated) not so difficult and very much in the spirit of classical algebraic topology. The first step is the constructions of a linear map

$$\mu : H_2(M) \longrightarrow H^2(\mathcal{A}_k^*/\mathcal{G}).$$

Here M is a simply-connected closed smooth oriented Riemannian manifold and $\mathcal{A}_k/\mathcal{G}$ is the space of gauge-equivalence classes of connections on an $SU(2)$-bundle P_k with $c_2 = k$ as described in Section 2. We denote by $\mathcal{A}_k^*$ the irreducible connections, where a connection is called reducible if it comes from a connection of a decomposition of the $SU(2)$-bundle into a sum of line bundles. For this one considers the subgroups $\mathcal{G}_0 \subset \mathcal{G}$ of all gauge transformations fixing a prescribed fibre of the $SU(2)$-bundle. The action of $\mathcal{G}_0$ on $\mathcal{A}_k^*$ is free and the same is true for the pull-back of P_k to $\mathcal{A}_k^* \times M$ under the second projection. We denote the quotient bundle by $P_k' \to \mathcal{A}_k^*/\mathcal{G}_0 \times M$.

Now we define $\mu' : H_2(M) \to H^2(\mathcal{A}_k^*/\mathcal{G}_0)$ by assigning to $\alpha \in H_2(M)$ the slant product of $c_2(P_k')$ with α:

$$\mu'(\alpha) = c_2(P_k')/\alpha \in H^2(\mathcal{A}_k^*/\mathcal{G}_0).$$

$\mathcal{A}_k^*/\mathcal{G}_0$ projects onto $\mathcal{A}_k^*/\mathcal{G}$ and one can factor μ' through $\mathcal{A}_k^*/G$ to obtain

$$\mu : H_2(M) \longrightarrow H^2(\mathcal{A}_k^*/\mathcal{G}).$$

Now, suppose that the moduli space $\mathcal{M}(M, k, g)$ is a smooth manifold of dimension $2d$, is contained in $\mathcal{A}_k^*/\mathcal{G}$, and that $\mu(\alpha)$ extends to its compactification. Then one can evaluate $\mu(\alpha)$ on the fundamental class to define

$$q_k(\alpha) = \langle \mu(\alpha)^d, [\overline{\mathcal{M}}(M, k, g)] \rangle.$$

Under appropriate conditions on M and k these assumptions can be fulfilled and it can be shown (a rather delicate point) that the invariant is independent of the choice of a generic Riemannian metric on M. It also depends on the choice of an orientation of $\mathcal{M}(M, k, g)$, another subtle point one has to take care of. For details of the definition of these invariants I refer to [DK]. These are Donaldson's celebrated polynomial invariants (under slightly weaker assumptions Donaldson constructs similar invariants which depend on additional information in homology and are actually numerical invariants).

Now, I want to continue with the description of the development. After the appearance of Donaldson's invariants the computation of their leading coefficient and sometimes a little bit more was carried out for most of the known simply-connected algebraic surfaces. The reason, why algebraic surfaces are the most studied examples, is that although Donaldson's invariants are differential topological invariants, most information which was so far obtained from them depends on a theorem of Donaldson which identifies the moduli space of asd-connections on an algebraic surface with the moduli space of stable vector bundles. These objects are not easy to handle but until now in most cases at least a partial successful computation could be achieved. A particular attractive case was the Barlow surface, a surface homeomorphic to $\mathbb{C}P^2 \# 8\overline{\mathbb{C}P^2}$. Kotschick [Ko] studied modified Donaldson invariants for bundles with structure group $SO(3)$ and proved among other things that these two surfaces are not diffeomorphic. The most delicate computations were carried out for elliptic surfaces. They lead for example to the result that two simply-connected elliptic surfaces of geometric genus 1 are diffeomorphic if and only if they are deformation equivale nt, i.e. there is a holomorphic family containing both [Ba2], [MO], and the same holds for non simply-connected surfaces of geometric genus 0 [Ba1]. Similar computations are the heart of the following qualitative result which I first heard in a talk by J. Morgan during the 1988 Oberwolfach meeting on 4-manifolds.

Theorem 9 [FM2]. *The natural map*

$$\begin{Bmatrix} \text{algebraic surfaces modulo} \\ \text{deformation equivalence} \end{Bmatrix} \longrightarrow \begin{Bmatrix} \text{smooth oriented 4-manifolds modulo} \\ \text{orientation preserving diffeomorphism} \end{Bmatrix}$$

is finite-to-one.

It is open if this map is one-to-one.
If one proceeds further to the topological category one can look at the map

$$\begin{Bmatrix} \text{smooth structures on} \\ \text{algebraic surfaces} \end{Bmatrix} \longrightarrow \begin{Bmatrix} \text{topological oriented 4-manifolds modulo} \\ \text{orientation preserving homeomorphisms} \end{Bmatrix}$$

and ask when does a fibre has order > 1, or equivalently which algebraic surfaces admit at least two smooth structures. The following result says that there are at least two smooth structures if the fundamental group is finite, the surface is minimal and if the Euler characteristic is large enough.

Theorem 10 [HK1]. *A minimal algebraic surface with finite fundamental group π and Euler characteristic greater than a constant $c(\pi)$ admits at least two smooth structures.*

These two smooth structures are smoothly s-cobordant and thus give counter-examples to the smooth s-cobordism theorem in dimension 4. Theorem 10 is based on the indecomposibility theorem above and a homeomorphism classification in rather simple terms following the lines outlined at the end of Section 1, which also leads sometimes to better results like for instance that every algebraic surface with non-trivial cyclic fundamental group admits at least two smooth structures [HK3]. The proof uses Freedman's topological s-cobordism theorem, surgery theory and the stable algebra of quadratic forms following A. Bak and H. Bass. Here is a particularly simple case.

Theorem 11 [HK2]. *Two odd oriented smooth closed 4-manifolds M_0 and M_1 with finite fundamental group π such that M_0 decomposes as $M_0' \sharp S^2 \times S^2$ are homeomorphic if and only if they have the same signature, Euler characteristic and image of fundamental class in $H_4(\pi; \mathbb{Z})$.*

To get a general picture we compare the three categories

$$\{\text{complex manifolds}\} \longrightarrow \{\text{smooth manifold}\} \longrightarrow \{\text{topological manifolds}\}.$$

In even real dimensions not equal to 4 the difference between the smooth and the topological category is comparatively small, the fibres of the forgetful map are always finite. The fibres of the first forgetful map is in general huge as one can already see in real dimension 2.

In dimension 4 the sketched results and also the construction of the Donaldson invariants in terms of differential equations indicates that the first two categories are more closely connected whereas the topological category is very different. This cannot be made precise but the following question indicates its flavour:

Is every simply-connected smooth 4-manifold different from the 4-sphere diffeomorphic to the connected sum of complex surfaces?

Instead of complex surfaces one could have said algebraic surfaces since every simply-connected complex surface is diffeomorphic to an algebraic surface.

For several years this question was one of the hopes to establish a new "normal" picture of 4-manifolds. But it turned out that this is not true. During a conference at McMaster University (Canada) in 1990 (by the way a conference which was planned as a follow up of the Oberwolfach 4-manifolds meeting in 1988) B. Gompf presented a construction of a smooth 4-manifold which is homeomorphic to the K3-surface and asked if it admits a complex structure. His construction was a modification of the logarithmic transformation and on the same conference T. Mrowka had reported on his more analytical way of computing Donaldson invariants for manifolds whose end is a 3-torus. It turned out that Mrowka had developed the main tools for proving that the manifold does not admit a complex structure.

Furthermore this manifold is smoothly indecomposable since Donaldson had generalized his result on definite intersection forms to show that a manifold with intersection form of the $K3$-surface is indecomposable [Do2].

Thus we obtain

Theorem 12 [GM]. *Not every simply-connected smooth 4-manifold different from S^4 is the connected sum of complex surfaces.*

I would like to finish this by no means complete list of results with a discussion of the progress on the generalized Thom conjecture. If M is a smooth oriented 4-manifold and α a 2-dimensional homology class, then there exists a smooth embedded oriented surface $F \subset M$ representing α, i.e. $\alpha = i_*[F]$, where $[F]$ is the fundamental class of F and i is the inclusion. For this consider the Poincaré dual of α, represent it by a map into $\mathbb{C}P^N$ and let F be a transversal preimage of $\mathbb{C}P^{N-1}$. The question is what is the minimal genus of such a surface for a given homology class α.

The generalized Thom conjecture is the following

Conjecture. *Let X be an algebraic surface and C a smooth algebraic curve in X representing a class $\alpha \in H_2(X; \mathbb{Z})$. Then for each smooth embedded surface F representing α the genus of F is greater or equal to the genus of C.*

If this conjecture is true it is one of the results which support the picture that complex or algebraic surfaces are closely related to smooth manifolds. In the last 3 years considerable progress was made on this important problem and the Mathematisches F orschungsinstitut Oberwolfach played a certain role in this.

Let me first make the conjecture more explicit by using the adjunction formula

$$2g(C) - 2 = C \cdot C + K \cdot C,$$

where $g(C)$ is the genus of C and K is the canonical class. Thus the conjecture says:

$$2g(F) - 2 \geq \alpha \cdot \alpha + K \cdot \alpha.$$

A related conjecture is the following:

$$2g(F) - 2 \geq \alpha \cdot \alpha,$$

if $\alpha \cdot \alpha \geq 0$. Here one does not assume that X is an algebraic surface but a simply-connected smooth 4-manifold with non-trivial Donaldson invariant. If $K = 0$ (like for instance the important case of the $K3$-surface), this second conjecture implies the generalized Thom conjecture.

A few years ago, Mathematisches Forschungsinstitut Oberwolfach created a prize for young European mathematicians. Part of this prize is that Oberwolfach finances a stay up to 3 months at Oberwolfach where the prize winner has the right to invite a colleague with whom he would like to cooperate.

The first prize winner was Peter Kronheimer from Oxford who received it in February 1991. During this ceremony, Kronheimer gave a talk in which he presented a program which should lead to a proof of the second conjecture. This program was actually a complete proof modulo a delicate analytic piece concerning the asd-connections on $X - F$ which have finite actions of the curvature F_A i.e.

$$\int_{X-F} |F_A|^2 < \infty$$

I remember when I talked to Kronheimer about his fascinating program on this occasion that he was a bit sceptical. Maybe the prize was a help to collect new energy. Kronheimer invited Tom Mrowka from Caltech, Pasadena, to Oberwolfach as his partner and in joint work they made a breakthrough leading to the following result.

Theorem 13 (Kronheimer-Mrowka [KM]). *Let M be a smooth, closed, oriented simply-connected 4-manifold with $b_2^+(M)$ odd and ≥ 3 and which has nontrivial polynominal invariants.*

Then the genus of any orientable surface F, other than a sphere of self-intersection -1 or an inessential sphere of self-intersection 0, satisfies the inequality

$$2g(F) - 2 \geq F \cdot F$$

As a corollary this implies the Thom conjecture for $K3$-surfaces. It also solves a long standing problem on algebraic knots or links L in S^3 saying that the genus of a smooth oriented surface in the 4-ball with oriented boundary L is not less than the genus of the Milnor fibre.

During the 1992 Oberwolfach meeting on 4-manifolds Kronheimer sketched another important step towards the Thom conjecture. The state of the art is now:

Theorem 14 (Kronheimer [Kro]). *The generalized Thom conjecture holds at least under the following assumptions:*

1. *X is a simply-connected algebraic surface,*
2. *C is an algebraic curve with $C \cdot C \geq 0$,*
3. *There is a class $w \in H_2(X; \mathbb{C})$ dual to a holomorphic 2-form on X, such that $q_k(w + \bar{w}) > 0$ for sufficiently large k.*

Do we have a good conception of 4-manifolds by now? I think it is too early to say yes. What we can see is that for simply-connected topological 4-manifolds the topological classification is complete. For smooth manifolds the difference to the topological category is enormous and the relation to the world of algebraic surfaces is close although not as close as we expected for some time. All this is based on the same phenomenon: the moduli space of asd-connections and derived invariants.

For non simply-connected manifolds we know much less. It is very likely that most of Donaldson's work goes through with appropriate modifications for arbitrary fundamental groups but we would not be surprised if the generalization of Freedman's work to non poly- (finite or cyclic) groups would be wrong.

4. What Would We Still Like to Know?

When Mathematisches Forschungsinstitut Oberwolfach was founded in 1944 none of the results mentioned in this article was known. With some right one can say that there was no knowledge of 4-manifolds. In these 50 years the picture has changed dramatically. It is difficult to estimate how large the percentage of what we know is compared to what we would like to know.

I think what we miss most is an effective method for proving that two 4-manifolds are actually diffeomorphic. To my mind the optimal result would be to find necessary and sufficient conditions under which a 5-dimensional smooth h- or s-cobordism is a product.

Problem 1: Prove a 5-dimensional smooth h- or s-cobordism theorem.

One can ask if this would be an effective way for getting a classification. The problem might be that there are too many smooth h- or s- cobordisms between two given 4-manifolds. At least for simply-connected manifolds this is not the case since I can rather easily show the following result using my modified surgery theory:

Theorem 15. *Let M_0 and M_1 be closed oriented smooth simply-connected 4-manifolds. Then the set of smooth h-cobordisms between M_0 and M_1 is isomorphic to the set of isomorphisms between the intersection forms of M_0 and M_1.*

The inclusions of M_0 and M_1 into W induce an isomorphism between the intersection forms and this is the correspondence. Thus if we fix an isomorphism between the intersection forms on M_0 and M_1, then there is a unique h-cobordism joining these manifolds and if we would have invariants deciding if an h-cobordism is a product, then we would have obstructions for the existence of a diffeomorphism between M_0 and M_1 inducing the given isomorphism.

To prove an h-cobordism theorem one has to look at the space of Morse functions on it and to search for obstructions for finding one without critical points. In principle, the Donaldson polynomials are such obstructions and it might be helpful to relate them more directly to the singularities of a Morse function. This might also be a useful step towards the solution of the next problem.

Problem 2: Find a combinatorial formula for the Donaldson polynomial invariants (if they are defined).

The background information for this is the fact that a 4-dimensional PL-manifold admits a smooth structure and that two PL-isomorphic smooth 4-manifolds are diffeomorphic [Ce]. Thus the Donaldson invariants are combinatorial invariants but we don't know a combinatorial formula. This problem is probably very hard as one can learn from the similar question of finding combinatorial formulas for the rational Pontrjagin classes.

The third problem is

Problem 3: Is there any topological 4-manifold with unique smooth structure? In particular, is this true for S^4?

The last question is the smooth 4-dimensional Poincaré conjecture. In dimension $\neq 4$ manifolds with unique smooth structure exist ([Moi] in dimension 3, [Kr] in dimension > 4).

Finally I would like to mention the outstanding problem for topological 4-manifolds.

Poblem 4: For which fundamental groups does the 5-dimensional topological s-cobordism theorem hold?

Here the problem is to find additional invariants or to prove that the s-cobordism theorem is always true.

At the moment all these problems look very hard. But, who knows: At the time we celebrate the next anniversary of Oberwolfach, we might know some or all the answers. In any case one can be sure that any substantial progress will be reported or discussed at one of the future Oberwolfach conferences.

References

[Ba1] S. Bauer: The diffeomorphism classification of non-simply connected Dolgatchev surfaces. J. reine angew. Math. (to appear)

[Ba2] S. Bauer: Diffeomorphism type of elliptic surfaces with $p_g = 1$. Preprint, 1992

[Ca] A. Casson: Lectures on new infinite constructions in 4-dimensional manifolds. Notes by L. Guillou, Orsay

[Ce] J. Cerf: Sur les difféomorphismes de la sphère de dimension trois ($\Gamma_4 = 0$). Lecture Notes in Mathematics, vol. 53. Springer 1968

[Do1] S. K. Donaldson: An application of gauge theory to the topology of 4-manifolds. J. Diff. Geom. **18**, 269–316 (1983)

[Do2] S. K. Donaldson: The orientation of Yang-Mills moduli spaces and 4-manifold topology. J. Diff. Geom. **26**, 397–428 (1987)

[Do3] S. K. Donaldson: Irrationality and the h-cobordism conjecture. J. Diff. Geom. **26**, 141–168 (1987)

[Do4] S. K. Donaldson: Polynomial invariants for smooth 4-manifolds. Topology **29**, 257–316 (1990)

[DK] S. K. Donaldson and P. Kronheimer: The geometry of four-manifolds. Oxford Science Publ., Oxford 1990

[Fr1] M. H. Freedman: The topology of four-dimensional manifolds. J. Diff. Geom. **17**, 357–453 (1982)

[Fr2] M. H. Freedman: The disk theorem for four-dimensional manifolds. In: Proc. Int. Conf. Warsaw 1984, pp. 647–663

[FQ] M. H. Freedman and F. Quinn: Topology of 4-manifolds. Princeton Univ. Press, Princeton 1990

[FM1] R. Friedman and J. Morgan: On the diffeomorphism type of certain algebraic surfaces I, II. J. Diff. Geom. **27**, 297–398 (1988)

[FM2] R. Friedman and J. Morgan: Complex versus differentiable classification of algebraic surfaces. Topology and Its Applications **32**, 135–139 (1989)

[GM] R. Gompf and T. Mrowka: A family of non-complex homotopy $K3$ surfaces. Preprint

[HK1] I. Hambleton and M. Kreck: Smooth structures on algebraic surfaces with finite fundamental group. Inv. Math. **102**, 109–114 (1990)

[HK2] I. Hambleton and M. Kreck: Cancellation of hyperbolic forms and topological four-manifolds. J. reine angew. Math. (to appear)

[HK3] I. Hambleton and M. Kreck: Cancellation, elliptic surfaces and the topology of certain four-manifolds. J. reine angew. Math. (to appear)

[Ke] M. Kervaire: Le théorème de Barden-Mazur-Stallings. Comment. Math. Helv. **40**, 31–42 (1965)

[Ko] D. Kotschick: On the geometry of certain 4-manifolds. D. Phil. thesis, Oxford 1990

[Kr] M. Kreck: Manifolds with unique differentiable structure. Topology **23**, 219–232 (1984)

[KM] P. Kronheimer and T. Mrowka: Gauge theory for embedded surfaces I, II. Preprint

[Kro] P. Kronheimer: The genus-minimizing property of algebraic curves. Bull. AMS (to appear)

[KS] R. Kirby and L. Siebenmann: Foundational essays on topological manifolds, smoothings and triangulations. Annals of Math. Studies No. 88, Princeton University Press, 1977

[Moi] E. Moise: Affine structures on 3-manifolds. Ann. Math. **56**, 96–114 (1952)

[Mi] J. Milnor: On simply-connected 4-manifolds. Symp. Int. de Topologia Algebraica, Mexico 1958, pp. 122–128

[MM] J. Morgan and T. Mrowka: A note on Donaldson's polynomial invariants. International Math. Res. Notices **10**, 223–230 (1992)

[MO] J. Morgan and K. O'Grady: The smooth classification of fake $K3$'s and similar surfaces. Preprint 1992

[OV] C. Okonek and A. Van de Ven: Stable vector bundles and differentiable structures on certain algebraic surfaces. Inv. Math. **86**, 357–370 (1989)

[Ro] V.A. Rohlin: A new result in the theory of 4-dimensional manifolds. Soviet Math. Doklady **8**, 221–224 (1952)

[Ta] C.H. Taubes: Gauge theory on assymptotically periodic 4-manifolds. J. Diff. Geom. **25**, 363–430 (1986)

[Wa1] C.T.C. Wall: On simply connected 4-manifolds. J. Lond. Math. Soc. Coc. **39**, 141–149 (1964)

[Wa2] C.T.C. Wall: Surgery on Compact Manifolds. Academic Press, New York 1970

[Wh] J.H.C. Whitehead: On simply connected 4-dimensional polyhedra. Comment. Math. Helv. **22**, 48–92 (1949)

Some Remarks on Geometric Mechanics

Jerrold E. Marsden

1. Introduction

This paper gives a few new developments in mechanics, as well as some remarks of a historical nature. To keep the discussion focussed, most of the paper is confined to equations of "rigid body", or "hydrodynamic" type on Lie algebras or their duals. In particular, we will develop the variational structure of these equations and will relate it to the standard variational principle of Hamilton.

Even this small area of mechanics is fascinating from the historical point of view. In fact, it is quite surprising how long it can sometimes take for fundamental results of the masters to be tied together and to filter into the main literature and to become "well-known". In particular, part of our story follows a few fragments of a thread through the works of Euler, Lagrange, Lie, Poincaré, Clebsch, Ehrenfest, Hamel, Arnold, and many others.

Although Newton's discoveries were directly motivated by planetary motion, the realm of mechanics expanded well beyond particle mechanics with the work of Euler, Lagrange, and others to include fluid and solid mechanics. Today we see its methods permeating large areas of physical phenomena besides these, including electromagnetism, plasma physics, classical field theories, general relativity, and quantum mechanics. Part of what makes this unified point of view possible is the abstraction, often in a geometric way, of the underlying structures in mechanics.

Two general points of view emerged early on concerning the basic structures in mechanics. One, which is commonly referred to as "Lagrangian mechanics" can be based in variational principles, and the other, "Hamiltonian mechanics", rests on symplectic and Poisson geometry. As we shall see shortly, the history of this development is actually quite complex.

How rigid body mechanics, fluid mechanics and their generalizations fit into this story is quite interesting because of the way their equations fit into the schemes of Lagrange and Hamilton. For example, the way the equations are normally presented (in body representation for the rigid body, and in spatial representation for ideal fluids), they do not *literally* fit in as written. However, through a process of *reduction*, whereby the quotient by a Lie group of symmetries is taken, one gets in either picture, a clear understanding of how the variational and symplectic (or Poisson) structures descend to the quotient space. Since the reduction of variational principles has received less attention in the literature than that of symplectic and

Poisson reduction, the paper spends more time on that aspect. Indeed, although the results here are very simple, they do appear to be new. A more general reduction procedure for Lagrangian systems that will also be sketched, is due to Marsden and Scheurle [1993].

A specific instance of this reduction procedure, which the paper will focus on, for both simplicity of exposition and its historical relevance, is that of equations on Lie algebras $\mathfrak{g}$ or their duals $\mathfrak{g}^*$. The equations on $\mathfrak{g}$ fit into the "Lagrangian mechanics" scheme, while those on $\mathfrak{g}^*$ fit into that of "Hamiltonian mechanics". The equations on $\mathfrak{g}$ will be called the **Euler-Poincare equations**, while those on $\mathfrak{g}^*$ will be called the **Lie-Poisson equations**.

Mechanics has not only undergone considerable internal maturation, but its links with other areas of science and mathematics have strengthened considerably. For example, in engineering and physics, we have come to a much deeper understanding of stability, bifurcation and pattern formation through the maturation of mechanics and concurrent developments in dynamical systems. Perhaps the best known example of how mechanics links with mathematics is the use of symplectic techniques in representation theory through the work of Kostant, Kirillov, Souriau, Guillemin, Sternberg, and many others. There are of course many other examples of deep links with mathematics and these mathematical bonds appear to be strengthening.

Acknowledgments. I would like to especially thank Hans Duistermaat, Tudor Ratiu, Jürgen Scheurle, Juan Simo, Alan Weinstein, and Norman Wildberger for helpful discussions and comments. Sections 3 and 4 are based on notes kindly supplied by Hans Duistermaat and are gratefully acknowledged. Some of the original research reported here was done jointly with Jürgen Scheurle, and is hereby acknowledged as well.

2. Some Basic Principles of Mechanics

Let Q be an n-manifold and TQ its tangent bundle. Coordinates $q^i, i = 1, \ldots, n$ on Q induce coordinates $(q^i, \dot{q}^i)$ on TQ, called **tangent coordinates**. A mapping $L : TQ \to \mathbb{R}$ is called a **Lagrangian**. Often we choose L to be $L = K - V$ where $K(v) = \frac{1}{2}\langle v, v \rangle$ is the **kinetic energy** of the given mechanical system, and that thus defines a Riemannian metric and where $V : Q \to \mathbb{R}$ is the **potential energy**.

The **variational principle of Hamilton** singles out particular curves $q(t) \in Q$ by the condition

$$\delta \int_b^a L(q(t), \dot{q}(t))dt = 0, \tag{2.1}$$

where the variation is over smooth curves in Q with fixed endpoints. Note that (2.1) is unchanged if we replace the integrand by $L(q, \dot{q}) - \frac{d}{dt}S(q, t)$ for any function $S(q, t)$. This reflects the **gauge invariance** of classical mechanics and is closely related to Hamilton-Jacobi theory.

If one prefers, Hamilton's variational principle states that the map I defined by $I(q(\cdot)) = \int_a^b L(q(t), \dot{q}(t))dt$ from the space of curves with prescribed endpoints in Q to $\mathbb{R}$ has a critical point at the curve in question. In any case, a basic, but elementary result of the calculus of variations is that Hamilton's variational principle for a curve $q(t)$ is equivalent to the condition that this curve satisfy the **Euler-Lagrange equations**:

$$\frac{d}{dt}\frac{\partial L}{\partial \dot{q}^i} - \frac{\partial L}{\partial q^i} = 0. \tag{2.2}$$

The **Maupertuis principle of critical action**, which is closely related to Hamilton's principle, states that the integral of the canonical one form be stationary relative to curves with the energy constrained to a fixed value and with temporal variations of the endpoints possible.

Let us recall a few other basic results about this formalism. Given $L : TQ \to \mathbb{R}$, let $\mathbb{F}L : TQ \to T^*Q$, called the **ber derivative**, be the derivative of L in the fiber direction. In coordinates,

$$(q^i, \dot{q}^j) \mapsto (q^i, p_j)$$

where $p_j = \partial L/\partial \dot{q}^j$. A Lagrangian L is called **hyperregular** if $\mathbb{F}L$ is a diffeomorphism. If L is a hyperregular Lagrangian, we define the corresponding **Hamiltonian** by

$$H(q^i, p_j) = p_i\dot{q}^i - L.$$

The change of data from L on TQ to H on T^*Q is called the **Legendre transform**.

For the hyperregular case, the Euler-Lagrange equations for L are equivalent to Hamilton's equations for H, namely,

$$\dot{q}^i = \frac{\partial H}{\partial p_i} \tag{2.3}$$

$$\dot{p}_i = -\frac{\partial H}{\partial q^i}. \tag{2.4}$$

These equations define a vector field X_H on T^*Q that is related to the canonical symplectic form $\Omega = \sum_{i=1}^n dq^i \wedge dp_i$ by $\mathbf{i}_{X_H}\Omega = dH$, where $\mathbf{i}$ denotes the interior product. They can also be written in Poisson bracket form $\dot{F} = \{F, H\}$ where

$$\{F, K\} = \sum_{i=1}^n \frac{\partial F}{\partial q_i}\frac{\partial K}{\partial p_i} - \frac{\partial K}{\partial q^i}\frac{\partial F}{\partial p_i}$$

is the canonical Poisson bracket. One can, as is well known, also cast Hamilton's equations directly into a variational form on phase space (unlike Hamilton's principle, which is presented on configuration space).

In a relativistic context one finds that the two conditions $p_j = \partial L/\partial \dot{q}^j$ and $H = p_i\dot{q}^i - L$, defining the Legendre transform, fit together as the spatial and temporal components of a single object. Suffice it to say that the formalism developed here is useful in the context of relativistic fields.

3. Some Early History of the Euler-Lagrange Equations and Symplectic Geometry

In this section we make a few remarks concerning the history of the Euler-Lagrange equations.[1] Naturally, much of the story focuses on Lagrange. Section V of Lagrange's *Mecanique Analytique* contains the equations of motion in Euler-Lagrange form (2.2). Lagrange writes $Z = T - V$ for what we would call the Lagrangian today. In the preceding section of *Mecanique Analytique*, Lagrange came to these equations by asking for a coordinate invariant (*i.e.,* a covariant) expression for mass times acceleration. His conclusion is that it is given (in abbreviated notation) by $(d/dt)(\partial T/\partial v) - \partial T/\partial q$, which transforms under changes of configuration variables as a 1-form. This approach is closely related to Lagrange's introduction of generalized coordinates, which we would today refer to by saying that the configuration space is a differentiable manifold.

Interestingly, Lagrange does *not* recognize the equations of motion as being equivalent to the variational principle $\delta \int L \, dt = 0$. In fact, this principle was observed only a few decades later by Hamilton. The peculiar fact about this is that Lagrange *did* know the general form of the differential equations for variational problems and he actually had commented on Euler's proof of this – his early work on this in 1759 was admired very much by Euler. He immediately applied it to give a proof of the Maupertuis principle of least action, as a consequence of Newton's equations of motion. This principle, apparently having its roots in early work of Leibnitz, is a less natural principle in the sense that the curves are only varied over those which have a constant energy. It is also Hamilton's principle that applies in the *time dependent* case, when H is *not* conserved and which also generalizes to allow for certain external forces as well.

This discussion in the *Mecanique Analytique precedes* the equations of motion in general coordinates, and correspondingly is written in the case that the kinetic energy is of the form $\sum_i m_i v_i^2$, with constant $m_i's$. Wintner [1941] is also amazed by the fact that the more complicated Maupertuis principle historically precedes Hamilton's principle. One possible explanation is that Lagrange did not consider L as an interesting physical quantity; for him it was only a convenient function for writing down the equations of motion in a coordinate-invariant fashion. The time span between his work on variational calculus and the *Mecanique Analytique* (1788, 1808) could also be part of the explanation; he may have not been thinking of the variational calculus at the time he addressed the question of a coordinate invariant formulation of the equations of motion.

Section V starts by discussing the fact that the position and velocity at time t depend on the initial position and velocity, which can be chosen freely. We write this as (suppressing the coordinate indices for simplicity): $q = q(t, q_0, v_0), v = v(t, q_0, v_0)$, and in modern terminology we would talk about the flow in $x = (q, v)$-

---

[1] Many of these interesting historical points were conveyed by Hans Duistermaat. The reader can profitably consult with the standard texts such as those of Whittaker, Wintner, and Lanczos listed in the bibliography for additional information.

space. One problem in reading Lagrange is that he does not explicitly write the variables on which his quantities depend. In any case, he then makes an infinitesimal variation in the initial conditions and looks at the corresponding variations of position and velocity at time t. In our notation we would write $\delta x = (\partial x/\partial x_0)(t, x_0)\delta x_0$ and we would say that he considers the tangent mapping of the flow on the tangent bundle of $X = TQ$. Now comes the first interesting result. He takes two such variations, one denoted by δx and the other by Δx, and he writes down a bilinear form $\omega(\delta x, \Delta x)$, in which we recognize ω as the pull-back of the canonical symplectic form on the cotangent bundle of Q, by means of the fiber derivative $\mathbb{F}L$. What he then shows is that this symplectic product is constant as a function of t. This is nothing else than the *invariance of the symplectic form ω under the flow in TQ*.

It is striking that Lagrange gets the invariance of the symplectic form in TQ and *not* in T^*Q. In fact, Lagrange does *not* look at the equations of motion in the cotangent bundle via the transformation $\mathbb{F}L$; again it is Hamilton who observes that these take the Hamiltonian form (2.3). This is retrospectively puzzling since, later on in section V, Lagrange states very explicitly that it useful to pass to the (q, p)-coordinates by means of the coordinate transformation $\mathbb{F}L$ and one even sees written down a system of ordinary differential equations *in Hamiltonian form*, but with the total energy function H replaced by some other mysterious function $-\Omega$. Lagrange does use the letter H for the constant value of the energy, apparently in honor of Huygens. He also knew about the conservation of momentum as a result of translational symmetry.

The part where he discusses the Hamiltonian form deals with the case in which he modifies the system by perturbing the potential from $V(q)$ to $V(q) - \Omega(q)$, leaving the kinetic energy unchanged. To this perturbation problem, he applies his famous method of variation of constants, which is presented here in a truly nonlinear framework! In our notation, he keeps $t \mapsto x(t, x_0)$ as the solution of the unperturbed system, and then looks at the differential equations for $x_0(t)$ that make $t \mapsto x(t, x_0(t))$ a solution of the perturbed system. The result is that, if V is the vector field of the unperturbed system and $V + W$ is the vector field of the perturbed system, then $dx_0/dt = ((e^{tV})^*W)(x_0)$. Thus, $x_0(t)$ is the solution of the time dependent system, the vector field of which is obtained by pulling back W by means of the flow of V after time t. In the case Lagrange considers, the dq/dt-component of the perturbation is equal to zero, and the dp/dt-component is equal $\partial\Omega/\partial q$. Thus, it is obviously in a Hamiltonian form; this discussion does not use anything about Legendre-transformations (which Lagrange does not seem to know). But Lagrange knows already that the flow of the unperturbed system preserves the symplectic form, and he shows that the pull-back of his W under such a transformation is a vector field in Hamiltonian form. This is a time-dependent vector field defined by the function $G(t, q_0, p_0) = -\Omega(q(t, q_0, p_0))$. A potential point of confusion is that Lagrange denotes this by just $-\Omega$, and writes down expressions like $d\Omega/dp$, and one might first think these are zero because Ω was assumed to depend only on q. Lagrange presumably means that $dq_0/dt = \partial G/\partial p_0$, $dp_0/dt = -\partial G/\partial q_0$.

Most classical textbooks on mechanics, for example Routh, correctly point out that Lagrange has the invariance of the symplectic form in (q, v) coordinates (rather than in the canonical (q, p) coordinates). Less attention is paid to the equations obtained by the method of variation of constants that he wrote in Hamiltonian form. We do note, however, that this point is discussed in Weinstein [1981]. In fact, we should point out that the whole question of linearizing the Euler-Lagrange and Hamilton equations and retaining the mechanical structure is remarkably subtle (see Marsden, Ratiu, and Raugel [1991], for example).

Lagrange continues by introducing the *Poisson brackets* for arbitrary functions, arguing that these are useful in writing the time derivative of arbitrary functions of arbitrary variables, along solutions of systems in Hamiltonian form. He also continues by saying that if Ω is small, then $x_0(t)$ in zero order approximation is a constant and he obtains the next order approximation by an integration over t; here Lagrange introduces the first steps of the so-called *method of averaging*. When Lagrange discovered (in 1808) the invariance of the symplectic form, the variations-of-constants equations in Hamiltonian form and the Poisson brackets, he was already 73 years old. It is quite probable that Lagrange shared some of his ideas on brackets with Poisson at this time. In any case, it is clear that Lagrange had a surprisingly large part of the symplectic picture of classical mechanics.

4. Some History of Poisson Structures

Following from the work of Lagrange and Poisson mentioned above, the general concept of Poisson manifold probably should be credited to Sophus Lie in his treatise on transformation groups about 1880 in the chapter on "function groups". As was pointed out in Weinstein [1983], he also defined quite explicitly, a Poisson structure on the dual of a general Lie algebra; because of this, Marsden and Weinstein [1983] coined the phrase "Lie-Poisson bracket" for this object, and this terminology is now in common use. We recall the definition at the start of the next section. However, it is not clear that Lie realized that the Lie-Poisson bracket is obtained by a simple reduction process, namely that it is induced from the canonical cotangent Poisson bracket on T^*G by passing to $\mathfrak{g}$ regarded as the quotient T^*G/G, as will be explained in the next section. (This fact seems to have been first noted for the corresponding symplectic context by Marsden and Weinstein [1974]).

As noted by Weinstein [1983], Lie seems to have come very close, and may have even understood implicitly, the general concepts of momentum map and coadjoint orbit. The link between the closedness of the symplectic form and the Jacobi identity is a little harder to trace explicitly; some comments in this direction are given in Souriau [1970].

Lie starts by taking functions $F_1, \ldots, F_r$ on a symplectic manifold M, with the property that there exist functions G_{ij} of r variables, such that

$$\{F_i, F_j\} = G_{ij}(F_1, \ldots, F_r).$$

In Lie's time, functions were implicitly assumed to be analytic. The collection of all functions φ of $F_1, \ldots, F_r$ is the "function group" and is provided with the bracket

$$[\varphi, \psi] = \sum_{ij} G_{ij} \varphi_i \psi_j , \qquad (4.5)$$

where

$$\varphi_i = \frac{\partial \varphi}{\partial F_i} \quad \text{and} \quad \psi_j = \frac{\partial \psi}{\partial F_j} .$$

Considering $F = (F_1, \ldots, F_r)$ as a map from M to an r-dimensional space P and φ and ψ as functions on P, one may formulate this as: $[\varphi, \psi]$ is a Poisson structure on P, with the property that

$$F^*[\varphi, \psi] = \{F^*\varphi, F^*\psi\} .$$

Lie writes down the equations for the G_{ij} that follow from the antisymmetry and the Jacobi identity for the bracket $\{,\}$ on M. He continues with the question: suppose we have given a system of functions G_{ij} in r variables that satisfy these equations, is it induced as above from a function group of functions of $2n$ variables? He shows that under suitable rank conditions the answer is yes. As we shall see below, this result is the precursor to many of the fundamental results about the geometry of Poisson manifolds.

It is obvious that if G_{ij} is a system that satisfies the equations that Lie writes down, then (4.5) is a Poisson structure in the r-dimensional space. Vice versa, for any Poisson structure $[\varphi, \psi]$, the functions

$$G_{ij} = [F_i, F_j]$$

satisfy Lie's equations.

Lie continues with more remarks on local normal forms of function groups (*i.e.,* of Poisson structures), under suitable rank conditions, which are not always stated as explicitly as one would like. These amount to the statement that a Poisson structure of constant rank is determined from a foliation by symplectic leaves. It is this characterization that Lie uses to get the symplectic form on coadjoint orbits. On the other hand, Lie does not apply the symplectic form on the coadjoint orbits to representation theory – representation theory of Lie groups started only later with Schur on GL_n, Elie Cartan on representations of semisimple Lie algebras and much later, in the 1930's by Weyl for compact Lie groups. The coadjoint orbit symplectic structure was connected with representation theory in the work of Kirillov and Kostant. On the other hand, Lie *did* apply the Poisson structure on the dual of the Lie algebra to prove that every abstract Lie algebra can be realized as a Lie algebra of Hamiltonian vector fields, or as a Lie subalgebra of the Poisson algebra of functions on some symplectic manifold. This is "Lie's third fundamental theorem" in the form as given by Lie.

Of course, in geometry, people like Engel, Study and in particular Elie Cartan studied Lie's work intensely and propagated it very actively. However, through the tainted glasses of retrospection, Lie's work on Poisson structures did not appear to receive as much attention in mechanics; for example, even though Cartan himself did very important work in mechanics, he did not seem to realize that the Lie-Poisson bracket was central to the Hamiltonian description of some of the rotating fluid systems he was himself studying. However, others, such as Hamel [1904, 1949] did study Lie intensively and used it to make substantial contributions and extensions (such as to the study of nonholonomic systems, including rolling constraints), but many other active schools seem to have missed it. Even more surprising in this context is the contribution of Poincaré [1901, 1910] to the Lagrangian side of the story, a tale that we shall come to shortly. But we are getting ahead of ourselves – before telling this part of the story let us study some of the theory of mechanics and Lie algebras from the Hamiltonian point of view.

5. Lie-Poisson Structures and the Rigid Body

We now summarize a few topics in the dynamics of systems associated with Lie groups from a modern point of view to put the preceding historical comments in perspective.

Let G be a Lie group and $\mathfrak{g} = T_e G$ its Lie algebra with $[\,,\,] : \mathfrak{g} \times \mathfrak{g} \to \mathfrak{g}$ the associated Lie bracket. The dual space $\mathfrak{g}^*$ is a Poisson manifold with either of the two brackets

$$\{f, k\}_{\pm}(\mu) = \pm \left\langle \mu, \left[\frac{\delta f}{\delta \mu}, \frac{\delta k}{\delta \mu} \right] \right\rangle . \tag{5.6}$$

Here $\delta f / \delta \mu \in \mathfrak{g}$ is defined by $\left\langle v, \frac{\delta f}{\delta \mu} \right\rangle = \mathbf{D} f(\mu) \cdot v$ for $v \in \mathfrak{g}^*$, where $\mathbf{D}$ denotes the Frechet derivative. (In the infinite dimensional case one needs to worry about the existence of $\delta f / \delta \mu$). See, for instance, Marsden and Weinstein [1982, 1983] for applications to plasma physics and fluid mechanics. The notation $\delta f / \delta \mu$ is used to conform to the functional derivative notation in classical field theory. In coordinates, $(\xi^1, \ldots, \xi^m)$ on $\mathfrak{g}$ and corresponding dual coordinates $(\mu_1, \ldots, \mu_m)$ on $\mathfrak{g}^*$, the **Lie-Poisson bracket** (5.6) is

$$\{f, k\}_{\pm}(\mu) = \pm \mu_a C^a_{bc} \frac{\partial f}{\partial \mu_b} \frac{\partial k}{\partial \mu_c} ; \tag{5.7}$$

here, C^a_{bc} are the structure constants of $\mathfrak{g}$ defined by $[e_a, e_b] = C^c_{ab} e_c$, where $(e_1, \ldots, e_m)$ is the coordinate basis of $\mathfrak{g}$ and where, for $\xi \in \mathfrak{g}$, we write $\xi = \xi^a e_a$, and for $\mu \in \mathfrak{g}^*$, $\mu = \mu_a e^a$, where $(e^1, \ldots, e^m)$ is the dual basis. As we mentioned earlier, formula (5.7) appears explicitly in Lie [1890] (see 75).

Which sign to take in (5.7) is determined by understanding **Lie-Poisson reduction**, which can be summarized as follows. Let

$$\lambda : T^*G \to \mathfrak{g}^* \quad \text{be defined by} \quad p_g \mapsto (T_e L_g)^* p_g \in T_e^* G \cong \mathfrak{g}^* , \tag{5.8}$$

and

$$\rho : T^*G \to \mathfrak{g}^* \quad \text{be defined by} \quad p_g \mapsto (T_e R_g)^* p_g \in T_e^* G \cong \mathfrak{g}^* . \tag{5.9}$$

Then λ *is a Poisson map if one takes the* $-$ *Lie-Poisson structure on* $\mathfrak{g}^*$ *and* ρ *is a Poisson map if one takes the* $+$ *Lie-Poisson structure on* $\mathfrak{g}^*$. This procedure uniquely characterizes the Lie-Poisson bracket and is a basic example of Poisson reduction.

Every left invariant Hamiltonian and Hamiltonian vector field is mapped by λ to a Hamiltonian and Hamiltonian vector field on $\mathfrak{g}^*$. There is a similar statement for right invariant systems on T^*G. One says that the original system on T^*G has been **reduced** to $\mathfrak{g}^*$. The reason λ and ρ are both Poisson maps is perhaps best understood by observing that they are both equivariant momentum maps generated by the action of G on itself by right and left translations, respectively together with the fact that equivariant momentum maps are always Poisson maps (see, for example, Marsden et. al. [1983]).

The Euler equations of motion for rigid body dynamics are given by

$$\dot{\Pi} = \Pi \times \Omega,\qquad(5.10)$$

where $\Pi = \mathbb{I}\Omega$ is the body angular momentum, $\mathbb{I}$ is the moment of inertia tensor, and Ω is the body angular velocity. Euler's equations are Hamiltonian relative to the minus Lie-Poisson structure. To see this, take $G = SO(3)$ to be the configuration space. Then $\mathfrak{g} \cong (\mathbb{R}^3, \times)$ and we identify $\mathfrak{g} \cong \mathfrak{g}^*$ using the standard inner product on Euclidean space. The corresponding (minus) Lie-Poisson structure on $\mathbb{R}^3$ is given by

$$\{f, k\}(\Pi) = -\Pi \cdot (\nabla f \times \nabla k).\qquad(5.11)$$

For the rigid body one chooses the minus sign in the Lie-Poisson bracket because the rigid body Lagrangian (and hence Hamiltonian) is *left* invariant and so its dynamics pushes to $\mathfrak{g}^*$ by the map λ in (5.8).

To understand the way the Hamiltonian function originates, it is helpful to recall some basic facts about rigid body dynamics. We regard an element $R \in SO(3)$ giving the configuration of the body as a map of a reference configuration $\mathcal{B} \subset \mathbb{R}^3$ to the current configuration $R(\mathcal{B})$; the map R takes a reference or label point $X \in \mathcal{B}$ to a current point $x = R(X) \in R(\mathcal{B})$. When the rigid body is in motion, the matrix R is time dependent and the velocity of a point of the body is $\dot{x} = \dot{R}X = \dot{R}R^{-1}x$. Since R is an orthogonal matrix, $R^{-1}\dot{R}$ and $\dot{R}R^{-1}$ are skew matrices, and so we can write

$$\dot{x} = \dot{R}R^{-1}x = \omega \times x,\qquad(5.12)$$

which defines the **spatial angular velocity vector** ω. The corresponding *body angular velocity* is defined by

$$\Omega = R^{-1}\omega,\quad i.e.,\quad R^{-1}\dot{R}v = \Omega \times v\qquad(5.13)$$

so that Ω is the angular velocity relative to a body fixed frame. The kinetic energy is

$$K = \frac{1}{2}\int_{\mathcal{B}} \rho(X)\|\dot{R}X\|^2\, d^3X,\qquad(5.14)$$

where ρ is a given mass density in the reference configuration. Since

$$\|\dot{R}X\| = \|\omega \times x\| = \|R^{-1}(\omega \times x)\| = \|\Omega \times X\|,$$

K is a quadratic function of Ω. Writing

$$K = \frac{1}{2}\Omega^T \mathbb{I}\Omega \tag{5.15}$$

defines the **moment of inertia tensor** $\mathbb{I}$, which, if the body does not degenerate to a line, is a positive definite 3×3 matrix, or equivalently, a quadratic form. This quadratic form can be diagonalized, and this defines the principal axes and moments of inertia. In this basis, we write $\mathbb{I} = \mathrm{diag}(I_1, I_2, I_3)$. The function K is taken to be the Lagrangian of the system on $TSO(3)$ and by means of the Legendre transformation we get the corresponding Hamiltonian description on $T^*SO(3)$. One observes that the Lagrangian and the Hamiltonian are left invariant functions and so can be expressed in body representation. In this way, we obtain the formula for the Hamiltonian in body representation $H(\Pi) = \frac{1}{2}\Pi \cdot (\mathbb{I}^{-1}\Pi)$. One can then verify directly from the chain rule and properties of the triple product that Euler's equations are equivalent to the following equation for all $f \in \mathcal{F}(\mathbb{R}^3)$:

$$\dot{f} = \{f, H\}. \tag{5.16}$$

If $(P, \{ , \})$ is a Poisson manifold, a function $C \in \mathcal{F}(P)$ satisfying $\{C, f\} = 0$ for all $f \in \mathcal{F}(P)$ is called a *Casimir function*. In the case of the rigid body, every function $C : \mathbb{R}^3 \to \mathbb{R}$ of the form $C(\Pi) = \Phi(\|\Pi\|^2)$, where $\Phi : \mathbb{R} \to \mathbb{R}$ is a differentiable function, is a Casimir function, as is readily checked. Casimir functions are constants of the motion for *any* Hamiltonian since $\dot{C} = \{C, H\} = 0$ for any H. In particular, for the rigid body, $\|\Pi\|^2$ is a constant of the motion. Casimir functions and momentum maps play a key role in the stability theory of relative equilibria (see Marsden [1992] and references therein and for references and a discussion of the relation between Casimir functions and momentum maps.

As we have remarked, the maps λ and ρ induce Poisson isomorphisms between $(T^*G)/G$ and $\mathfrak{g}^*$ (with the $-$ and $+$ brackets respectively) and this is a special instance of Poisson reduction. The following result is one useful way of formulating the general relation between T^*G and $\mathfrak{g}^*$. We treat the left invariant case for simplicity.

Theorem 5.1. *Let G be a Lie group and $H : T^*G \to \mathbb{R}$ be a left invariant Hamiltonian. Let $h : \mathfrak{g}^* \to \mathbb{R}$ be the restriction of H to the identity. For a curve $p(t) \in T^*_{g(t)}G$, let $\mu(t) = (T^*_{g(t)}L) \cdot p(t) = \lambda(p(t))$ be the induced curve in $\mathfrak{g}^*$. Assuming that $\dot{g}(t) = \partial H/\partial p$, the following are equivalent:*

(i) *$p(t)$ is an integral curve of X_H; i.e., Hamilton's equations on T^*G hold,*
(ii) *for any smooth function $F \in \mathcal{F}(T^*G)$, $\dot{F} = \{F, H\}$, where $\{ , \}$ is the canonical bracket on T^*G*
(iii) *$\mu(t)$ satisfies the **Lie-Poisson equations***

$$\frac{d\mu}{dt} = \mathrm{ad}^*_{\delta h/\delta \mu}\mu \tag{5.17}$$

where $\mathrm{ad}_\xi : \mathfrak{g} \to \mathfrak{g}$ is defined by $\mathrm{ad}_\xi \eta = [\xi, \eta]$ and ad^*_ξ is its dual, i.e.,

$$\dot{\mu}_a = C^d_{ba}\frac{\delta h}{\delta \mu_b}\mu_d \tag{5.18}$$

(iv) *for any $f \in \mathcal{F}(\mathfrak{g}^*)$, we have*

$$\dot{f} = \{f, h\}_- \tag{5.19}$$

where $\{\,,\,\}_-$ *is the minus Lie-Poisson bracket.*

We now make some remarks about the proof. First of all, the equivalence of (i) and (ii) is general for any cotangent bundle, as is well known. The equivalence of (ii) and (iv) follows from the fact that λ is a Poisson map and $H = h \circ \lambda$. Finally, we establish the equivalence of (iii) and (iv). Indeed, $\dot{f} = \{f, h\}_-$ means

$$\left\langle \dot{\mu}, \frac{\delta f}{\delta \mu} \right\rangle = -\left\langle \mu, \left[\frac{\delta f}{\delta \mu}, \frac{\delta h}{\delta \mu} \right] \right\rangle$$

$$= \left\langle \mu, \mathrm{ad}_{\delta h / \delta \mu} \frac{\delta f}{\delta \mu} \right\rangle$$

$$= \left\langle \mathrm{ad}^*_{\delta h / \delta \mu} \mu, \frac{\delta f}{\delta \mu} \right\rangle.$$

Since f is arbitrary, this is equivalent to (iii).

6. A Little History of the Equations of Mechanics on Lie Algebras and Their Duals

The above theory describes the adaptation of the concepts of Hamiltonian mechanics to the context of the duals of Lie algebras. This theory could easily have been given shortly after Lie's work, but evidently it was not observed for the rigid body or ideal fluids until the work of Pauli [1953], Martin [1959], Arnold [1966], Ebin and Marsden [1970], Nambu [1973], and Sudarshan and Mukunda [1974], all of whom were, it seems, unaware of Lie's work on the Lie-Poisson bracket. It would appear that even Elie Cartan was unaware of this aspect of Lie's work, which does seem surprising. Perhaps it is less surprising when one thinks for a moment about how many other things Cartan was involved in at the time. Nevertheless, one is struck by the amount of rediscovery and confusion in this subject. Evidently this situation is not unique to mechanics.

One can also write the equations directly on the Lie algebra, bypassing the Lie-Poisson equations on the dual. The resulting equations were first written down on a general Lie algebra by Poincaré [1901]; we refer to these as the **Euler-Poincare equations**. Arnold [1988] and Chetaev [1989] emphasized these equations as important in the recent literature. We shall develop them from a modern point of view in the next section. Poincaré [1910] goes on to study the effects of the deformation of the earth on its precession – he apparently recognizes the equations as Euler equations on a semi-direct product Lie algebra. In general, the command that Poincaré had of the subject is most impressive, and is hard to match in his near contemporaries, except perhaps Riemann and Routh. It is noteworthy that Poincaré

[1901] has no bibliographic references and is only three pages in length, so it is rather hard to trace his train of thought or his sources – compare this style with that of Hamel [1904]! In particular, he gives no hints that he understood the work of Lie on the Lie-Poisson structure, but of course Poincaré understood the Lie group and Lie algebra concepts very well indeed.

Our derivation of the Euler-Poincaré equations in the next section is based on a reduction of variational principles, not on a reduction of the symplectic or Poisson structure, which is natural for the dual. We also show that the Lie-Poisson equations are related to the Euler-Poincaré equations by the "fiber derivative" in the same way as one gets from the ordinary Euler-Lagrange equations to the Hamilton equations. Even though this is relatively trivial in the present context, it does not appear to have been written down before.

In the dynamics of ideal fluids, the resulting variational principle is essentially what has been known as "Lin constraints". (See Cendra and Marsden [1987] for a discussion of this theory and for further references; that paper introduced a constrained variational principle closely related to that given here, but using Lagrange multipliers rather than the direct and simpler approach in this paper). Variational principles in fluid mechanics itself has an interesting history, going back to Ehrenfest, Boltzman, and Clebsch, but again, there was little if any contact with the heritage of Lie and Poincaré on the subject. Even as recently as Seliger and Witham [1968] it was remarked that "Lin's device still remains somewhat mysterious from a strictly mathematical view". It is our hope that the methods of the present paper remove some of this mystery.

One person who was well aware of the work of both Lie and Poincaré was Hamel. However, despite making excellent contributions, he seemed to miss the true simplicity of the situation, and instead got tangled up in the concept of "quasi-coordinates".

How does Lagrange fit into this story? In *Mecanique Analytique*, volume 2, equations A on page 212 are the Euler-Poincaré equations for the rotation group written out explicitly for a reasonably general Lagrangian. Of course, he must have been thinking of the rigid body equations as his main example. We should remember that Lagrange also developed the key concept of the Lagrangian representation of fluid motion, but it is not clear that he understood that both systems are special instances of one theory. Lagrange spends a large number of pages on his derivation of the Euler-Poincaré equations for $SO(3)$, in fact, a good chunk of volume 2 of *Mecanique Analytique*. His derivation is not as clean as we would give today, but it seems to have the right spirit of a reduction method. That is, he tries to get the equations from the Euler-Lagrange equations on $TSO(3)$ by passing to the Lie algebra.

Because of the above facts, one might argue that the term "Euler-Lagrange-Poincaré" equations is the right nomenclature for these equations. Since Poincaré noted the generalization to arbitrary Lie algebras, and applied it to interesting fluid problems, it is clear that his name belongs, but in light of other uses of the term "Euler-Lagrange", it seems that "Euler-Poincaré" is a reasonable choice.

Marsden and Scheurle [1992], [1993] and Weinstein [1993] have studied a more general version of Lagrangian reduction whereby one drops the Euler-Lagrange equations from TQ to TQ/G. This is a nonabelian generalization of the classical Routh method, and leads to a very interesting coupling of the Euler-Lagrange and Euler-Poincaré equations. This problem was also studied by Hamel [1904] in connection with his work on nonholonomic systems (see Koiller [1992] and Bloch, Krishnaprasad, Marsden and Murray [1993] for more information).

7. The Euler-Poincaré Equations

Above, we saw how to write the Lagrangian of rigid body motion as a function $L : TSO(3) \to \mathbb{R}$ and that the Lagrangian can be written entirely in terms of the body angular velocity. From the Lagrangian point of view, the relation between the motion in R space and that in body angular velocity (or Ω) space is as follows.

Theorem 7.1. *The curve $R(t) \in SO(3)$ satisfies the Euler-Lagrange equations for*

$$L(R, \dot{R}) = \frac{1}{2} \int_B \rho(X) \|\dot{R} X\|^2 \, d^3 X \qquad (7.20)$$

if and only if $\Omega(t)$ defined by $R^{-1} \dot{R} v = \Omega \times v$ for all $v \in \mathbb{R}^3$ satisfies Euler's equations:

$$\mathbb{I} \dot{\Omega} = \mathbb{I} \Omega \times \Omega . \qquad (7.21)$$

Moreover, this equation is equivalent to conservation of the spatial angular momentum:

$$\frac{d}{dt} \pi = 0 ,$$

where $\pi = R \mathbb{I} \Omega$.

One instructive way to prove this *indirectly* is to pass to the Hamiltonian formulation and use Lie-Poisson reduction, as outlined above. One way to do it *directly* is to use variational principles. By Hamilton's principle, $R(t)$ satisfies the Euler-Lagrange equations if and only if

$$\delta \int L \, dt = 0 .$$

Let $l(\Omega) = \frac{1}{2}(\mathbb{I}\Omega) \cdot \Omega$ so that $l(\Omega) = L(R, \dot{R})$ if R and Ω are related as above. To see how we should transform Hamilton's principle, we differentiate the relation $R^{-1} \dot{R} v = \Omega \times v$ with respect to R to get

$$-R^{-1}(\delta R) R^{-1} \dot{R} v + R^{-1}(\delta \dot{R}) v = \delta \Omega \times v . \qquad (7.22)$$

Let the skew matrix $\hat{\Sigma}$ be defined by

$$\hat{\Sigma} = R^{-1}\delta R \tag{7.23}$$

and define the vector Σ by

$$\hat{\Sigma}v = \Sigma \times v. \tag{7.24}$$

Note that

$$\dot{\hat{\Sigma}} = -R^{-1}\dot{R}R^{-1}\delta R + R^{-1}\delta\dot{R},$$

so

$$R^{-1}\delta\dot{R} = \dot{\hat{\Sigma}} + R^{-1}\dot{R}\hat{\Sigma} \tag{7.25}$$

substituting (7.25) and (7.23) into (7.22) gives

$$-\hat{\Sigma}\hat{\Omega}v + \dot{\hat{\Sigma}}v + \hat{\Omega}\hat{\Sigma}v = \widehat{\delta\Omega}v$$

i.e.,

$$\widehat{\delta\Omega} = \dot{\hat{\Sigma}} + [\hat{\Omega}, \hat{\Sigma}]. \tag{7.26}$$

The identity $[\hat{\Omega}, \hat{\Sigma}] = (\Omega \times \Sigma)\hat{\ }$ holds by Jacobi's identity for the cross product, and so

$$\delta\Omega = \dot{\Sigma} + \Omega \times \Sigma. \tag{7.27}$$

These calculations prove the following

Theorem 7.2. *Hamilton's variational principle*

$$\delta\int_a^b L\,dt = 0 \tag{7.28}$$

*on $SO(3)$ is equivalent to the **reduced variational principle***

$$\delta\int_a^b l\,dt = 0 \tag{7.29}$$

on $\mathbb{R}^3$ where the variations $\delta\Omega$ are of the form (7.27) with $\Sigma(a) = \Sigma(b) = 0$.

To complete the proof of Theorem 7.1, it suffices to work out the equations equivalent to the reduced variational principle (7.29). Since $l(\Omega) = \frac{1}{2}\langle\mathbb{I}\Omega, \Omega\rangle$, and $\mathbb{I}$ is symmetric, we get

$$\delta\int_a^b l\,dt = \int_a^b \langle\mathbb{I}\Omega, \delta\Omega\rangle dt$$

$$= \int_a^b \langle\mathbb{I}\Omega, \dot{\Sigma} + \Omega \times \Sigma\rangle dt$$

$$= \int_a^b \left[\left\langle -\frac{d}{dt}\mathbb{I}\Omega, \Sigma\right\rangle + \langle\mathbb{I}\Omega, \Omega \times \Sigma\rangle\right]$$

$$= \int_a^b \left\langle -\frac{d}{dt}\mathbb{I}\Omega + \mathbb{I}\Omega \times \Omega, \Sigma\right\rangle dt,$$

where we have integrated by parts and used the boundary conditions $\Sigma(b) = \Sigma(a) = 0$. Since Σ is otherwise arbitrary, (7.29) is equivalent to

$$-\frac{d}{dt}(\mathbb{I}\Omega) + \mathbb{I}\Omega \times \Omega = 0,$$

which are Euler's equations. That these are equivalent to the conservation of spatial angular momentum is a straightforward calculation. Note that alternatively, one can use Noether's theorem to prove conservation of spatial angular momentum, and from this one can derive the Euler equations. □

We now generalize this procedure to an arbitrary Lie group and later will make the direct link with the Lie-Poisson equations.

Theorem 7.3. *Let G be a Lie group and $L : TG \to \mathbb{R}$ a left invariant Lagrangian. Let $l : \mathfrak{g} \to \mathbb{R}$ be its restriction to the identity. For a curve $g(t) \in G$, let $\xi(t) = g(t)^{-1} \cdot \dot{g}(t)$; i.e., $\xi(t) = T_{g(t)}L_{g(t)^{-1}}\dot{g}(t)$. Then the following are equivalent*

(i) *$g(t)$ satisfies the Euler-Lagrange equations for L on G,*
(ii) *the variational principle*

$$\delta \int L(g(t), \dot{g}(t))dt = 0 \tag{7.30}$$

holds, for variations with fixed endpoints,
(iii) *the **Euler-Poincare equations** hold:*

$$\frac{d}{dt}\frac{\delta l}{\delta \xi} = \mathrm{ad}_\xi^* \frac{\delta l}{\delta \xi}, \tag{7.31}$$

(iv) *the variational principle*

$$\delta \int l(\xi(t))dt = 0 \tag{7.32}$$

holds on $\mathfrak{g}$, using variations of the form

$$\delta \xi = \dot{\eta} + [\xi, \eta], \tag{7.33}$$

where η vanishes at the endpoints,
(v) *conservation of spatial angular momentum holds:*

$$\frac{d}{dt}\pi = 0,$$

where π is defined by

$$\pi = \mathrm{Ad}_{g^{-1}}^* \frac{\partial l}{\partial \xi}.$$

We comment on the proof. First of all, the equivalence of (i) and (ii) holds on the tangent bundle of any configuration manifold Q. Secondly, (ii) and (iv) are equivalent. To see this, one needs to compute the variations $\delta\xi$ induced on $\xi = g^{-1}\dot{g} = TL_{g^{-1}}\dot{g}$ by a variation of g. To calculate this, we need to differentiate $g^{-1}\dot{g}$ in the direction of a variation δg. If $\delta g = dg/d\varepsilon$ at $\varepsilon = 0$, where g is extended to a curve g_ε, then, roughly speaking,

$$\delta\xi = \frac{d}{d\varepsilon}g^{-1}\frac{d}{dt}g$$

while if $\eta = g^{-1}\delta g$, then

$$\dot{\eta} = \frac{d}{dt}g^{-1}\frac{d}{d\varepsilon}g\,.$$

It is thus plausible that the difference $\delta\xi - \dot{\eta}$ is the commutator, $[\xi, \eta]$. Above, we saw the explicit verification of this for the rigid body, and the same proof works for any matrix group. For a complete proof for the general case, see Bloch, Krishnaprasad, Marsden and Ratiu [1993b] (it also follows from formulas in Marsden, Ratiu and Raugel [1991]).

The proof that (iii) and (v) are equivalent is a straightforward verification. We also note that conservation of the spatial angular momentum follows from Noether's theorem (indeed the spatial angular momentum is the value of the momentum map for the left action of the group), so this can be used to give another derivation of the Euler-Poincaré equations.

To complete the proof, we show the equivalence of (iii) and (iv). Indeed, using the definitions and integrating by parts,

$$\delta\int l(\xi)dt = \int \frac{\delta l}{\delta\xi}\delta\xi\,dt$$
$$= \int \frac{\delta l}{\delta\xi}(\dot{\eta} + \mathrm{ad}_\xi\eta)dt$$
$$= \int\left[-\frac{d}{dt}\left(\frac{\delta l}{\delta\xi}\right) + \mathrm{ad}_\xi^*\frac{\delta l}{\delta\xi}\right]\eta\,dt$$

so the result follows. $\qquad\qquad\qquad\qquad\qquad\qquad\qquad\qquad\qquad\qquad\square$

Since the Euler-Lagrange and Hamilton equations on TQ and T^*Q are equivalent, it follows that the Lie-Poisson and Euler-Poincaré equations are also equivalent. To see this *directly*, we make the following Legendre transformation from $\mathfrak{g}$ to $\mathfrak{g}^*$:

$$\mu = \frac{\delta l}{\delta\xi}, \quad h(\mu) = \langle\mu, \xi\rangle - l(\xi)\,.$$

Note that

$$\frac{\delta h}{\delta\mu} = \xi + \left\langle\mu, \frac{\delta\xi}{\delta\mu}\right\rangle - \left\langle\frac{\delta l}{\delta\xi}, \frac{\delta\xi}{\delta\mu}\right\rangle = \xi$$

and so it is now clear that the Lie-Poisson and Euler-Poincaré equations are equivalent.

8. The Reduced Euler-Lagrange Equations

As we have mentioned, the Lie-Poisson and Euler-Poincaré equations occur for many systems besides the rigid body equations. They include the equations of fluid and plasma dynamics, for example. For many other systems, such as a rotating molecule or a spacecraft with movable internal parts, one can use a combination of equations of Euler-Poincaré type and Euler-Lagrange type. Indeed, on the Hamiltonian side, this process has undergone development for quite some time, and is discussed briefly below. On the Lagrangian side, this process is also very interesting, and has been recently developed by, amongst others, Marsden and Scheurle [1993]. The general problem is to drop Euler-Lagrange equations and variational principles from a general velocity phase space TQ to the quotient TQ/G by a Lie group action of G on Q. If L is a G-invariant Lagrangian on TQ, it induces a reduced Lagrangian l on TQ/G.

An important ingredient in this work is to introduce a connection A on the principal bundle $Q \rightarrow S = Q/G$, assuming that this quotient is nonsingular. For example, the mechanical connection (see Kummer [1981], Marsden [1992] and references therein), may be chosen for A. This connection allows one to split the variables into a horizontal and vertical part.

We let x^α, also called "internal variables", be coordinates for shape space Q/G, η^a be coordinates for the Lie algebra $\mathfrak{g}$ relative to a chosen basis, l be the Lagrangian regarded as a function of the variables x^α, $\dot{x}^\alpha$, η^a, and let C_{db}^a be the structure constants of the Lie algebra $\mathfrak{g}$ of G.

If one writes the Euler-Lagrange equations on TQ in a local principal bundle trivialization, using the coordinates x^α introduced on the base and η^a in the fiber, then one gets the following system of **Hamel equations**

$$\frac{d}{dt}\frac{\partial l}{\partial \dot{x}^\alpha} - \frac{\partial l}{\partial x^\alpha} = 0 \tag{8.34}$$

$$\frac{d}{dt}\frac{\partial l}{\partial \eta^b} - \frac{\partial l}{\partial \eta^a}C_{db}^a\eta^d = 0. \tag{8.37}$$

However, this representation of the equations does not make global intrinsic sense (unless $Q \rightarrow S$ admits a global flat connection). The introduction of a connection allows one to intrinsically and globally split the original variational principle relative to horizontal and vertical variations. One gets from one form to the other by means of the velocity shift given by replacing η by the vertical part relative to the connection:

$$\xi^a = A_\alpha^a \dot{x}^\alpha + \eta^a .$$

Here, A_α^d are the local coordinates of the connection A. This change of coordinates is well motivated from the mechanical point of view since the variables ξ have the interpretation of the locked angular velocity and they often complete the square (help to diagonalize) the kinetic energy expression. The resulting **reduced Euler-Lagrange equations** have the following form:

$$\frac{d}{dt}\frac{\partial l}{\partial \dot{x}^\alpha} - \frac{\partial l}{\partial x^\alpha} = \frac{\partial l}{\partial \xi^a}\left(B^a_{\alpha\beta}\dot{x}^\beta + B^a_{\alpha d}\xi^d\right) \tag{8.36}$$

$$\frac{d}{dt}\frac{\partial l}{\partial \xi^b} = \frac{\partial l}{\partial \xi^a}(B^a_{b\alpha}\dot{x}^\alpha + C^a_{db}\xi^d). \tag{8.37}$$

In these equations, $B^a_{\alpha\beta}$ are the coordinates of the curvature B of A, $B^a_{\alpha d} = C^a_{bd}A^b_\alpha$ and $B^a_{d\alpha} = -B^a_{\alpha d}$.

It is interesting to note that the matrix

$$\begin{bmatrix} B^a_{\alpha\beta} & B^a_{\alpha d} \\ B^a_{d\alpha} & c^a_{bd} \end{bmatrix}$$

is itself the curvature of the connection regarded as residing on the bundle $TQ \to TQ/G$.

The variables ξ^a may be regarded as the rigid part of the variables on the original configuration space, while x^α are the internal variables. As in Simo, Lewis, and Marsden [1991], the division of variables into internal and rigid parts has deep implications for both stability theory and for bifurcation theory, again, continuing along lines developed originally by Riemann, Poincaré and others. The main way this new insight is achieved is through a careful split of the variables, using the (mechanical) connection as one of the main ingredients. This split puts the second variation of the augmented Hamiltonian at a relative equilibrium as well as the symplectic form into "normal form". It is somewhat remarkable that they are *simultaneously* put into a simple form. This link helps considerably with an eigenvalue analysis of the linearized equations, and in Hamiltonian bifurcation theory–see for example, Bloch, Krishnaprasad, Marsden and Ratiu [1993a].

One of the key results in Hamiltonian reduction theory says that the reduction of a cotangent bundle T^*Q by a symmetry group G is a bundle over T^*S, where $S = Q/G$ is shape space, and where the fiber is either $\mathfrak{g}^*$, the dual of the Lie algebra of G, or is a coadjoint orbit, depending on whether one is doing Poisson or symplectic reduction. We refer to Montgomery, Marsden, and Ratiu [1984] and Marsden [1992] for details and references. The reduced Euler-Lagrange equations give the analogue of this structure on the tangent bundle.

Remarkably, equations (8.36) are formally identical to the equations for a mechanical system with classical nonholonomic velocity constraints (see Neimark and Fufaev [1972] and Koiller [1992].) The connection chosen in that case is the one-form that determines the constraints. This link is made precise in Bloch, Krishnaprasad, Marsden and Murray [1993]. In addition, this structure appears in several control problems, especially the problem of stabilizing controls considered by Bloch, Krishnaprasad, Marsden, and Sanchez [1992].

For systems with a momentum map $\mathbf{J}$ constrained to a specific value μ, the key to the construction of a reduced Lagrangian system is the modification of the Lagrangian L to the Routhian R^μ, which is obtained from the Lagrangian by subtracting off the mechanical connection paired with the constraining value μ of the momentum map. On the other hand, a basic ingredient needed for the reduced Euler-Lagrange equations is a velocity shift in the Lagrangian, the shift being determined by the connection, so this velocity shifted Lagrangian plays the role that the Routhian does in the constrained theory.

Conclusions

The current vitality of mechanics, including the investigation of fundamental questions, is quite remarkable, given its long history and development. This vitality comes about through rich interactions with both pure mathematics (from topology and geometry to group representation theory) and through new and exciting applications to areas such as control theory. It is perhaps even more remarkable that absolutely fundamental points, such as a clear and unambiguous linking of Lie's work on the Lie-Poisson bracket on the dual of a Lie algebra and Poincaré's work on the Euler-Poincaré equations on the Lie algebra itself, with the most basic of examples in mechanics, such as the rigid body and the motion of ideal fluids, took nearly a century to complete. The attendant lessons to be learned about communication between mathematics and the other mathematical sciences are, hopefully, obvious.

References

Abraham, R. and J. Marsden (1978): Foundations of mechanics. Addison-Wesley Publishing Co., Reading, Mass.

Arms, J.M., J.E. Marsden and V. Moncrief (1981): Symmetry and bifurcations of momentum mappings. Comm. Math. Phys. **78**, 455–478

Arnold, V.I. (1966): Sur la géometrie differentielle des groupes de Lie de dimenson infinie et ses applications à l'hydrodynamique des fluids parfaits. Ann. Inst. Fourier, Grenoble **16**, 319–361

Arnold, V.I. (1988): Dynamical systems III. Encyclopaedia of mathematics, vol. 3. Springer, Berlin Heidelberg New York

Arnold, V.I. (1989): Mathematical methods of classical mechanics, 2nd edn. Graduate Texts in Mathematics, vol. 60. Springer, Berlin Heidelberg New York

Bloch, A.M., P.S. Krishnaprasad, J.E. Marsden and R. Murray (1993): Nonholonomic mechanical systems with symmetry (in preparation)

Bloch, A.M., P.S. Krishnaprasad, J.E. Marsden and T.S. Ratiu (1993a): Dissipation induced instabilities. Ann. Inst. H. Poincaré, Analyse Nonlineaire (to appear)

Bloch, A.M., P.S. Krishnaprasad, J.E. Marsden and T.S. Ratiu (1993b): The Euler-Poincaré equations and double bracket dissipation (preprint)

Bloch, A.M., P.S. Krishnaprasad, J.E. Marsden and G. Sánchez de Alvarez (1992): Stabilization of rigid body dynamics by internal and external torques. Automatica **28**, 745–756

Cendra, H., A. Ibort and J.E. Marsden (1987): Variational principles on principal fiber bundles: a geometric theory of Clebsch potentials and Lin constraints. J. Geom. Phys. **4**, 183–206

Cendra, H. and J.E. Marsden (1987): Lin constraints, Clebsch potentials and variational principles. Physica D **27**, 63–89

Chetayev, N.G. (1989): Theoretical Mechanics. Springer, New York

Ebin, D.G. and J.E. Marsden (1970): Groups of diffeomorphisms and the motion of an incompressible fluid. Ann. Math. **92**, 102–163

Hamel, G (1904): Die Lagrange-Eulerschen Gleichungen der Mechanik. Z. Math. Phys. **50**, 1–57

Hamel, G (1949): Theoretische Mechanik. Springer, Berlin Heidelberg New York

Koiller, J. (1992): Reduction of some classical nonholonomic systems with symmetry. Arch. Rat. Mech. Ann. **118**, 113–148

Kummer, M. (1981): On the construction of the reduced phase space of a Hamiltonian system with symmetry. Indiana Univ. Math. J. **30**, 281–291

Lanczos, C. (1949): The variational principles of mechanics. University of Toronto Press

Lewis, D., T.S. Ratiu, J.C. Simo and J.E. Marsden (1992): The heavy top, a geometric treatment. Nonlinearity **5**, 1–48

Lie, S. (1890): Theorie der Transformationsgruppen, three volumes. Teubner, Leipzig (reprinted by Chelsea, N.Y.)

Marsden, J.E. (1992): Lectures on Mechanics. London Mathematical Society Lecture note series, vol. 174. Cambridge University Press

Marsden, J.E., R. Montgomery and T. Ratiu (1990): Reduction, symmetry, and phases in mechanics. Memoirs AMS vol. 436, p. 1–110

Marsden, J.E., T.S. Ratiu and G. Raugel (1991): Symplectic connections and the linearization of Hamiltonian systems. Proc. Roy. Soc. Ed. A **117**, 329–380

Marsden, J.E. and J. Scheurle (1993): Lagrangian reduction and the double spherical pendulum, ZAMP **44**, 17–43

Marsden, J.E. and J. Scheurle (1993): The reduced Euler-Lagrange equations. Fields Institute Communications **1**, 139–164

Marsden, J.E., and A. Weinstein (1974): Reduction of symplectic manifolds with symmetry. Rep. Math. Phys. **5**, 121–130

Marsden, J.E. and A. Weinstein (1982): The Hamiltonian structure of the Maxwell-Vlasov equations. Physica D **4**, 394–406

Marsden, J.E. and A. Weinstein (1983): Coadjoint orbits, vortices and Clebsch variables for incompressible fluids. Physica D **7**, 305–323

Marsden, J.E., A. Weinstein, T.S. Ratiu, R. Schmid and R.G. Spencer (1983): Hamiltonian systems with symmetry, coadjoint orbits and plasma physics. In: Proc. IUTAM-IS1MM Symposium on Modern Developments in Analytical Mechanics, Torino 1982. Atti della Acad. della Sc. di Torino **117**, 289–340

Martin, J.L. (1959): Generalized classical dynamics and the "classical analogue" of a Fermi oscillation. Proc. Roy. Soc. A **251**, 536

Montgomery, R. (1984): Canonical formulations of a particle in a Yang-Mills field. Lett. Math. Phys. **8**, 59–67

Montgomery, R., J.E. Marsden and T.S. Ratiu (1984): Gauged Lie-Poisson structures. Cont. Math. AMS **28**, 101–114

Naimark, Ju. I. and N.A. Fufaev (1972): Dynamics of nonholonomic systems. Translations of Mathematical Monographs, AMS, vol. 33

Nambu, Y. (1973): Generalized Hamiltonian dynamics. Phys. Rev. D **7**, 2405–2412

Pauli, W. (1953): On the Hamiltonian structure of non-local field theories. Il Nuovo Cimento **X**, 648–667

Poincaré, H. (1901): Sur une forme nouvelle des equations de la mecanique. Compt. Rend. Acad. Sci. **132**, 369–371

Poincaré, H. (1910): Sur la precession des corps deformables. Bull Astron.

Routh, E.J. (1877): Stability of a given state of motion. Reprinted in Stability of Motion, ed. A.T. Fuller. Halsted Press, New York 1975

Seliger, R.L. and G.B. Whitham (1968): Variational principles in continuum mechanics. Proc. Roy. Soc. Lond. **305**, 1–25

Simo, J.C., D. Lewis and J.E. Marsden (1991): Stability of relative equilibria I: The reduced energy momentum method. Arch. Rat. Mech. Anal. **115**, 15–59

Sjamaar, R. and E. Lerman (1991): Stratified symplectic spaces and reduction. Ann. Math. **134**, 375–422

Smale, S. (1970): Topology and Mechanics. Invent. Math. **10**, 305–331, **11**, 45–64

Souriau, J.M. (1970): Structure des Systemes Dynamiques. Dunod, Paris

Sudarshan, E.C.G. and N.Mukunda (1974): Classical mechanics: a modern perspective. Wiley, New York 1974; 2nd edn. Krieber, Melbourne (Florida) 1983

Weinstein, A. (1983): Sophus Lie and symplectic geometry. Expo. Math. **1**, 95–96

Weinstein, A. (1981): Symplectic geometry. Bulletin A.M.S. (new series) **5**, 1–13

Weinstein, A. (1993): Lagrangian mechanics and groupoids. Preprint

Wintner, A. (1941): The Analytical Foundations of Celestial Mechanics. Princeton University Press

Neuere Entwicklungen in der Komplexen Geometrie

Thomas Peternell und Michael Schneider

Ziel dieses Artikel ist es, über wichtige Entwicklungen der **Komplexen Geometrie** der letzten 15–20 Jahre zu berichten. Unter Komplexer Geometrie verstehen wir die Untersuchung der komplexen Mannigfaltigkeiten mit Methoden der algebraischen Geometrie, der komplexen Differentialgeometrie, der reellen und komplexen Analysis sowie der (Differential-) Topologie. Wir beschränken uns hier auf **kompakte** algebraische und Kählersche Mannigfaltigkeiten. Aufgrund der Weite des Gebietes klammern wir auch die projektive Geometrie (d.h. Untervarietäten des projektiven Raumes) aus.

Wir behandeln insbesondere folgende Teilaspekte

— Grundlegende Begriffe der Klassifikationstheorie
— Birationale Klassifikationstheorie („Mori-Theorie")
— Fanomannigfaltigkeiten
— Mannigfaltigkeiten semi-positiver Krümmung
— Vektorbündelmethoden.

Dieser Artikel richtet sich in erster Linie an Nichtspezialisten. Um die Darstellung leichter lesbar zu machen, haben wir Unvollständigkeit beim Zitieren der Literatur angestrebt.

1. Was ist Klassifikationstheorie algebraischer Mannigfaltigkeiten?

Gewisse Objekte zu klassifizieren bedeutet, sich eine Übersicht zu verschaffen, die Objekte einzuteilen nach gewissen Merkmalen. Die schönste (oder schlimmste?) Welt ist die der Botaniker oder Zoologen, die dieses Problem einfach dadurch lösen, daß sie eine vollständige Liste der Arten angeben. Obwohl man sich nach M. Reid algebraische Mannigfaltigkeiten am besten als Lebewesen („alive") vorstellt, so ist doch die Klassifikation (kompakter) algebraischer Mannigfaltigkeiten nicht so „einfach". Schon ein Blick auf den eindimensionalen Fall zeigt dies: Klassifikationsparameter ist das Geschlecht $g = \frac{1}{2}\dim H^1(X, \mathbb{R})$. Ist $g = 0$, so ist X der 1-dimensionale projektive Raum, ist $g = 1$, so ist X ein Torus, und ist $g \geq 2$, dann ist eine so explizite Beschreibung nicht mehr möglich. Topologisch (oder diffeomorph) gesehen sind alle Riemannschen Flächen vom Geschlecht g gleich, die komplexe Struktur variiert jedoch für $g \geq 2$ in $3g - 3$ Parametern, wie schon Riemann erkannte. Eine Liste ist also nicht möglich.

Auf der anderen Seite verbreiten $\mathbb{P}_1$ und Tori Hoffnung; daß es nämlich auch im höherdimensionalen im Sinne von Listen klassifizierbare Mannigfaltigkeiten gibt. Dies ist tatsächlich nur sehr eingeschränkt der Fall (Beispiel: Fanomannigfaltigkeiten).

Bei jedem Klassifikationsproblem ist natürlich zunächst der Isomorphiebegriff festzulegen. Für algebraische (oder allgemeiner kompakte komplexe) Mannigfaltigkeiten gibt es verschiedene Möglichkeiten:

a) Homöomorphie oder Diffeomorphie
b) birationale Äquivalenz
c) biholomorphe Äquivalenz
d) Deformationen.

Die Fragestellungen zu a) lauten dann:

a_1) Welche topologischen (differenzierbaren) Mannigfaltigkeiten besitzen eine algebraische (oder komplexe) Struktur?
a_2) Wieviele?

Schon für sehr einfache topologische Mannigfaltigkeiten können diese Fragen sehr schwierig sein. So ist zum Beispiel bis heute nicht bekannt, ob S^6 eine komplexe Struktur besitzt (sie kann jedenfalls nicht algebraisch sein) oder wieviele algebraische Strukturen die n-dimensionale Quadrik ($n \geq 4$, n gerade) trägt.

Oft führt a_2) auf sog. Modulräume; z.B. bilden die Riemannschen Flächen vom Geschlecht $g \geq 2$ einen $(3g - 3)$-dimensionalen Raum, der dann näher studiert werden muß. Modulraumprobleme sind allerdings nicht Thema dieses Artikels.

Völlig anderer Natur ist der birationale Äquivalenzbegriff. Hier soll nicht die Topologie bei äquivalenten Objekten, sondern der Funktionenkörper der rationalen (= meromorphen) Funktionen invariant sein. Zum Beispiel kann man in einer 2-dimensionalen Mannigfaltigkeit einen Punkt ersetzen durch einen $\mathbb{P}_1$, diese neue Fläche hat dann denselben Funktionenkörper, die Topologie jedoch hat sich geändert (siehe Abschnitt 2). Klassifikation modulo birationaler Äquivalenz hieße dann im Idealfall in jeder birationalen Äquivalenzklasse einen „schönen" Vertreter zu finden (siehe Abschnitt 2, Theorie der minimalen Modelle).

Der prominenteste Spezialfall: man entscheide von einer vorgegebenen Mannigfaltigkeit, ob sie rational ist, d.h. ob deren Funktionenkörper der des projektiven Raums $\mathbb{P}_n$ ist. Punkt d) soll bedeuten, daß man zwei Mannigfaltigkeiten als gleich ansieht, wenn sie durch Deformation der komplexen Struktur auseinander hervorgehen. Insbesondere sind solche Mannigfaltigkeiten dann diffeomorph (Satz von Ehresmann).

Essentiell für die Klassifikationstheorie sind natürlich Invarianten, deren wichtigste hier besprochen werden sollen.

Zunächst betrachten wir

$$q_i(X) = \dim_{\mathbb{C}} H^0(X, \Omega_X^i),$$

die Maximalzahl linear unabhängiger holomorpher i-Formen, $q(X) = q_1(X)$ heißt auch die Irregularität von X. Die Zahlen $q_i(X)$ sind birationale Invarianten von X. Aufgrund der Hodge-Zerlegung gilt

$$\dim_{\mathbb{C}} H^1(X, \mathbb{C}) = 2q(X),$$

also ist $q(X)$ sogar eine topologische Variante. Es ist unbekannt, ob $q_i(X)$ auch für $i \geq 2$ eine topologische Invariante ist. Für kompakte Riemannsche Flächen ist $q(X)$ nichts anderes als das Geschlecht.

Als nächstes besprechen wir die sogenannten Plurigeschlechter.

Es sei $\omega_X = \Omega_X^n$ die Garbe der holomorphen n-Formen auf der n-dimensionalen projektiven Mannigfaltigkeit X. Ein globaler Schnitt in $\omega_X^{\otimes m}$ hat lokal die Form

$$s = f(dz_1 \wedge \ldots \wedge dz_n)^{\otimes m},$$

f eine lokale holomorphe Funktion. Wir setzen: $P_m(X) = \dim_{\mathbb{C}} H^0(X, \omega_X^{\otimes m})$.

Die Plurigeschlechter $P_m(X)$ sind birationale Invarianten. Sie werden verwendet, um die Kodairadimension $\kappa(X)$ in folgender Weise zu definieren.

Ist $P_m(X) = 0$ für alle $m > 0$, so setzen wir $\kappa(X) = -\infty$. Ist $P_m(X) \leq 1$, jedoch $P_{m_0}(X) = 1$ für ein m_0, so definieren wir $\kappa(X) = 0$. Ansonsten wächst $P_m(X)$ etwa wie m^k, und man setzt $\kappa(X) = k$. Stets gilt $\kappa(X) \leq n = \dim X$. Die wichtigste Grobeinteilung projektiver Mannigfaltigkeiten geschieht nach der Kodairadimension. Im Fall Riemannscher Flächen ($n = 1$) gilt:

$$\kappa(X) = -\infty \iff X = \mathbb{P}_1$$

$$\kappa(X) = 0 \iff X \text{ ist ein Torus}$$

$$\kappa(X) = 1 \iff g(X) \geq 2.$$

Mannigfaltigkeiten mit $\kappa(X) = \dim X$ heißen vom allgemeinen Typ. Sie entziehen sich weitgehend einer Klassifikation. Im Abschnitt 2, 3 und 4 dieses Artikels werden ausführlich die einzelnen Klassen behandelt mit Ausnahme der Mannigfaltigkeiten vom allgemeinen Typ.

Für birationale Fragen ist meist das kanonische Bündel $K_X = \Lambda^r T_X^*$ entscheidend, dessen Garbe der holomorphen Schnitte genau $\omega_X = \Omega_X^n$ ist, mehr Information trägt aber natürlich das holomorphe Tangentialbündel T_X selbst, ist dafür aber sehr viel unhandlicher.

Das Tangentialbündel betreffend stellen wir folgende Punkte in den Vordergrund:

- Chernklassen
- Positivität
- Stabilität.

Allgemein kann man jedem holomorphen Vektorbündel oder komplexem C^∞-Bündel E Chernsche Klassen $c_i(E) \in H^{2i}(X, \mathbb{R})$ zuordnen. Es gibt bekanntlich verschiedene Methoden, diese zu definieren: topologische, differentialgeometrische und algebraische. Die Chernschen Klassen sind topologische Invarianten von E, und es ist $c_i(E) = 0$, falls $i > r$. Hier ist r der Rang von E.

Ist speziell $E = T_X$, so heißen $c_i(X) = c_i(T_X)$ die Chernschen Klassen der Mannigfaltigkeit X. Sie sind biholomorphe Invarianten von X; homöomorphe komplexe Mannigfaltigkeiten haben i.a. voneinander verschiedene Chernsche Klassen. Da für ein Bündel E vom Rang r gilt:

$$c_1(E) = c_1(\Lambda^r E),$$

ist die Kenntnis von $c_1(X)$ im wesentlichen äquivalent zur (numerischen) Kenntnis von K_X. Die höheren Chernklasssen $c_i(X)$ enthalten daher Informationen, die im kanonischen Bündel nicht mehr enthalten sind.

Es gibt noch mehr „charakteristische" Klassen von X, z.B. die Pontrjaginschen Klassen $p_i \in H^{4i}(X, \mathbb{R})$; erwähnt sei hier die Stiefel-Whitney-Klasse

$$w_2 \in H^2(X, \mathbb{Z}_2),$$

die das Bild von $c_1(X)$ unter $H^2(X, \mathbb{Z}) \to H^2(X, \mathbb{Z}_2)$ ist (tatsächlich ist $c_1(X) \in H^2(X, \mathbb{Z})$). Die Klasse $w_2(X)$ ist eine topologische Invariante von X, ebenso wie die p_i. Wir verweisen auf [Hi 66].

Der Unterschied zwischen T_X und $\det T_X = -K_X$ wird durch die höheren Chernklassen realisiert. Sie sind i.a. noch viel schwieriger zu berechnen als $c_1(X)$. Eine Ausnahme macht die höchste Chernklasse $c_n(X)$; es gilt nämlich der Satz von Hopf:

$$c_n(X) = \chi_{top}(X),$$

für jede n-dimensionale kompakte komplexe Mannigfaltigkeit X, wobei

$$\chi_{top}(X) = \sum_{i=0}^{2n} (-1)^i b_i(X)$$

die topologische Eulercharakteristik von X ist.

Eine Sonderrolle spielt auch $c_2(X)$, wie wir bei der Diskussion der Stabilität sehen werden.

Die überragende Bedeutung der Chernklassen wird am deutlichsten beim Satz von Riemann-Roch (in der modernen Form von Hirzebruch, Grothendieck, Atiyah-Singer,...). Sei dazu X eine algebraische oder nicht algebraische kompakte komplexe Mannigfaltigkeit, E ein holomorphes Vektorbündel und $\mathcal{O}(E)$ die Garbe der holomorphen Schnitte. Die Kohomologievektorräume $H^q(X, \mathcal{O}(E))$ sind endlich-dimensional, und mit

$$\chi(X, \mathcal{O}(E)) = \sum_{i=0}^{n} (-1)^i \dim H^i(X, \mathcal{O}(E))$$

bezeichnet man die holomorphe Eulercharakteristik von E. Der Satz von Riemann-Roch berechnet nun $\chi(X, \mathcal{O}(E))$ in Abhängigkeit der $c_i(E)$ und $c_j(X)$. Zum Beispiel ist (E trivial vom Rang 1):

$$\chi(X, \mathcal{O}_X) = \frac{c_1(X) \cdot c_2(X)}{24} ,$$

wenn $\dim X = 3$. Bemerkenswert ist, daß in dieser Formel $c_3(X)$ nicht auftritt. Dies ist eine Spezialität der Dimension 3 und von großer Tragweite. In höherer Dimension muß man leider mit dem Auftauchen aller Chernscher Klassen in der Formel für die holomorphe Eulercharakteristik rechnen. Der Punkt symbolisiert das Schnittprodukt

$$H^2(X, \mathbb{R}) \times H^4(X, \mathbb{R}) \rightarrow H^6(X, \mathbb{R}) ,$$

und $H^6(X, \mathbb{R})$ wird mit $\mathbb{R}$ identifiziert.

Zum Satz von Riemann-Roch siehe z.B. [Hi 66], [Ful 84].

Im allgemeinen gibt es zwischen den Chernschen Klassen eines holomorphen Bündels E keine Beziehungen. Ist jedoch E positiv oder stabil, so ändert sich die Situation. Zur Stabilität verweisen wir auf Abschnitt 5.

Wir erörtern nun den Begriff der Positivität. Positivität (von Vektorbündeln) ist im Grunde – von der Konzeption her – ein eher analytisch-differentialgeometrischer als ein algebraischer Begriff, obwohl er auch algebraisch gefaßt werden kann. Des öfteren wird von der Positivität jedoch nur eine eher algebraische Konsequenz benötigt: Chernklassenungleichungen. Diese bilden jedoch sozusagen nur die Spitze des Eisberges: sie tragen keinesfalls genügend Informationen, um auf Positivität zurückschließen zu können. Damit ist es in der algebraischen Geometrie oft nicht viel einfacher, die Positivität oder Negativität eines Bündels nachzuweisen, als in der Differentialgeometrie eine Metrik mit bestimmter Krümmung zu konstruieren (obwohl, wie wir gleich sehen werden, der Positivitätsnachweis in der algebraischen Geometrie den Vorteil hat , daß vorher kein globales Objekt (die Metrik) zu konstruieren ist, sondern daß sofort losgerechnet werden kann).

Wir definieren zunächst den Begriff des amplen (positiven) Geradenbündels. Es sei L ein holomorphes Geradenbündel auf einer kompakten Mannigfaltigkeit X, $\mathcal{L} = \mathcal{O}(L)$ die Garbe der holomorphen Schnitte. L (oder $\mathcal{L}$) heißt ample (positiv), wenn gilt:

$$H^q(X, \mathcal{F} \otimes \mathcal{L}^{\otimes \mu}) = 0 ,$$

für $q > 0, \mu \geq \mu_0(\mathcal{F})$ und jede kohärente Garbe $\mathcal{F}$. Es würde genügen, für $\mathcal{F}$ Idealgarben der Form $\mathcal{I}_p, \mathcal{I}_{pq}, \mathcal{I}_p^2, p, q \in X$, zu nehmen.

Wichtige äquivalente Aussagen sind

(1) Es gibt ein μ, so daß der Schnittvektorraum $H^0(X, \mathcal{L}^{\otimes \mu})$ eine Einbettung von X in den projektiven Raum $\mathbb{P}(H^0(X, \mathcal{L}^{\mu}))$ definiert.

(2) (X sei als projektiv vorausgesetzt) Für alle s-codimensionalen irreduziblen kompakten algebraischen Mengen $Y \subset X$ gilt: $(c_1(L)^{n-s} \cdot Y) > 0$.

(3) Es gibt eine Hermitesche Metrik auf L mit positiver Krümmung.

Besonders wichtig für unsere Betrachtungen ist die Charakterisierung (2), während (3) die Verbindung zur Differentialgeometrie herstellt. Um dies näher auszuführen, betrachten wir den reellen Vektorraum $N^1(X)$, der von den numerischen Äquivalenzklassen irreduzibler reduzierter Hyperflächen erzeugt wird. Dabei heißen Hyperflächen (oder Divisoren) D_1, D_2 numerisch äquivalent, in Zeichen

$$D_1 \equiv D_2 \,,$$

wenn $D_1 \cdot C = D_2 \cdot C$ für jede irreduzible Kurve $C \subset X$ gilt; der Punkt ist wie üblich das Schnittprodukt. Man kann zeigen, daß $D_1 \equiv D_2$ genau dann, wenn die Kohomologieklassen $D_i \in H^2(X, \mathbb{R})$ übereinstimmen. Insbesondere ist $N^1(X)$ endlichdimensional. Der von den amplen Divisoren (=Geradenbündeln) erzeugte Kegel heißt ampler Kegel K. Analog definiert man den Vektorraum $N_1(X)$, der von den numerischen Äquivalenzklassen irreduzibler Kurven erzeugt wird. In $N_1(X)$ definieren wir den von den effektiven Kurven $\sum n_i C_i, n_i \in \mathbb{N}$, erzeugten Kegel $NE(X)$. Diese Kegel sind erstmals von Kleiman [Kl 68] betrachtet worden. Er zeigte, daß $\overline{K}$ der zu $\overline{NE}(X)$ duale Kegel ist. Insbesondere gilt für einen Divisor D:

$$D \in \overline{K} \text{ genau dann, wenn } D \cdot \alpha \geq 0$$

für alle $\alpha \in \overline{NE}(X)$. Dies ist wiederum äquivalent zu folgender Aussage: es gibt $\varepsilon > 0$, so daß $D \cdot C \geq \varepsilon \, \| [C] \|$ für alle irreduziblen Kurven $C \subset X$.

Hierbei ist $\| \cdot \|$ irgendeine Norm auf dem endlichdimensionalen Raum $N_1(X)$ und $[C]$ bezeichnet die Klasse von C.

Ein Divisor ist also ample, wenn die durch ihn definierte Linearform auf $\overline{NE}(X)$ positiv ist. Ein Divisor $D \in \overline{K}$ heißt *nef* (dies ist eine Abkürzung für „numerically eventually free", was bedeutet, daß das numerische Verhalten das eines Geradenbündels ist, welches durch globale Schnitte erzeugt ist). D ist somit nef, wenn D als Limes von amplen Divisoren mit rationalen Koeffizienten, sogenannten $\mathbb{Q}$-Divisoren geschrieben werden kann. Ein $\mathbb{Q}$-Divisor hat per definitionem die Gestalt $\sum r_i Y_i$ mit irreduziblen Hyperflächen $Y_i \subset X$ und rationalen Zahlen r_i. Die Geometrie des Kegels $\overline{NE}(X)$ ist ein Schlüssel zum Verständnis projektiver Mannigfaltigkeiten (zumindest solcher, für die das kanonische Bündel K_X nicht in $\overline{K}$ liegt.) Dieser Aspekt wird eingehend im nächsten Abschnitt besprochen.

Der Begriff der „Ampleheit" (Positivität) kann auf Vektorbündel E vom Rang $r \geq 2$ ausgedehnt werden. Dies geschieht durch ein Verfahren von grundsätzlicher Bedeutung, das in gewissem Umfang erlaubt, Fragen über Vektorbündel auf Probleme für Geradenbündel zu reduzieren. Dazu betrachten wir das zu E assoziierte $\mathbb{P}_{r-1}$-Bündel $\mathbb{P}(E)$ über X, das entsteht, indem wir in jeder Faser E_x durch Betrachten der Hyperebenen zum projektiven Raum übergehen. $\mathbb{P}(E)$ heißt das zu E assoziierte projektive Bündel. Es trägt ein „tautologisches" Geradenbündel $\mathcal{O}_{\mathbb{P}(E)}(1)$. Das Bündel E heißt nun ample, wenn das zugehörige tautologische Geradenbündel ample ist. Grauert [Gr 62] hat eine geometrische Charakterisierung gegeben: E ist ample genau dann, wenn man den Nullschnitt im dualen Bündel E^* (in der analytischen oder algebraischen Kategorie) auf einen Punkt kontrahieren kann. Es gibt auch eine kohomologische Charakterisierung. Des weiteren existieren a priori viel natürlichere differentialgeometrische Versionen, besonders wichtig ist die auf Griffiths zurückgehende: E heißt (Griffiths-)positiv, wenn es eine Metrik h auf E gibt, für deren Krümmung(smatrix) in lokalen Koordinaten für X und E

$$(\Theta_{ijkl}), 1 \le i, j \le n = \dim X, 1 \le k, l \le r = rg\, E$$

gilt:

$$\sum \Theta_{ijkl} \zeta_i \bar{\zeta}_j v_k \bar{v}_l > 0$$

für alle von Null verschiedenen $\zeta = (\zeta_i) \in \mathbb{C}^n$, $v = (v_k) \in \mathbb{C}^r$.

Ist ein Vektorbündel positiv im Sinne von Griffiths, so auch ample. Während für Geradenbündel auch die Umkehrung gilt, ist dies im Fall höheren Ranges unbekannt (und vielleicht auch gar nicht richtig) und ein interessantes Problem. Übrigens: Ist E nur nef, was bedeuten soll, daß das zugehörige tautologische Bündel nef ist, so hat E im allgemeinen keine Metrik semipositiver Krümmung. Dies wurde in [DPS 92] gezeigt. Die „entartete" Version des obigen Problems ist also negativ beantwortet.

Ist speziell $E = T_X$, so ist die Krümmung von E bzgl. einer Metrik h nichts anderes als die holomorphe Bischnittkrümmung der komplexen Differentialgeometrie. Die Krümmung von $\Lambda^n T_X = \det T_X$, des antikanonischen Bündels, ist hingegen die Ricci-Krümmung. Auf diese Gesichtspunkte werden wir im 4. Abschnitt zurückkommen.

Oft ist es schwierig, die Positivität eines Vektorbündels nachzuweisen. Jedoch gibt es einfache notwendige numerische Kriterien: Chernklassenungleichungen. Sei also X eine projektive Mannigfaltigkeit und E ein positives oder amples holomorphes Vektorbündel auf X. Dann gibt es gewisse „positive" Polynome P in den Chernklassen $c_j(E)$, so daß

$$P(c_1(E), \ldots, c_r(E)) > 0.$$

Diese Polynome sind von der Form

$$P = \sum a_\lambda S_\lambda,$$

wobei $a_\lambda \ge 0$, nicht alle 0 und S_λ das Schurpolynom zur Partition λ ist (vgl. [Ful 84]). Die einfachsten Beispiele sind die Chernklassen selbst sowie die inversen Chernklassen, die sog. Segreklassen. Konkret gilt: $c_i(E) > 0$, des weiteren $(c_1^2 - c_2)(E) > 0$, $(c_1^3 - 2c_1 c_2 + c_3)(E) > 0, \ldots$.

Da aber Chernklassen und deren Produkte i.a. keine Zahlen sind, sondern je nach Bedarf Zykeln oder Kohomologieklassen, muß erklärt werden, was $P(c_1, \ldots, c_r) > 0$ überhaupt bedeuten soll, nämlich:

$$P(c_1, \ldots, c_r) \cdot Y > 0$$

für alle s-codimensionalen irreduziblen algebraischen Varietäten $Y \subset X$, wenn $\deg P = s$ ist und somit $P(c_1, \ldots, c_r) \in H^{2s}(X, \mathbb{R})$.

Es gibt noch eine weitere Art von Chernklassenungleichungen, solche nämlich, die von der Stabilität (oder Kähler (Hermite) -Einstein Eigenschaft) herkommen.

Es ist dies hauptsächlich die sogenannte Bogomolov-Miyaoka-Yau-Ungleichung. Siehe hierzu Abschnitt 5.

2. Birationale Klassifikationstheorie

Ziel der birationalen Klassifikationstheorie algebraischer Mannigfaltigkeiten ist es, in jeder birationalen Äquivalenzklasse einen ausgezeichneten Vertreter, ein sogenanntes minimales Modell, zu konstruieren. In Dimension 2, für (algebraische Flächen), ist es sehr einfach, ein solches Modell zu konstruieren, und es ist im wesentlichen eindeutig bestimmt.

Sei dazu X eine kompakte algebraische Fläche. Eine glatte rationale Kurve $C \subset X$ heißt (-1)-Kurve, wenn gilt: $C^2 = -1$. Es gibt dann eine holomorphe Abbildung $\pi : X \to Y$ auf eine andere algebraische Fläche Y, so daß $\pi(C)$ ein Punkt ist und $\pi : X \setminus C \to Y \setminus \{p\}$ biholomorph ist. Man sagt, daß X längs C niedergeblasen wird, π heißt (Hopfscher) σ-Prozeß. Nach endlich vielen Niederblasungen erhält man aus X eine neue Fläche X', die dann keine (-1)-Kurven mehr enthält. Eine solche Fläche nennt man minimal. Bis auf wenige explizite Ausnahmen gibt es in jeder birationalen Äquivalenzklasse genau ein minimales Modell (siehe [BPV 84]).

Dies war die Betrachtungsweise bis zum Beginn der 80er Jahre.

Heute würde man die Situation jedoch differenzierter beschreiben. Dazu führen wir allgemein auf einer n-dimensionalen Mannigfaltigkeit X das kanonische Bündel $K_X = \Omega_X^n$ (Garbe der holomorphen n-Formen) ein. Die 1. Chernsche Klasse von X ist $c_1(X) = c_1(K_X^*) \in H^2(X, \mathbb{R})$. Ist C in der Fläche X eine (-1)-Kurve, so stellen wir fest: $(K_X \cdot C) = -(c_1(X) \cdot C) = -1$. Ist X nun minimal, $\neq \mathbb{P}_2$, so ist X entweder ein $\mathbb{P}_1$-Bündel über eine Kurve, und es gilt für jede Faser F: $(K_X \cdot F) = -1$, oder K_X ist nef: $(K_X \cdot C) \geq 0$ für jede Kurve $C \subset X$.

Ungefähr bedeutet die Aussage „K_X nef" für X, Ricci-seminegativ gekrümmt zu sein. Ist $X = \mathbb{P}_2$ und $l \subset \mathbb{P}_2$ eine Gerade, so gilt $(K_X \cdot l) = -2$.

Was kann also das kanonische Bündel K_X einer Fläche daran hindern, nef zu sein? Drei Dinge: eine (-1)-Kurve, eine $\mathbb{P}_1$-Bündelstruktur, oder X ist einfach $\mathbb{P}_2$. Ein ähnliches Bild erhält man im Grunde auch in höherer Dimension. Nur fragt sich: was sind dort „(-1)-Kurven"; wie verallgemeinert man $\mathbb{P}_1$-Bündelstrukturen; was ist das Analogon für $\mathbb{P}_2$ (nur $\mathbb{P}_n$?)?

In höherer Dimension war lange Zeit unklar, was ein minimales Modell und was das Analogon von (-1)-Kurven sein sollte. Es wurde versucht, minimale Modelle abbildungstheoretisch zu definieren, siehe Fujiki [Fu 80]. Danach heißt eine kompakte Mannigfaltigkeit X minimal in seiner birationalen (bimeromorphen) Äquivalenzklasse, wenn jede bimeromorphe Abbildung $f : Y \longrightarrow X$ (Y wieder eine kompakte Mannigfaltigkeit) automatisch holomorph ist. Fujiki zeigt dann: X ist minimal genau dann, wenn X keine rationale Kurve enthält (es sei dim $X \geq 3$). Mit diesem Begriff der Minimalität kann man jedoch nicht in jeder bimeromorphen Äquivalenzklasse ein minimales Modell finden. Die grundlegende Idee der Entwicklung in den 80er Jahren ist, nicht alle rationalen Kurven zu verbieten, sondern nur „geometrisch relevante". Zum Beispiel werden wir sehen, daß sehr singuläre rationale Kurven oder glatte rationale Kurven etwa mit Normalenbündel $\mathcal{O}(-2) \oplus \mathcal{O}(-1)$ oder noch negativerem Normalenbündel nicht geometrisch relevant sind.

Der Durchbruch wurde schließlich 1979/82 von S. Mori erzielt. Der Ausgangspunkt war 1979 Moris Beweis der sogenannten Hartshorne-Frankel-Vermutung, daß jede kompakte komplexe Mannigfaltigkeit mit „positivem" Tangentialbündel schon biholomorph zum projektiven Raum $\mathbb{P}_n$ sein muß [Mo 79]. Insbesondere entwickelte Mori eine spektakuläre Methode durch Reduktion auf Charakteristik p rationale Kurven zu konstruieren. 1982 untersuchte Mori in [Mo 82] dann die Geometrie rationaler Kurven; nicht aller, aber solcher rationaler Kurven C mit

$$c_1(X) \cdot C > 0, \quad \text{d.h.} \ (K_X \cdot C) < 0.$$

Diese Theorie wurde dann weiterentwickelt von Y. Kawamata, M. Reid, J. Kollár, V. Shokurov, S. Mori u.a. Grundlegende Artikel, z.T. übersichtsartig, sind [KMM 87], [Mo 87], [CKM 88], [Ko 87,92], [Wi 87a]. Zentral ist der sogenannte

Kegelsatz. *Sei X eine projektive Mannigfaltigkeit, $N_1(X)$ der reelle Vektorraum der numerischen Äquivalenzklassen effektiver 1-Zykel $\sum n_i C_i$, $C_i \subset X$ irreduzible Kurven. Sei $\overline{NE}(X) \subset N_1(X)$ der abgeschlossene Kegel der effektiven 1-Zykel. Dann ist $\overline{NE}(X)$ lokal polyhedral im Bereich $\{[C] \mid c_1(X) \cdot C \geq 0\}$. Genauer gilt folgendes.*

Sei L ein festes amples Geradenbündel, $\varepsilon > 0$.

Dann gibt es Halbstrahlen $R_1, \ldots, R_k \subset \overline{NE}(X)$, die geometrisch extremal in $\overline{NE}(X) = \sum_i^k R_i + \{\alpha \in N_1(X) \mid (c_1(X) \cdot C) \leq \varepsilon(c_1(L) \cdot C)\}$ sind und $(c_1(X) \cdot R_i) > 0$ erfüllen.

Jeder Halbstrahl R_i ist von der Form $\mathbb{R}_+[C_i]$, mit einer rationalen Kurve C_i. Dies ist alles andere als trivial, der einzig bekannte Beweis benützt wiederum Reduktion auf Charakteristik p.

Die Zahl k hängt von ε ab, und „i.a." ist $\lim_{\varepsilon \to 0} k(\varepsilon) = \infty$!

R_i wird als extremaler Strahl und C_i als extremale rationale Kurve bezeichnet. Ist z.B. X ein $\mathbb{P}_r$-Bündel über Y und l eine Gerade in einer Faser, so ist l eine extremale Kurve. Man kann sagen, daß die extremalen rationalen Kurven die geometrisch relevanten rationalen Kurven im Sinne der Theorie der minimalen Modelle sind. In Dimension 2 sind diese genau die oben beschriebenen: (-1)-Kurven, Fasern von $\mathbb{P}_1$-Bündeln und Geraden in der projektiven Ebene. In Dimension 3 ist die Liste bereits deutlich komplizierter, wir verweisen auf [Mo 82]. Man bemerkt, daß im Fall der Dimension 2 die extremalen rationalen Kurven als Fasern besonders einfacher Abbildungen auftauchen (im Fall der projektiven Ebene ist dies die konstante Abbildung). Wirklich wichtig wird der Kegelsatz durch seine geometrischen Implikationen, die besagen, daß eine analoge Aussage auch in höherer Dimension gilt falls K_X nicht nef ist. Sei dazu $R = \mathbb{R}_+[C_0]$ ein extremaler Strahl. Es gibt dann ein holomorphes Geradenbündel H auf X mit

(1) $(H.C) \geq 0$ für jede Kurve $C \subset X$ („H ist nef")
(2) $(H.C) = 0 \Leftrightarrow [C] \in R$.

Man kann zeigen, daß mH durch globale Schnitte erzeugt wird, und man erhält eine holomorphe Abbildung $f : X \to Y$, so daß gilt

(a) Y ist normal und projektiv
(b) $f(C)$ ist ein Punkt $\Leftrightarrow [C] \in R$.
(c) $-K_X$ ist f-ample.

f wird als Kontraktion des extremalen Strahls R bezeichnet und kontrahiert genau diejenigen Kurven, die homolog zu R sind. Wichtiges technisches Hilfsmittel zur Konstruktion von f ist das sogenannte

Base Point Free Theorem (Kawamata, Shokurov). *Ist H ein Geradenbündel auf der n-dimensionalen projektiven Mannigfaltigkeit X, so daß gilt:*

(1) $H^n > 0$
(2) H *ist* nef
(3) $H + K_X$ *ist* nef.

Dann ist mH global erzeugt für $m \gg 0$.

Dieser Satz wiederum fußt auf einem Kohomologie-Verschwindungssatz von Kawamata-Viehweg, der den berühmten Kodairaschen Verschwindungssatz „numerisch" verallgemeinert. Der Kegelsatz kann auch aus dem Base Point Free Theorem abgeleitet werden. Wichtig sind ferner gewisse technische Varianten, die $\mathbb{Q}$-Divisoren zulassen, also Divisoren (Geradenbündel) mit rationalen Koeffizienten. Obwohl diese geometrisch eigentlich gar keinen Sinn machen, erlauben sie doch völlig neue technische Freiheiten.

Wir wollen nun einen Blick auf die Struktur von f werfen. Ist man in einer Faserungssituation, d.h. $\dim Y < \dim X$, so ist die allgemeine Faser von F eine algebraische Mannigfaltigkeit F mit $-K_F$ positiv. Eine solche Mannigfaltigkeit wird als Fano-Mannigfaltigkeit bezeichnet. Diese können äquivalent auch dadurch definiert werden, daß F eine Kählermetrik positiver Ricci-Krümmung trägt. Fano-Mannigfaltigkeiten werden intensiv in Abschnitt 3 behandelt. In diesem Fall ($\dim(Y) < \dim(X)$) ist stets $\kappa(X) = -\infty$. Ist X eine Fläche, so definiert f eine $\mathbb{P}_1$-Bündelstruktur auf X.

Ist $\dim Y = \dim X$, so ist f eine „Modifikation". Eine algebraische Menge $E \subset X$ wird abgebildet auf eine niederdimensionale algebraische Menge E'. Im allgemeinen entstehen Singularitäten in Y längs $f(E) = E'$. Jedoch sind diese relativ harmlos, falls E eins-codimensional ist, die auftretenden Singularitäten werden als terminal bezeichnet und verhalten sich in vieler Hinsicht nicht anders als glatte Punkte. Insbesondere ist ein Vielfaches des kanonischen (Weil)-Divisors ein Cartierdivisor, definiert also ein holomorphes Geradenbündel; deswegen ist $\kappa(X)$ definiert für Varietäten mit nur terminalen Singularitäten, und es gilt darüber hinaus $\kappa(X) = \kappa(\hat{X})$ für jede Desingularisation $\hat{X}$ von X.

In Dimension 2 tauchen terminale Singularitäten noch nicht auf, und es gibt nur eine Möglichkeit für E, nämlich $E = \mathbb{P}_1$; dies ist ein wesentlicher Grund (aus höherer Warte), warum die Flächentheorie so viel einfacher ist.

Wir geben nun die Definition von terminalen Singularitäten an (siehe etwa [KMM 87], [Re 87]). Es sei X ein normaler komplexer Raum (eine normale algebraische Varietät). X sei $\mathbb{Q}$-Gorenstein (d.h. für ein $m \in \mathbb{N}$ ist der „kanonische" Weil-Divisor mK_X ein Cartier-Divisor und X ist Cohen-Macaulay). Es sei $\pi : \hat{X} \to X$ eine Desingularisation, $E_1, \ldots, E_n$ die π-exzeptionellen Divisoren. Dann hat X nur *terminale* Singularitäten, wenn gilt:

$$K_{\hat{X}} = \pi^*(K_X) + \sum_{i=1}^{n} d_i E_i \,,$$

mit $d_i > 0$.

Sind nur alle $d_i \geq 0$, so hat X kanonische Singularitäten. Ist speziell X eine normale Fläche, so sind terminale Singularitäten automatisch glatt, während die kanonischen Singularitäten genau die rationalen Doppelpunkte sind. Echte terminale Singularitäten tauchen erst in Dimension 3 auf. Die Bedeutung terminaler oder kanonischer Singularitäten kann man aus der Formel

$$\pi_*(K_{\hat{X}}) = K_X$$

ablesen, die impliziert, daß für projektive Varietäten X mit nur kanonischen Singularitäten und einer Desingularisation $\hat{X}$ stets gilt: $\kappa(X) = \kappa(\hat{X})$. Dies ist für beliebige Varietäten völlig falsch.

Wir fahren nun fort in der Beschreibung der Struktur von f. Ist in unserer Situation nun $\dim E \leq \dim X - 2$, so besitzt Y „unangenehme" Singularitäten, ein Phänomen, das erstmals in Dimension 3 auftaucht.

Dieses Phänomen hat schwerwiegende Konsequenzen. Man möchte nämlich, wenn $f : X \to Y$ eine Modifikation ist, das Verfahren auf Y wiederholen. Das geht natürlich nur, wenn es auf Y einen extremalen Strahl gibt, d.h. nach dem Kegelsatz, wenn K_Y nicht nef ist. Im Fall $\dim E = \dim X - 1$, hat, wie bereits erwähnt, Y nur terminale Singularitäten. Dann gilt der Kegelsatz und Kontraktionssatz noch immer, und ist also K_Y nicht nef, so kann man induktiv fortfahren. Ist aber $\dim E \leq \dim X - 2$, so kann man keine extremalen Kontraktionen mehr konstruieren, weil Y zu singulär ist, und es scheint, als ob das Verfahren hier abbricht. An diesem Punkt kommt die Theorie der sogenannten flips ins Spiel; hierbei handelt es sich um eine Art von „surgery" auf komplexen Mannigfaltigkeiten: eine höchstens 2-codimensionale algebraische Menge A in Y wird herausgenommen und eine andere (A') eingesetzt, so daß insgesamt eine neue Varietät Y' entsteht, die nun wieder nur terminale Singularitäten besitzt. Eine solche Operation wird „flip" genannt. Der entscheidende Unterschied zwischen A und A' ist, daß K_Y auf A negativ, K_Y' auf A' jedoch positiv ist. Es handelt sich also um eine höchst unsymmetrische Situation. Zwei essentielle Probleme sind nun:

(a) die Existenz von flips

(b) eine Endlichkeitsaussage, die besagt, daß man nach höchstens endlich vielen flips eine Kontraktion findet, die entweder auf etwas niederdimensionales abbildet oder einen Divisor kontrahiert.

(b) ist deutlich einfacher als (a) und wurde von Shokurov [Sh 86] in Dimension 3 und in [KMM 87] in Dimension 4 bewiesen.

(a) ist ein sehr tiefliegendes Problem und wurde von S. Mori [Mo 88] in Dimension 3 gelöst. Die Dimension 3 geht im Beweis ganz entscheidend ein.

Zusammenfassend erhält man

Satz. *Jede 3-dimensionale projektive Mannigfaltigkeit ist birational äquivalent zu einer normalen $\mathbb{Q}$-faktoriellen projektiven Varietät X' mit nur terminalen Singularitäten, so daß entweder*

(a) $K_{X'}$ nef *ist*

oder

(b) *auf X' eine extremale Kontraktion $f : X' \to Y'$ mit $\dim Y' < \dim X'$ existiert.*

Insbesondere ist im Fall (b) $\kappa(X) = -\infty$ und X wird durch rationale Kurven überdeckt. Im Fall (a) sagt man, daß X' ein minimales Modell von X ist. Eine birationale Abbildung $X \to X'$ kann durch eine Sequenz von „divisoriellen Kontraktionen" und flips gewonnen werden. Ein X' wie in (b) wird *nicht* als minimal bezeichnet (im Gegensatz zur Flächentheorie).

Im Fall höherer Dimension wird jener Satz vermutet, man spricht von der Vermutung über die Existenz minimaler Modelle (minimal model conjecture).

Zumindest in Dimension 3 können wir jedoch festhalten: Eine projektive Varietät mit höchstens terminalen Singularitäten hat Kodairadimension $-\infty$, wenn sie eine überdeckende Familie rationaler Kurven besitzt (solche Varietäten heißen „uniruled").

Im Zusammenhang mit minimalen Modellen ist der kanonische Ring

$$R = \bigoplus_{\nu=0}^{\infty} H^0(X, \nu K_X)$$

wichtig. Die zentrale Frage ist, ob R endlich erzeugt ist. Für ein beliebiges Geradenbündel L statt K_X ist dies Hilberts 14. Problem, und im allgemeinen ist die Antwort negativ. Zariski hat jedoch gezeigt, daß R endlich erzeugt ist, falls mL global erzeugt ist für ein m. Insbesondere ist also der kanonische Ring R endlich erzeugt, falls X minimal ist mit $\kappa(X) = \dim X$ (also X vom allgemeinen Typ ist) (base point free theorem). Ist nun allgemeiner X eine beliebige projektive Mannigfaltigkeit mit $\kappa(X) = \dim X$, so daß der kanonische Ring R von X global erzeugt ist, so ist $\mathrm{Proj}(R)$ fast ein minimales Modell bis auf die Tatsache, daß die Singularitäten von $\mathrm{Proj}(R)$ nur kanonisch, nicht aber terminal sind. Man sagt, daß $\mathrm{Proj}(R)$

ein kanonisches Modell von X ist. Die Voraussetzung „X vom allgemeinen Typ" wird verwendet, um sicherzustellen, daß die rationale Strukturabbildung

$$f : X \to \mathrm{Proj}(R)$$

birational ist, d.h. $\dim \mathrm{Proj}(R) = \dim X$.

Wir fassen zusammen:

Ist X eine projektive Mannigfaltigkeit vom allgemeinen Typ, so hat X ein kanonisches Modell genau dann, wenn der kanonische Ring endlich erzeugt ist.

Übrigens ist es typisch in der Klassifikationstheorie, daß das kanonische Bündel K_X viele Eigenschaften besitzt, die beliebige Geradenbündel nicht teilen.

Es stellt sich nunmehr das Problem, die Struktur minimaler Varietäten X zu untersuchen. Zentral ist hier die sogenannte

„Abundance Conjecture" (Kawamata). *Ist X minimal, so gilt $\kappa(X) \geq 0$, und für ein genügend großes m ist mK_X global erzeugt.*

Diese Vermutung besagt also, daß aus einer numerischen Bedingung (nämlich $(K_X \cdot C) \geq 0$ für alle Kurven $C \subset X$) die Existenz von vielen Schnitten in mK_X, also globalen algebraischen Objekten, folgt. Eine Konsequenz wäre, falls stets minimale Modelle existieren, daß $\kappa(X) = -\infty$ schon die Existenz einer X überdeckenden Familie rationaler Kurven nach sich zieht.

Eine solche Familie existiert jedenfalls dann, wenn $(K_X \cdot C) < 0$ für „genügend viele" Kurven C gilt (Mori-Mukai), eine Aussage, die mittels Reduktion mod p bewiesen wird.

Die Abundance Conjecture ist bewiesen in Dimension 3 durch Arbeiten von Miyaoka [Mi 88] und Kawamata [Ka 85a,92]; in höherer Dimension sind nur Spezialfälle bekannt.

Um einen etwas genaueren Einblick in die Struktur von Varietäten X mit $\kappa(X) \geq 0$ zu erhalten, betrachten wir minimale Varietäten X, so daß für ein $m > 0$ das „plurikanonische" Bündel mK_X durch globale Schnitte erzeugt ist, der kanonische Ring sei also endlich erzeugt. In jeder birationalen Äquivalenzklasse algebraischer Mannigfaltigkeiten mit $\kappa \geq 0$ sollte (s.o.) ein solches Modell existieren. Wir setzen voraus, daß $mK_X \neq \mathcal{O}_X$, also $\kappa(X) > 0$ und auch, daß X nicht vom allgemeinen Typ ist. Zur Struktur von Mannigfaltigkeiten mit $\kappa(X) = 0$ vgl. Abschnitt 4. Es sei $f : X \to Y$ die durch die globalen Schnitte von mK_X definierte Abbildung; f hat zusammenhängende Fasern, und Y ist dann eine normale projektive Varietät mit $\kappa(X) = \dim Y$. Wüßten wir nicht, daß mK_X global erzeugt ist, so wäre f nur eine rationale Abbildung. f heißt Iitaka-Faserung und es gilt: $Y = \mathrm{Proj}(\bigoplus H^0(mK_X))$. Sei F die allgemeine Faser von f. Dann ist F eine normale Gorensteinvarietät mit nur terminalen Singularitäten, es gilt $mK_X|F = \mathcal{O}_F$ und folglich ist K_F numerisch trivial. Daher ist $\kappa(F)$ (=Kodairadimension einer beliebigen Desingularisation) $= 0$. Über $\kappa(Y)$ kann fast keine Aussage gemacht werden bis auf eine Abschätzung, die aus folgender Vermutung, die man mit $\mathbf{C_{n,m}}$ bezeichnet, folgt.

Vermutung $C_{n,m}$. *Sei $f : X \to Y$ eine surjektive holomorphe Abbildung normaler projektiver Varietäten mit zusammenhängenden Fasern. Sei $\dim X = n$, $\dim Y = m$. Dann gilt:*

$$\kappa(X) \geq \kappa(Y) + \kappa(F),$$

F die „generische geometrische Faser ".

Diese Vermutung – eines der zentralen Probleme in der Klassifikationstheorie – ist in vielen Fällen bestätigt worden (Viehweg, siehe etwa [Vi 80], Kawamata u.a.), jedoch allgemein noch unbekannt (siehe [Mo 87] für eine Übersicht).

Kawamata [Ka 85b] hat gezeigt, daß sie aus der Vermutung über minimale Modelle folgt.

In unserer Situation folgt, daß $\kappa(Y) \leq \kappa(X)$.

Es ist durchaus möglich, daß $\kappa(Y) < \kappa(X)$, z.B. gibt es elliptische Flächen X über $\mathbb{P}_1$ mit $\kappa(X) = 1$, d.h. also minimale Flächen X mit einer Abbildung $f : X \to \mathbb{P}_1$, so daß die allgemeine Faser elliptisch ist.

3. Fano-Mannigfaltigkeiten

Wir wollen einen detaillierten Blick auf Mannigfaltigkeiten X mit $\kappa(X) = -\infty$ werfen. Nehmen wir einmal an, daß die Vermutung über minimale Modelle (Abschnitt 2) stimmt. Dann ist X birational zu einer Varietät X', welche eine extremale Kontraktion $f : X' \to Y'$ mit $\dim Y' < \dim X'$ besitzt. Die allgemeine glatte Faser F von f hat dann negatives kanonisches Bündel K_F, ist also per definitionem eine *Fano-Mannigfaltigkeit*. Die Basis Y' kann beliebig sein. Man kann also sagen, daß die Fano-Mannigfaltigkeiten die elementaren Bausteine der Varietäten negativer Kodairadimension sind. Daher ist deren Studium zentral für das Verständnis der gesamten Klasse.

Eine Fano-Mannigfaltigkeit X ist also dadurch definiert, daß das antikanonische Bündel ample ist. Nach Aubin und Yau ist dies äquivalent zur Existenz einer Kählermetrik positiver Ricci-Krümmung auf X (siehe Abschnitt 4). Fano-Mannigfaltigkeiten sind also auch aus der Sicht der Differentialgeometrie interessante Objekte. Dieser Aspekt wird in Abschnitt 4 zum Tragen kommen.

Zunächst sollen einige Beispiele ein Gefühl für diese Klasse vermitteln. In Dimension 1 existiert natürlich nur $\mathbb{P}_1$, im Fall von Flächen haben wir $\mathbb{P}_2$, $\mathbb{P}_1 \times \mathbb{P}_1$ und Aufblasungen von $\mathbb{P}_2$ in bis zu 8 Punkten in genügend allgemeiner Lage (sogenannte del Pezzo-Flächen). Alle diese sind klassische Objekte der algebraischen Geometrie und rational. In höherer Dimension wird die Situation komplizierter. Prominenteste Beispiele sind $\mathbb{P}_n$, die Quadrik Q_n, allgemeiner homogen-rationale Mannigfaltigkeiten und Hyperflächen in $\mathbb{P}_n$ vom Grad $\leq n$.

Natürlich ist nicht jede rationale Mannigfaltigkeit eine Fanomannigfaltigkeit, denn die Fano-Eigenschaft wird durch Aufblasen von allgemeinen Unterräumen zerstört. Andererseits ist bei weitem nicht jede Fanomannigfaltigkeit rational, das vielleicht prominenteste Beispiel ist die 3-dimensionale Kubik im $\mathbb{P}_4$ (Clemens-Griffiths [CG 72]).

Nun einige grundlegende Aussagen zur Struktur einer Fano-Mannigfaltigkeit X. X ist einfach zusammenhängend (vgl. Abschnitt 4) und mit Hilfe des Kodaira-schen Verschwindungssatzes kann man recht viele analytische Kohomologiegrup-pen berechnen, z.B. ist $H^q(X, \mathcal{O}) = 0$ für $q > 0$. Im weiteren wollen wir uns der Einfachheit halber auf den wichtigsten Fall, daß $b_2(X) = \dim_{\mathbb{C}} H^2(X, \mathbb{C}) = 1$, beschränken. Es ist dann Pic(X), die Gruppe der holomorphen Geradenbündel, isomorph zu $\mathbb{Z}$, so daß es einen eindeutig bestimmten amplen Erzeuger H von Pic(X) gibt. Folglich existiert ein $r \in \mathbb{N}$, so daß

$$-K_X = rH .$$

Diese Zahl r heißt der Index von X und ist eine wichtige Invariante. Der klassische Satz von Kobayashi-Ochiai [KO 73] besagt, daß

(1) $r \leq \dim X + 1$
(2) $r = \dim X + 1$ genau dann, wenn $X \cong \mathbb{P}_n$
(3) $r = \dim X$ genau dann, wenn $X \cong Q_n$.

Alle Werte zwischen 1 und $n + 1$ tauchen als Index einer n-dimensionalen Fano-Mannigfaltigkeit auf. Man kann in etwa sagen, daß die Mannigfaltigkeit um so komplizierter ist, je kleiner der Index ist.

Wichtig für das Verständnis der Fano-Mannigfaltigkeiten sind die rationalen Kurven. Aufgrund von Moris Charakteristik p Trick und dessen Verfeinerungen weiß man, daß Fano-Mannigfaltigkeiten überdeckende Familien rationaler Kurven besitzen. 1991 haben jedoch Kollár-Miyaoka-Mori [KMM 91] sowie Campana [Ca 92] noch mehr bewiesen: Man kann je zwei (allgemeine) Punkte stets durch eine rationale Kurve verbinden; man spricht daher von rationalem Zusammenhang.

Darüber hinaus kann man diese rationalen Kurven C auch noch dem Grad nach beschränken, d.h. $(-K_X \cdot C) \leq f(n)$ mit einer nur von der Dimension n abhängen-den Funktion f. Daraus folgt mit allgemeinen Sätzen, daß die n-dimensionalen Fano-Mannigfaltigkeiten (mit $b_2 = 1$) eine beschränkte Familie bilden, d.h. es gibt nur endlich viele modulo Deformation. Prinzipiell sind Fano-Mannigfaltigkeiten einer festen Dimension also klassifizierbar. Tatsächlich existiert eine solche Klassi-fikation in Dimension 3; begonnen wurde sie von Fano und später Roth, fortgesetzt und vollendet von Iskovskih, Shokurov, Mukai (Mori-Mukai für $b_2 \geq 2$) u.a.. Als Übersichtsartikel sei [Mr 81] empfohlen, als neuere Literatur verweisen wir auf [Mu 89] und [Is 90].

Von grundsätzlicher Bedeutung ist das Linearsystem $|H|$. Meist ist es „sehr ample", d.h. definiert eine Einbettung in einen projektiven Raum, stets jedoch gibt es sehr viele Elemente in $|H|$ und jedenfalls immer glatte Mitglieder. Dies ist jedoch nicht so einfach zu zeigen; tatsächlich gelang Iskovskih seine Klassifikation modulo zweier Annahmen:

(1) $|H|$ enthält ein glattes Element
(2) es gibt eine Gerade auf X.

Eine „Gerade" ist per definitionem eine rationale Kurve $C \subset X$ (evtl. singulär) mit $(H \cdot C) = 1$.

Sowohl (1) als auch (2) wurden für $\dim X = 3$ von Shokurov [Sh 80a,b] hergeleitet und sind unbewiesen im höherdimensionalen (für $n = 4, r = 2$ wurde (1) von Wilson [Wi 87b] gezeigt).

Übrigens: definiert $|H|$ eine Einbettung $X \hookrightarrow \mathbb{P}_N$, so ist das Bild einer Geraden C eine „echte" Gerade in $\mathbb{P}_n$, daher der Name.

Die Geraden spielen in der Geometrie 3-dimensionaler Fano-Mannigfaltigkeiten eine wesentliche Rolle, abhängig vom Index. Zum Beispiel betrachten wir die Familie $(C_t)_{t \in T}$ der Geraden durch einen festen (allgemeinen) Punkt $x_0 \in X$.

Dann wird $\dim T$ durch folgende Tabelle gegeben.

$r = $ Index von X	$\dim T$
4 $(X = \mathbb{P}_3)$	2
3 $(X = Q_3)$	1
2	0 (endlich viele Geraden)
1	$-\infty$ $(T = \emptyset)$

Im Fall $r = 1$ bilden alle Geraden wenigstens noch eine Fläche in X, die man berechnen kann.

Während im Fall $\dim X = 3$ das Klassifikationsproblem befriedigend gelöst ist, ist das im höherdimensionalen natürlich nicht erreichbar. Ist jedoch der Index r groß, so gibt es eine Klassifikation für $r = \dim X - 1$ (Fujita) und $r = \dim X - 2$ (Mukai, falls $|H|$ genügend glatte Elemente besitzt). Glatte Hyperflächen $Y \in |H|$ sind von Bedeutung, weil sie eine Induktion über $\dim X$ erlauben: hat X den Index r, so ist Y wieder Fano mit Index $r - 1$. Ist etwa $r = n - 2$, so kann man diesen Schritt $(n - 3)$-mal durchführen, erhält eine 3-dimensionale Fano-Mannigfaltigkeit Z mit Index 1, und kann dann deren Klassifikation verwenden. Schlecht in diesem Sinne sind z.B. 4-dimensionale Fano-Mannigfaltigkeiten vom Index 1: dann ist Y nicht mehr Fano, sondern $K_Y = \mathcal{O}_Y$; Y ist eine Calabi-Yau-Mannigfaltigkeit, die viel schwieriger zu behandeln ist.

Fanomannigfaltigkeiten, zumindestens solche von großem Index, können durch die Existenz ampler Vektorbündel charakterisiert werden, Als Prototyp sollte folgender, schon erwähnter Satz von Mori angesehen werden, der die bekannte Hartshorne-Frankel Vermutung löst.

Satz ([Mo 79]). *Jede kompakte Mannigfaltigkeit mit amplem Tangentialbündel ist isomorph zum projektiven Raum.*

Eine differentialgeometrische Version stammt von Siu-Yau [SY 80]; wir werden auf diesen Komplex noch im folgenden Abschnitt ausführlich eingehen.

Moris Satz kann wie folgt verallgemeinert werden:

Satz. *Sei X eine n-dimensionale projektive Mannigfaltigkeit, E ein amples Vektorbündel auf X vom Rang r mit $c_1(E) = c_1(X)$.*

(1) *Stets gilt $r \le n+1$ und $r = n+1$ genau dann, wenn $X \simeq \mathbb{P}_n$ und $E \simeq \mathcal{O}(1)^{n+1}$.*

(2) *Es ist $r = n$ genau dann, wenn (X, E) eines der folgenden Paare ist: $(\mathbb{P}_n, T_{\mathbb{P}_n})$, $(\mathbb{P}_n, \mathcal{O}(2) \oplus \mathcal{O}(1)^{n-1})$, $(Q_n, \mathcal{O}(1)^n)$.*

(3) *Es ist $r = n - 1$ genau dann, wenn für den Index i gilt: $i \ge n - 1$, sodann ist X der projektive Raum, die Quadrik oder eine del Pezzo Mannigfaltigkeit, mit gewissen klassifizierbaren Bündeln E.*

Einen Beweis des schwierigsten letzten Falles findet man in [PSW 92] und dort auch die Literatur zu den anderen Fällen.

Im allgemeinen könnte man einen Zusammenhang folgender Art vermuten (es sei $b_2(X) = 1$):

Index$(X) \ge r$ gilt genau dann, wenn es ein amples Vektorbündel vom Rang mindestens r auf X gibt, so daß $c_1(E) = c_1(X)$. Dies wäre leicht zu beweisen, wenn die Existenz von Geraden auf X gesichert wäre. Interessant wäre es auch, diejenigen Fanomannigfaltigkeiten zu charakterisieren, für die eine äussere Potenz $\Lambda^q T_X$ ample ist für ein q. Im Falle $q = 2$ erwartet man:

$\Lambda^2 T_X$ is ample genau dann, wenn X der projektive Raum oder die Quadrik ist. Hierzu ist in den letzten Monaten ein preprint von Cho und Sato erschienen.

Beide Vermutungen würden, falls sie positiv bestätigt werden könnten, jeweils eine Hierarchie von Fanomannigfaltigkeiten begründen, an deren Spitze der projektive Raum steht und am Ende im gewissen Sinn die allgemeinsten Fanomannigfaltigkeiten. Im ersten Fall würde diese Hierarchie genau die durch den Index gegebene sein.

Eine weitere naheliegende Frage ist es, Mannigfaltigkeiten mit „semipositivem" Tangentialbündel zu charakterisieren. Da dies eng mit differentialgeometrischen Fragestellungen zu tun hat, verschieben wir die Diskussion auf den nächsten Abschnitt, auch wenn es sich teilweise um eine algebraische Problematik handelt.

4. Mannigfaltigkeiten semi-positiver Krümmung

Wie mehrfach angedeutet, ist eine Klassifikation der Mannigfaltigkeiten allgemeinen Typs ziemlich hoffnungslos. Diese Mannigfaltigkeiten haben die Tendenz, negativ gekrümmt zu sein. Im vorliegenden Abschnitt wenden wir uns der entgegengesetzten Klasse zu – den Mannigfaltigkeiten semi-positiver Krümmung.

Dazu müssen wir zunächst den Krümmungsbegriff, den wir verwenden wollen, präzisieren.

Sei X eine komplexe Mannigfaltigkeit und E ein holomorphes Vektorbündel über X mit einer hermiteschen Metrik h. Es gibt genau einen Zusammenhang $D_{E,h}$, der mit h und der holomorphen Struktur verträglich ist. Der Chernsche Krümmungstensor ist gegeben durch

$$\Theta_h(E) = \frac{i}{2\pi} D_{E,h}^2 = \sum_{j,k,\lambda,\mu} \Theta_{jk\lambda\mu} dz_j \wedge d\bar{z}_k \otimes e_\lambda^* \otimes e_\mu \, ,$$

und ist eine hermitesche $(1, 1)$-Form mit Werten in $\text{End}(E)$. Hierbei sind $(z_1, \ldots, z_n)$ lokale Koordinaten in X und $(e_1, \ldots, e_r)$ ein lokaler holomorpher Rahmen von E.

Wir bezeichnen (E, h) als *Griffiths-semipositiv*, falls für $x \in X, \xi = (\xi_1, \ldots, \xi_n)$ $\in T_x(X), v = (v_1, \ldots, v_r) \in E_x$ gilt:

$$\Theta_h(E)(\xi \otimes v) = \sum_{j,k,\lambda,\mu} \Theta_{jk\lambda\mu} \xi_j \overline{\xi}_k v_\lambda \overline{v}_\mu \geq 0 \,.$$

Das Paar (E, h) ist *Griffiths-positiv gekrümmt*, falls für $\xi \neq 0, v \neq 0$ sogar

$$\Theta_h(E)(\xi \otimes v) > 0 \,.$$

Typische Beispiele: $T_{\mathbb{P}_n}$ ist versehen mit der Fubini-Studymetrik positiv; für jede homogene kompakte Mannigfaltigkeit X ist das Tangentialbündel T_X semipositiv, allgemeiner ist jedes global – erzeugte Vektorbündel versehen mit der Quotientenmetrik, semipositiv.

In der komplexen Differentialgeometrie ist der Begriff der *holomorphen Bischnittkrümmung* von Bedeutung. Eine kompakte Kählermannigfaltigkeit (X, g) hat semipositive holomorphe Bischnittkrümmung, falls (T_X, g) semi-positiv im Sinne von Griffiths ist. Analog wird positive holomorphe Bischnittkrümmung definiert. Es kann sehr wohl T_X semi-positiv sein, ohne daß X semi-positive holomorphe Bischnittkrümmung besitzt. Ein Beispiel ist etwa $X = \mathbb{P}(T_{\mathbb{P}_2})$. Hier existiert lediglich eine nicht Kählersche Hermitesche Metrik semi-positiver Krümmung.

Ist (E, h) ein hermitesches Vektorbündel, so hat man auf dem Geradenbündel $\det E$ die Metrik $\det h$, und für die Krümmung gilt

$$\Theta_{\det h}(\det E) = \text{Spur}(\Theta_h(E)) \,.$$

Ist insbesondere g eine Kählermetrik auf X, so ist die *Ricci-Krümmung* durch

$$\text{Ricci}(g) = \text{Spur}(\Theta_g(T_X))$$

gegeben. Die $(1, 1)$-Form $\text{Ricci}(g)$ ist eine repräsentierende Form von $c_1(X)$.

Die Calabi-Vermutung und ihre Folgen

Ein Meilenstein in der Anwendung differentialgeometrischer Methoden war die Lösung der Calabi-Vermutung [Ca 57] durch Yau [Yau 78] und Aubin [Au 76].

Calabi-Vermutung. *Ist X eine kompakte Kählermannigfaltigkeit und α eine geschlossene reelle $(1, 1)$-Form, die $c_1(X)$ repräsentiert, so gibt es genau eine Kählermetrik g auf X mit*

$$\text{Ricci}(g) = \alpha \,.$$

Ein Spezialfall der Vermutung besagt, daß die kompakten Kählermannigfaltigkeiten X mit $c_1(X) = 0$ genau die Ricciflächen sind.

Eng damit verknüpft ist folgendes Ergebnis, das von Aubin [Au 76] und Yau [Yau 77,78] erzielt wurde.

Satz. *Sei X eine kompakte Kählermannigfaltigkeit mit K_X positiv. Dann gibt es eine (eindeutig bestimmte) Kähler-Einstein-Metrik auf X.*

Hierbei heißt die Kählermetrik g auf X *Kähler-Einstein*, falls es eine reelle Konstante c gibt mit
$$\mathrm{Ricci}(g) = c\,\omega_g.$$
Hierbei bezeichnet ω_g die zu g gehörige (1, 1)-Form.

Zur Existenz einer Kähler-Einstein-Metrik ist folgende Monge-Ampère Gleichung zu lösen:
$$\det(g_{jk} + \frac{\partial^2\varphi}{\partial z_j \partial \overline{z}_k}) = \exp(\lambda\varphi + f).$$

Hierbei ist $g = \sum g_{jk} dz_j \otimes d\overline{z}_k$ die gegebene Kählermetrik, $\lambda \in \{-1, 0, 1\}$, je nachdem, ob $c_1(X) < 0, c_1(X) = 0$ oder $c_1(X) > 0$ und f eine differenzierbare Funktion. Gesucht ist eine glatte Lösung φ obiger Monge-Ampère Gleichung, so daß $(g_{jk} + \frac{\partial^2\varphi}{\partial z_j \partial \overline{z}_k})$ noch positiv definit ist. Im Fall $\lambda = 0$ muß die gegebene Funktion f noch die Bedingung $\int_X \exp(f) = vol(X)$ erfüllen.

Zusammenfassend besitzen also kompakte Kählermannigfaltigkeiten X mit $c_1(X) = 0$ oder $c_1(X) > 0$ (eindeutig bestimmte) Kähler-Einstein-Metriken.

Mannigfaltigkeiten mit $c_1(X) < 0$, d.h. also Fanomannigfaltigkeiten, besitzen im allgemeinen keine Kähler-Einstein-Metrik und es ist daher eine wichtige Herausforderung, die Teilklasse zu bestimmen, für die dies richtig ist.

Als nächstes sollen einige wichtige Folgerungen und Anwendungen der Lösung der Calabi-Vermutung gegeben werden.

Zerlegungssatz (vgl. [Kob 81], [Be 83]). *Sei X eine kompakte Kählermannigfaltigkeit mit $c_1(X) = 0$. Dann gibt es eine endliche unverzweigte Überlagerung $Y \longrightarrow X$ von der Form*
$$Y \simeq T \times Y_1 \times \ldots \times Y_r,$$
wobei die Y_i irreduzible Kählermannigfaltigkeiten mit Holonomiegruppe $SU(n_i)$ oder $Sp(m_i)$ sind.

Kobayashi [Kob 61] hatte schon vor der Lösung der Calabi-Vermutung gezeigt, daß sie folgendes impliziert.

Satz. *Jede Fanomannigfaltigkeit ist einfach zusammenhängend.*

Er zeigte nämlich, daß jede kompakte Kählermannigfaltigkeit (X, g) mit $\mathrm{Ricci}(g) > 0$ einfach zusammenhängend ist.

Lange vor der Lösung der Calabi-Vermutung war schon bekannt, daß die Existenz einer Kähler-Einstein-Metrik weitreichende Konsequenzen hat. Guggenheimer [Gu 52] hat für kompakte Kähler-Einsteinflächen X die Ungleichung

$$c_1(X)^2 \le 3c_2(X)$$

hergeleitet. Die Gleichheit tritt genau dann ein, wenn X konstante holomorphe Schnittkrümmung besitzt.

Die Erweiterung dieses Resultats auf höhere Dimension gelang Chen und Ogiue [CO 75]. Zusammen mit dem Resultat von Aubin und Yau ergibt sich speziell folgender

Satz. *Sei X eine n-dimensionale kompakte Mannigfaltigkeit mit positivem kanonischen Bündel, so gilt*

$$(nc_1(X)^2 - 2(n+1)c_2(X)) \cdot K_X^{n-2} \le 0.$$

Miyaoka [Mi 77] hatte für algebraische Flächen allgemeinen Typs unabhängig mit rein algebraischen Methoden die Ungleichung

$$c_1(X)^2 \le 3c_2(X)$$

hergeleitet. Dies verschärfte die schon vorher bekannte Bogomolovungleichung [Bo 78] $c_1(X)^2 \le 4c_2(X)$, die aus der Theorie der stabilen Vektorbündel folgt und auf die wir im Abschnitt 5 eingehen werden.

Weitere spektakuläre Anwendungen der Lösung der Calabivermutung betreffen die „Starrheit" des projektiven Raumes.

Satz (Hirzebruch-Kodaira-Yau). *Eine kompakte Kählermannigfaltigkeit, die homöomorph zu $\mathbb{P}_n$ ist, ist schon biholomorph äquivalent zu $\mathbb{P}_n$.*

Yau brachte damit die Untersuchungen von Hirzebruch und Kodaira [HK 57] zum Abschluß.

Im zweidimensionalen gelang es Yau, die Severivermutung zu lösen.

Satz (Yau). *Jede kompakte komplexe Fläche, die homöomorph zu $\mathbb{P}_2$ ist, ist schon biholomorph äquivalent zu $\mathbb{P}_2$.*

Die analoge Fragestellung in höherer Dimension ist ungeklärt, aber von Siu [Siu 89] wurde kürzlich das schwächere, lange Jahre offene Starrheitsproblem gelöst.

Satz (Siu). *Eine kompakte komplexe Mannigfaltigkeit, die aus $\mathbb{P}_n$ durch (globale) Deformation entsteht, ist biholomorph äquivalent zu $\mathbb{P}_n'$.*

Der 3-dimensionale Fall ist in wesentlich allgemeinerer Form (jede 3-dimensionale Moishezonmannigfaltigkeit, die homöomorph zu $\mathbb{P}_3$ ist, ist biholomorph zu $\mathbb{P}_3$) schon in [Pe 86], [Pe 93], [Na 87], [Ko 91] gelöst worden. Ein leichter zu verstehender Beweis des Satzes von Siu wäre sehr wünschenswert.

Ähnliche Starrheitsresultate werden für Quadriken, Grassmannsche und andere homogenrationale Mannigfaltigkeiten erwartet.

Mannigfaltigkeiten semipositiver Krümmung

Im Rest dieses Paragraphen wollen wir uns den semipositiv gekrümmten Mannigfaltigkeiten zuwenden.

Die Kenntnis der Riemannschen Flächen und die Enriquesklassifikation der algebraischen Flächen haben schon früh dazu geführt, den projektiven Raum $\mathbb{P}_n$ durch Krümmungseigenschaften zu charakterisieren.

Frankel [Fr 61] hat die Vermutung aufgestellt, daß $\mathbb{P}_n$ die einzige kompakte Kählermannigfaltigkeit mit positiver holomorpher Bischnittkrümmung ist. Diese Vermutung wurde 1970 von Hartshorne dahingehend verschärft, daß $\mathbb{P}_n$ die einzige kompakte komplexe Mannigfaltigkeit mit amplem Tangentialbündel ist. Für Flächen folgt dies relativ leicht aus der Klassifikation und in Dimension 3 wurde die Vermutung durch Sumihiro, Mori und Mabuchi gelöst.

In beliebiger Dimension gelang der Durchbruch Ende der 70er Jahre. Mori [Mo 79] löste die Hartshornevermutung mit Hilfe seines spektakulären Charakteristik p-Tricks, mit dem er rationale Kurven konstruierte. Dies war der Ausgangspunkt für die sogenannte Moritheorie (vgl. Abschnitt 2). Kurz darauf gelang Siu und Yau [SY 80] ein (analytischer) Beweis der Frankelvermutung. Hier spielen harmonische Abbildungen eine wesentliche Rolle. Dieser Zugang zur Starrheit von $\mathbb{P}_n$ wurde später von Siu erfolgreich auf andere Modellmannigfaltigkeiten (Quadriken,$\cdots$) ausgedehnt.

Nach den Arbeiten von Mori und Siu-Yau lag es nahe, etwas weiter in das Land der Mannigfaltigkeiten vorzudringen, die schwächeren Krümmungsvoraussetzungen genügen als der projektive Raum.

Durch eine ingeniöse Vermischung algebraischer (à la Mori) und analytischer Methoden gelang es Mok [Mk 88], die kompakten Kählermannigfaltigkeiten mit semipositiver holomorpher Bischnittkrümmung zu bestimmen.

Satz (Mok). *Sei X eine kompakte Kählermannigfaltigkeit mit semipositiver holomorpher Bischnittkrümmung. Dann ist die universelle Überlagerung $\tilde{X}$ von X von der Form*

$$\tilde{X} \simeq \mathbb{C}^k \times \mathbb{P}_{n_1} \times \ldots \times \mathbb{P}_{n_k} \times M_1 \times \ldots \times M_l,$$

wobei die M_i irreduzible kompakte Hermitesch-symmetrische Räume vom Rang > 1 sind.

Durch die Arbeiten von Mori trat ein sehr schwacher und rein algebraischer Semipositivitätsbegriff in den Vordergrund. Ein Geradenbündel L heißt nef, wenn $L \cdot C \geq 0$ für alle irreduziblen, reduzierten Kurven C. Ein Vektorbündel E heißt nef, wenn das Geradenbündel $\mathcal{O}_{\mathbb{P}(E)}(1)$ auf $\mathbb{P}(E)$ nef ist. Es bestehen folgende Implikationen

$$\text{global erzeugt} \Rightarrow \text{semipositive Krümmung} \Rightarrow \text{nef}.$$

Es gilt für keine dieser Implikationen die Umkehrung, nicht einmal für $E = T_X$, vgl. [DPS 92].

Der erste Schritt zur Klassifikation aller projektiven Mannigfaltigkeiten X mit T_X nef wurde in Dimension 3 (der Flächenfall ist mit Klassifikation nicht allzu schwer) unabhängig von Campana, Peternell [CP 91] und Zheng [Zh 90] gemacht. In [CP 91] wurde folgende allgemeine Vermutung aufgestellt.

Vermutung. *Ist X eine projektive Mannigfaltigkeit mit T_X nef, so ist nach evtl. endlicher unverzweigter Überlagerung die Albaneseabbildung*

$$\alpha : X \longrightarrow \mathrm{Alb}(X)$$

ein Faserbündel mit homogenrationaler Faser.

Die Lösung dieser Vermutung impliziert weitgehend die Klassifikation der Mannigfaltigkeiten, deren Tangentialbündel nef ist.

Ein Teil dieser Vermutung wurde in [DPS 92] gezeigt.

Satz. *Sei X eine kompakte Kählermannigfaltigkeit mit T_X nef. Dann ist die Albaneseabbildung $\alpha : X \longrightarrow \mathrm{Alb}(X)$ eine surjektive Submersion. Evtl. nach unverzweigter endlicher Überlagerung von X sind die Fasern F von α Fanomannigfaltigkeiten mit T_F nef.*

Stets ist die Fundamentalgruppe $\pi_1(X)$ von X eine Erweiterung von $\mathbb{Z}^q$ und einer endlichen Gruppe. Hierbei ist $q = q(X)$ die Irregularität von X.

Zum vollen Verständnis der Mannigfaltigkeiten, deren Tangentialbündel nef ist, fehlt im wesentlichen nur noch die Lösung folgender wichtiger

Vermutung. *Ist X eine Fanomannigfaltigkeit mit T_X nef, so ist X homogenrational.*

In Dimension drei ist diese Vermutung bestätigt.

Die Voraussetzung semi-positiver Krümmung ist natürlich sehr stark. Auch für spezielle Mannigfaltigkeiten wird die Krümmung in einigen Richtungen (semi-) positiv, in anderen Richtungen (semi-) negativ sein. Eine wesentlich schwächere Krümmungsvoraussetzung an X ist, daß $\det T_X$ nef ist. In [DPS 93] ist schon ein grundlegendes Resultat über die Struktur solcher Mannigfaltigkeiten enthalten.

Satz. *Sei X eine kompakte Kählermannigfaltigkeit mit $\det T_X$ nef. Dann hat $\pi_1(X)$ höchstens subexponentielles Wachstum.*

Hat $\det T_X$ sogar eine Metrik semipositiver Krümmung, so ist $\pi_1(X)$ von polynomialem Wachstum.

Dieses Ergebnis wurde für den Struktursatz [DPS 92] über Mannigfaltigkeiten mit T_X nef wesentlich benutzt. Hierbei heißt eine endlich erzeugte Gruppe $G = < g_1, \ldots, g_p >$ von subexponentiellem Wachstum, falls es zu jedem $\varepsilon > 0$ eine Konstante $C = C(\varepsilon)$ gibt, so daß für alle $k \geq 0$ gilt

$$N(k) \le C(\varepsilon)e^{\varepsilon k},$$

wobei $N(k)$ die Anzahl der Elemente aus G bezeichnet, die als Wörter der Länge $\le k$ in den Erzeugern $g_1, \dots, g_p$ geschrieben werden können.

Nach neueren Ergebnissen [DPS 94] ist $\pi_1(X)$ sogar fast-abelsch, falls $\det T_X$ semi-positiv gekrümmt ist.

Vermutung. *Sei X eine n-dimensionale kompakte Kählermannigfaltigkeit, so daß $\det T_X$ nef ist. Dann ist $\pi_1(X)$ fast abelsch.*

Zur Struktur der Albaneseabbildung wird in [DPS 93, 94] vermutet

Vermutung. *Sei X eine kompakte Kählermannigfaltigkeit, für die $\det T_X$ nef ist. Dann ist $\alpha : X \longrightarrow \mathrm{Alb}(X)$ eine surjektive Submersion, und nach endlicher unverzweigter Überlagerung ist X ein Produkt von Tori, Calabi-Yau-Mannigfaltigkeiten und rational zusammenhängenden Mannigfaltigkeiten.*

In [DPS 93, 94] wird dies bis auf rationalen Zusammenhang unter der stärkeren Voraussetzung gezeigt, daß $\det T_X$ semi-positiv gekrümmt ist.

Das genauere Verständnis der Mannigfaltigkeiten mit $\det T_X$ nef ist schwierig und interessant.

5. Vektorbündel und Anwendungen

Holomorphe Vektorbündel sind ein wichtiges Hilfsmittel beim Studium komplexer Mannigfaltigkeiten. Sie treten einmal geometrisch auf – als Tangentialbündel oder Normalenbündel einer Einbettung, zum anderen indirekt und verborgen etwa bei dem Diffeomorphieproblem komplexer Flächen.

Stabilität

Welche Klassen von Vektorbündeln sind nun besonders wichtig? Will man untersuchen, welche holomorphen Strukturen es auf einem gegebenen differenzierbaren komplexen Vektorbündel gibt, so stößt man auf ein Modulproblem. Dies ist im allgemeinen nicht sinnvoll lösbar und erfordert den Begriff der *Stabilität*, der von Mumford auf Kurven und Takemoto in höherer Dimension eingeführt wurde.

Ein holomorphes Vektorbündel E auf einer n-dimensionalen kompakten Kählermannigfaltigkeit (X, ω) heißt *stabil*, falls für jede kohärente Untergarbe $F \subset E$ mit $0 < rkF < rkE$ gilt

$$\mu(F) < \mu(E).$$

Hierbei ist $\mu(F) = \frac{\deg(F)}{rk(F)}$ die *Steigung* von F, $rk(F)$ der Rang von F und $\deg F = c_1(F) \cdot [\omega]^{n-1}$ der Grad von F bezüglich der Kählerform ω von X. Ist X

projektiv, so wird üblicherweise für ω die Kählerform genommen, die auf X durch eine Einbettung $X \hookrightarrow \mathbb{P}_N$ induziert wird.

Läßt man oben die Gleichheit zu, so heißt E *semi-stabil*.

Der Begriff der (Semi-) Stabilität hängt im allgemeinen von der Wahl der Polarisierung ω ab.

Für stabile bzw. semi-stabile Vektorbündel gibt es eine befriedigende Theorie der „Moduli" (vgl. [Ma 77,78]). Die Moduläume stabiler Bündel sind bei projektivem X selbst quasiprojektiv. Die Moduläume semi-stabiler Bündel sind projektiv, falls man noch zuläßt, daß die Bündel zu torsionsfreien Garben ausarten können. Zur Warnung sei gesagt, daß die Moduläume i.a. nicht singularitätenfrei sind, ja sogar nilpotente Elemente enthalten können.

Eine erste wichtige topologische Einschränkung ist durch die *Bogomolov-Ungleichung* gegeben:

$$((r - 1)c_1(E)^2 - 2rc_2(E)) \cdot [\omega]^{n-2} \leq 0 \,.$$

Hierbei ist $n = \dim X$ und $r = rkE$. Bogomolov hatte diese Ungleichung in [Bo 78] für algebraische Flächen gezeigt. Der Fall beliebiger Dimension ergibt sich durch Herunterschneiden mit generischen Hyperflächen hohen Grades auf eine Fläche. Daß dabei E semi-stabil bleibt, ist ein Resultat von Mehta und Ramanathan [MR 82].

Gelegentlich kann man für stabile Vektorbündel auf geeigneten Basisräumen eine bessere Ungleichung erzielen. So gilt z.B. für ein stabiles Bündel E auf $\mathbb{P}_n$

$$(r - 1)c_1(E)^2 - 2rc_2(E) < 0 \,.$$

Es sei hier auch noch erwähnt, daß es weitere Varianten der Stabilität gibt, die auf Bogomolov [Bo 78] und Gieseker [Gi 77] zurückgehen. Für einen genauen Vergleich der verschiedenen Stabilitätsbegriffe sei dem Leser das Buch [Kob 87] empfohlen.

Bogomolov hatte seine Ungleichung primär zur Anwendung auf Flächen allgemeinen Typs hergeleitet. Er bewies in [Bo 78], daß für eine Fläche allgemeinen Typs X das Cotangentialbündel Ω_X^1 semi-stabil bzgl. K_X ist. Daraus ergibt sich die Ungleichung

$$c_1(X)^2 \leq 4c_2(X) \,.$$

In beliebiger Dimension lautet sie für n-dimensionale Mannigfaltigkeiten mit K_X ample

$$((n - 1)c_1(X)^2 - 2nc_2(X)) \cdot K_X^{n-2} \leq 0$$

In Abschnitt 4 haben wir gesehen, daß die Lösung der Calabi-Vermutung durch Aubin und Yau mit Hilfe von [CO 75] zu der besseren Ungleichung

$$(nc_1(X)^2 - 2(n + 1)c_2(X)) \cdot K_X^{n-2} \leq 0$$

geführt hat. Diese sieht aus wie die Bogomolov-Ungleichung für ein geeignetes Vektorbündel von Rang $n + 1$. Es wäre interessant zu wissen, ob in der Tat die Yau-Ungleichung aus der von Bogomolov gefolgert werden kann. Außerdem stellt sich hier die Frage, ob es für Mannigfaltigkeiten allgemeinen Typs Ungleichungen à la Bogomolov oder Yau gibt, in denen die höheren Chernklassen vorkommen. Es ist jedenfalls so, daß die höheren Chernklassen von X durch polynomiale Ausdrücke in $c_1(X)$ und $c_2(X)$ abgeschätzt werden können (vgl. [CS 92]).

Vektorbündel und Diffeomorphie algebraischer Flächen

Waren anfangs Vektorbündel und ihre Modulräume *per se* untersucht worden, so hat sich zunehmend herausgestellt, daß sie wichtige Invarianten zum Verständnis der Mannigfaltigkeiten, auf denen sie leben, bilden.

In den bahnbrechenden Arbeiten von Donaldson über die Differentialtopologie kompakter reeller 4-dimensionaler orientierter Mannigfaltigkeiten wurden neue Invarianten gefunden *(Donaldsoninvarianten, Donaldsonpolynome)*. Dieses sind „Instantoneninvarianten" der differenzierbaren Struktur, die mit den Yang-Millsgleichungen der mathematischen Physik zu tun haben.

Wir wollen jetzt grob beschreiben, wie die Donaldsonpolynome erklärt sind. Für mehr Details sei auf die Übersichtsartikel [OV 90], [Ok 91] und die Bücher [DK 90], [FM 93] verwiesen. Sei dazu (M, g) eine 4-dimensionale kompakte orientierte Riemannsche Mannigfaltigkeit. Sei G eine kompakte Liegruppe (etwa $G = SU(2)$ oder $G = PU(2) \simeq SO(3)$) und P ein G-Prinzipalbündel auf M. Ist $\mathrm{ad}P$ das zu P assoziierte Bündel bezüglich der adjungierten Darstellung, so ist für jeden Zusammenhang D auf P die Krümmung D^2 eine 2-Form mit Werten in $\mathrm{ad}P$. *Instantonen* sind Extrema des Yang-Millsfunktionals

$$\|D\|_{YM} = \int_M \|D\|^2 dv \,,$$

wobei v die Volumenform und $\|D\|^2 = -\mathrm{Spur}(D^2 \wedge *D^2)$. Hierbei ist $*$ der Hodge $*$-Operator bezüglich der orientierten Riemannschen Mannigfaltigkeit (M, g). Wir wollen hier nur anti-selbstduale Instantonen betrachten, d.h. solche Zusammenhänge D mit

$$D^2 = - * D^2 \,.$$

Die Gruppe $\mathcal{G}(P)$ der Prinzipalbündelautomorphismen heißt *Gruppe der Eichtransformationen*, sie wirkt auf dem Raum $\mathcal{D}(P)$ aller Zusammenhänge auf P.

Der Modulraum

$$\mathcal{B}(P) = \mathcal{D}(P)/\mathcal{G}(P)$$

aller Zusammenhänge auf P (modulo Eichtransformationen) ist unendlich dimensional.

Sei ab jetzt $G = SU(2)$ oder $G = PU(2) \simeq SO(3)$ und ausserdem M einfach zusammenhängend mit $b_+(M)$ ungerade und $b_+(M) \geq 3$. Hierbei ist $b_+(M)$ die Anzahl der positiven Eigenwerte der Schnittform von M. Der für Anwendungen interessante Fall $b_+(M) = 1$ ist komplizierter und soll hier außer acht gelassen werden. Der Modulraum

$$\mathcal{M}(P, g) = \mathcal{D}_{asd}(P)/\mathcal{G}(P)$$

der anti-selbstdualen Zusammenhänge auf P (modulo Eichtransformationen) ist für generische Wahl von g eine orientierbare Untermannigfaltigkeit gerader Dimension $2d$ von $\mathcal{B}(P)$ (vgl. [Uh 82], [Do 87b]).

Auf $M \times \mathcal{B}(P)$ hat man ein universelles Bündel $\mathcal{P}$, dessen erste Pontrjaginklasse $p_1(\mathcal{P})$ in $H^4(M \times \mathcal{B}(P), \mathbb{Z})$ liegt. Der $(2, 2)$-Anteil $p_1^{2,2}(\mathcal{P})$ ist ein Element von $H^2(M, \mathbb{Z}) \otimes H^2(\mathcal{B}(P), \mathbb{Z})$, definiert also einen Homomorphismus

$$\mu : H_2(M, \mathbb{Z}) \longrightarrow H^2(\mathcal{B}(P), \mathbb{Z}) .$$

Die Donaldsonpolynome

$$\gamma_M^g(P) : S^d H_2(M, \mathbb{Z}) \longrightarrow \mathbb{Z}$$

können durch

$$\gamma_M^g(P)(u_1, \ldots, u_d) = (\mu(u_1) \cup \ldots \cup \mu(u_d))[\mathcal{M}(P, g)]$$

definiert werden.

Hierbei können geeignete Repräsentanten der u_i gewählt werden, so daß $\mu(u_1) \cup \ldots \cup \mu(u_d)$ kompakten Träger hat und auf $\mathcal{M}(P, g)$ ausgewertet werden kann. Dazu ist im Fall $G = SU(2)$ noch vorauszusetzen, daß $c_2(P) \geq (3b_+(M) + 5)/4$.

Die Donaldsonpolynome hängen a priori von der Riemannschen Metrik g ab. Donaldson [Do 90] zeigt, daß die Polynome $\gamma_M^g(P)$ unabhängig von g sind und daher nur von der differenzierbaren Struktur abhängen.

Kommt M von einer algebraischen Fläche, so können die Instantonenräume $\mathcal{M}(P, g)$ mit Hilfe der weiter unten beschriebenen Atiyah-Hitchin-Korrespondenz zu Modulräumen stabiler Vektorbündel in Beziehung gesetzt werden.

Modulräume stabiler Vektorbündel werden aber von den algebraischen Geometern seit 20 Jahren untersucht und erfahren hiermit tiefliegende Anwendungen.

Insbesondere wurden auf der topologischen 4-Mannigfaltigkeit, die aus $\mathbb{P}_2$ durch Aufblasen von 9 Punkten entsteht, unendlich viele verschiedene differenzierbare Strukturen entdeckt [Do 87a], [OV 86], [FM 88a], nämlich die Dolgachevflächen [Dol 81]. Diese Beispiele machen klar, daß in der Differentialtopologie die Dimension 4 eine Sonderrolle einnimmt. Für Dimension ≥ 5 gibt es höchstens endlich viele differenzierbare Strukturen [Sm 61].

Welche algebraischen Flächen sind nun diffeomorph? Van de Ven vermutet, daß die Kodairadimension eine Diffeomorphieinvariante ist (vgl. [Ve 86]).

Die Vermutung von Van de Ven scheint inzwischen gelöst (vgl. [PT 92], [FM 93]) bis auf die Möglichkeit, daß rationale Flächen diffeomorph zu nicht minimalen Flächen allgemeinen Typs sein könnten.

Wir haben früher bemerkt, daß zwei algebraische Flächen, die ineinander deformiert werden können, diffeomorph sind. Friedman und Morgan [FM 93] haben gezeigt, daß es zu festem Diffeomorphietyp höchstens endlich viele Deformationsklassen gibt. Sie vermuten sogar

Vermutung (Friedman, Morgan). *Zwei algebraische Flächen sind genau dann diffeomorph, wenn sie ineinander deformiert werden können.*

Da die Kodaira-Dimension eine Deformationsinvariante ist, ist die Vermutung von Friedman-Morgan stärker als die von Van de Ven.

Hermite-Einstein-Bündel

Wir kehren zum Beginn dieses Abschnitts zurück und wollen Stabilität differentialgeometrisch interpretieren.

In Analogie zu Kähler-Einstein-Metriken auf komplexen Mannigfaltigkeiten hat Kobayashi für holomorphe Vektorbündel Hermite-Einstein-Metriken eingeführt [Kob 82].

Sei dazu E ein holomorphes Vektorbündel auf der kompakten Kählermannigfaltigkeit (X, g). Eine hermitesche Metrik h auf E heißt Hermite-Einstein-Metrik, falls

$$\mathrm{Ricci}_g(h) = c\,h\,,$$

mit einer reellen Konstanten c.

Eine äquivalente Bedingung ist

$$\Lambda\Theta_h(E) = c\,id_E\,.$$

Hierbei ist Λ der zu $L = \omega \wedge \cdot$ adjungierte Operator.

Kobayashi formulierte die folgende Vermutung, die meist als Kobayashi-Hitchin-Korrespondenz bezeichnet wird.

Kobayashi-Hitchin-Korrespondenz. *Sei X eine kompakte Kählermannigfaltigkeit und E ein holomorphes Vektorbündel auf X. Das Bündel besitzt genau dann eine Hermite-Einstein-Metrik, wenn E direkte Summe stabiler Bündel mit gleicher Steigung ist.*

Lübke [Lu 82] tastete sich an diese Vermutung dadurch heran, daß er für Hermite-Einstein-Bündel die Bogomolov-Ungleichung für semi-stabile Bündel herleitete.

Sei E eine Hermite-Einstein-Bündel vom Rang r der kompakten Kählermannigfaltigkeit X. Dann gelten die *Lübkeungleichungen*

$$((r-1)c_1(E)^2 - 2rc_2(E)) \cdot \omega^{n-2} \le 0\,.$$

Gleichheit tritt genau dann auf, wenn E projektiv flach ist.

Die leichtere Richtung der Kobayashi-Hitchin-Vermutung wurde unabhängig von Kobayashi [Kob 82] und Lübke [Lu 83] bewiesen: Jedes Hermite-Einstein-Bündel ist direkte Summe stabiler Bündel gleicher Steigung.

Daß ein stabiles Bündel eine Hermite-Einstein-Metrik trägt, folgt für eindimensionales X aus den Arbeiten von Narasimhan und Seshadri über stabile Bündel auf Kurven [NS 65]. Den ersten Schritt in Richtung höherer Dimension tat Donaldson [Do 85] für algebraische Flächen indem er zeigte, daß stabile Bündel Hermite-Einstein-Metriken besitzen. Er benutzte das Resultat über Kurven und den Einschränkungssatz von Mehta und Ramanathan.

Für beliebige n-dimensionale kompakte Mannigfaltigkeiten haben schließlich Donaldson [Do 87c] im projektiven Fall und Uhlenbeck und Yau [UY 86] im kompakten Kählerfall gezeigt, daß stabile Bündel eine Hermite-Einstein-Metrik besitzen. Hier wird harte Analysis getrieben.

Viele Fragen über stabile Bündel lassen sich also differentialgeometrisch angreifen. Der Vorteil ist hier, daß konkret gerechnet werden kann. So folgt z.B. ganz einfach, daß alle symmetrischen Potenzen $S^m T_{\mathbb{P}_n}$ stabil sind.

Durch den Zusammenhang zwischen Stabilität und Hermite-Einstein-Metriken wurde wieder einmal eine Brücke zwischen differentialgeometrisch-analytischem und algebraisch-geometrischem Standpunkt geschlagen. Dieses Wechselspiel macht die Komplexe Geometrie so reizvoll.

Wir können unseren Ausflug in die Komplexe Geometrie nicht beenden, ohne kurz auf eine neue faszinierende Entwicklung hinzuweisen, die von Hitchin und Simpson eingeleitet wurde (vgl. [Hit 87], [Si 92]). Ein *Higgs-Bündel* (E, θ) auf einer kompakten Kählermannigfaltigkeit X ist ein holomorphes Vektorbündel E auf X zusammen mit einem Homomorphismus

$$\theta : E \longrightarrow E \otimes \Omega_X^1,$$

so daß $\theta \wedge \theta = 0$ in $End(E) \otimes \Omega_X^1$.

Der Begriff der Stabilität ist für Higgs-Bündel analog erklärt; die Untergarben müssen unter θ in sich abgebildet werden.

Mit Hilfe der Kobayashi-Hitchin-Korrespondenz kann der Modulraum der polystabilen Higgs-Bündel mit verschwindenden Chernklassen mit dem Modulraum der halbeinfachen Darstellungen der Fundamentalgruppe $\pi_1(X)$ identifiziert werden. Die multiplikative Gruppe $\mathbb{C}^*$ operiert durch

$$t \cdot (E, \theta) = (E, t\theta)$$

auf den Higgs-Bündeln und daher auf den Darstellungen von $\pi_1(X)$. Simpson zeigt, daß die Fixpunkte der $\mathbb{C}^*$-Aktion genau die Darstellungen von $\pi_1(X)$ sind, die von einer Variation der Hodgestruktur kommen.

Als Anwendung ergeben sich starke Einschränkungen für die Fundamentalgruppe einer Kählermannigfaltigkeit. So kann etwa $Sl(n, \mathbb{R})$ für $n \geq 3$ nicht als Fundamentalgruppe einer kompakten Kählermannigfaltigkeit auftreten.

References

[Au 76] Aubin, T.: Equations du type Monge-Ampère sur les variétés kählériennes compactes. CRAS A **283**, 119–121 (1976)
[Be 83] Beauville, A.: Variétés kählériennes dont la première classe de Chern est nulle. J. Diff. Geom. **18**, 755–782 (1983)
[Bo 78] Bogomolov, F. A.: Holomorphic tensors and vector bundles on projective varieties. Math. USSR Izv. **13**, 499–555 (1978)
[BPV 84] Barth, W., Peters, C., Van de Ven, A.: Compact complex surfaces. Ergebnisse der Mathematik (3) Band 4, Springer 1984

[Ca 54]   Calabi, E.: The space of Kähler metrics. Proc. Int. Congr. Math. Amsterdam 2, 206–207 (1954)

[Ca 57]   Calabi, E.: On Kähler manifolds with vanishing canonical class. Alg. Geometry and Topology (in honor of S. Lefschetz). Princeton Univ. Press 1957, pp. 78–89

[Ca 92]   Campana, F.: Connexité rationnelle des variétés de Fano. Ann. scient. Ec. Norm. Sup. **25**, 539–545 (1992)

[CG 72]   Clemens, H., Griffiths, P.: The intermediate Jacobian of the cubic threefold. Ann. Math. **95**, 281–356 (1972)

[CKM 88]   Clemens, H., Kollár, J., Mori, S.: Higher dimensional complex geometry. Astérisque vol. 166, 1988

[CO 75]   Chen, B. Y., Ogiue, K.: Some characterizations of complex space forms in terms of Chern classes. Quart. J. Math. Oxford **26**, 459–464, (1975)

[CP 91]   Campana, F., Peternell, T.: Projective manifolds whose tangent bundles are numerically effective. Math. Ann. **289**, 169–187 (1991)

[CS 92]   Catanese, F., Schneider, M.: Bounds for stable bundles and degrees of Weierstraß schemes. Math. Ann. **293**, 579–594 (1992)

[Do 85]   Donaldson, S.K.: Anti-self-dual Yang-Mills connections over complex algebraic surfaces and stable vector bundles. Proc. London Math. Soc. **50**, 1-26 (1985)

[Do 87a]   Donaldson, S. K.: Irrationality and the h-cobordism conjecture. J. Diff. Geom. **26**, 141–168 (1987)

[Do 87b]   Donaldson, S. K.: The orientation of Yang-Mills moduli spaces and 4-manifold topology. J. Diff. Geom. **26**, 397–428 (1987)

[Do 87c]   Donaldson, S. K.: Infinite determinants, stable bundles and curvature. Duke math. J. **54**, 231–247 (1987)

[Do 90]   Donaldson, S. K.: Polynomial invariants for smooth 4-manifolds. Topology **29**, 257–315 (1990)

[DK 90]   Donaldson, S. K., Kronheimer, P. B.: The geometry of four-manifolds. Clarendon Press, Oxford 1990

[Dol 81]   Dolgachev, I.: Algebraic surfaces with $p_g = q = 0$. In: Algebraic surfaces. CIME 1977, 1977, Cortona, Liguori, Napoli 1981, pp. 97–215

[DPS 92]   Demailly, J. P., Peternell, T., Schneider M.: Compact complex manifolds with numerically effective tangent bundles. To appear in J. Alg. Geometry (1993)

[DPS 93]   Demailly, J.P., Peternell, T., Schneider, M.: Kähler manifolds with numerically effective Ricci class. To appear in Comp. Math.

[DPS 94]   Demailly, J.P., Peternell, T., Schneider, M.: Compact Kähler manifolds with semi-positive Ricci class. In Vorbereitung

[FM 88a]   Friedman, R., Morgan, J.: On the diffeomorphism types of certain algebraic surfaces I, II. J. Diff. Geom. **27**, 297–369, 371–398 (1988)

[FM 88b]   Friedman, R., Morgan, J.: Algebraic surfaces and 4-manifolds: some conjectures and speculations. Bull. AMS **188**, 1–15 (1988)

[FM 93]   Friedman, R., Morgan, J.: Smooth four-manifolds and complex surfaces. Forthcoming book

[Fr 61]   Frankel, T.: Manifolds with positive curvature. Pacific J. Math. **11**, 165–174 (1961)

[Fu 80]   Fujiki, A.: On the minimal models of complex manifolds. Math. Ann. **253**, 111–128 (1980)

[Ful 84]   Fulton, W.: Intersection theory. Ergebnisse der Mathematik (3) Band 2, Springer 1984

[FU 84]   Freed, D., Uhlenbeck, K.: Instantons and 4-manifolds. M.S.R.I. publ. no. 1. Springer 1984

[Gi 77]   Gieseker, D.: On moduli of vector bundles on an algebraic surface. Ann. of Math. **106**, 45–60 (1977)

[Gr 62]   Grauert, H.: Über Modifikationen und exzeptionelle analytische Mengen. Math. Ann. **146**, 331–368 (1962)

[Gu 52]   Guggenheimer, H.: Über vierdimensionale Einsteinräume. Experientia **8**, 420–421 (1952)

[Hi 66]   Hirzebruch, F.: Topological methods in algebraic geometry. Springer 1966

[HK 57]   Hirzebruch, F., Kodaira, K.: On the complex projective spaces. J. Math. pure et appl. **36**, 201–216 (1957)

[Hit 87] Hitchin, N.: The self-duality equations on a Riemann surface. Proc. London Math. Soc. **55**, 59–126 (1987)

[Is 90]   Iskovskih, V. A.: Double projection from a line on Fano threefolds of the first kind. Math. USSR Sb. **66**, 265–284 (1990)

[Ka 85a] Kawamata, Y.: Pluricanonical systems on minimal algebraic varieties. Inv. Math. **79**, 567–588 (1985)

[Ka 85b] Kawamata, Y.: Minimal models and the Kodaira dimension of algebraic fiber spaces. Crelles J. **363**, 1–46 (1985)

[Ka 92]   Kawamata, Y.: Abundance Theorem for minimal threefolds. Inv. Math. **108**, 229–246 (1992)

[Kl 68]   Kleiman, S.: Towards a numerical theory of ampleness. Ann. Math. **84**, 293–344 (1968)

[KMM 87] Kawamata, Y., Matsuda, K., Matsuki, K.: Introduction to the minimal model program. Adv. Studies Pure Math. **10**, 283–360 (1987)

[KMM 91] Kollár, J., Miyaoka, Y., Mori, S.: Rationally connected varieties. Preprint 1991

[Ko 87]   Kollár, J.: The structure of algebraic 3-folds: an introduction to Mori's program. Bull. Amer. Math. Soc. **17**, 211–273 (1987)

[Ko 91]   Kollár, J.: Flips, flops and minimal models. Surveys Diff. Geom. **1**, 113–1991 (1991)

[Ko 92]   Kollár, J. et al.: Flips and abundance for algebraic 3-folds. Preprint 1992

[Kob 61] Kobayashi, S.: On compact Kähler manifolds with positive Ricci tensor. Ann. of Math. **74**, 570–574 (1961)

[Kob 81] Kobayashi, S.: Recent results in complex differential geometry. Jber. dt. Math.-Verein **83**, 147–158 (1981)

[Kob 82] Kobayashi, S.: Curvature and stability of vector bundles. Proc. Japan Acad. **58**, 158–162 (1982)

[Kob 87] Kobayashi, S.: Differential geometry of vector bundles. Princeton University Press 1987

[KO 73]   Kobayashi, S., Ochiai,T.: Characterization of complex projective spaces and hyperquadrics. J. Math. Kyoto Univ. **13**, 31–47 (1973)

[Lu 82]   Lübke, M.: Chernklassen von Hermite-Einstein Vektorbündeln. Math. Ann. **260**, 133–141 (1982)

[Lu 83]   Lübke, M.: Stability of Einstein-Hermitian vector bundles. Manuscripta math. **42**, 245–257 (1983)

[Ma 77]   Maruyama, M.: Moduli of stable sheaves I. J. Math. Kyoto Univ. **17**, 91–126 (1977)

[Ma 78]   Maruyama, M.: Moduli of stable sheaves II. J. Math. Kyoto Univ. **18**, 557–614 (1978)

[Mi 77]   Miyaoka, Y.: On the Chern numbers of surfaces of general type. Inv. Math. **32**, 225–237 (1977)

[Mi 88]   Miyaoka, Y.: Abundance conjecture for 3-folds: case $\nu = 1$. Comp. Math. **68**, 203–220 (1988)

[Mo 79]   Mori, S.: Projective manifolds with ample tangent bundles. Ann. Math. **110**, 593–606 (1979)

[Mo 82]   Mori, S.: Threefolds whose canonical bundles are not numerically effective. Ann. Math. **116**, 133–176 (1982)

[Mo 87]   Mori, S.: Classification of higher-dimensional varieties. Algebraic Geometry, Bowdoin 1985, Symp. Pure Math. **46**, 269–331 (1987)

[Mo 88]   Mori, S.: Flip conjecture and the existence of minimal models for 3-folds. J. Am. Math. Soc. **1**, 117–253 (1988)

[Mk 88]   Mok, N.: The uniformization theorem for compact Kähler manifolds of nonnegative holomorphic bisectional curvature. J. Diff. Geom. **27**, 179–214 (1988)

[Mr 81]   Murre, J.: Classification of Fano threefolds according to Fano and Iskovskih. Lecture Notes in Mathematics, vol. 947, 35–92 (1981)

[MR 82]   Mehta, V.B., Ramanathan, A.: Semistable sheaves on projective varieties and their restriction to curves. Math. Ann. **258**, 213–224 (1982)

[Mu 89]   Mukai, S.: New classification of Fano threefolds and Fano manifolds of coindex 3. Proc. Natl. Acad. Sci. USA **86**, 3000–3002 (1989)

[Na 87]   Nakamura, I.: Moishezon threefolds homeomorphic to $\mathbb{P}_3$. J. Math. Soc. Japan **39**, 521–535 (1987)

[NS 65]   Narasimhan, M.S., Seshadri, C.S.: Stable and unitary vector bundles on compact Riemann surfaces. Ann. Math. **82**, 540–567 (1965)

[Ok 91]   Okonek, C.: Instanton invariants and algebraic surfaces. In: Geometric Topology: Recent Developments, Montecatini Terme, 1990. LNM 1504, 138–186, Springer 1991

[OSS 80] Okonek, C., Schneider, M., Spindler, H.: Vector bundles over complex projective spaces. Progr. in Math. 3. Birkhäuser 1980

[OV 86]   Okonek, C., Van de Ven, A.: Stable bundles and differentiable structures on certain elliptic surfaces. Inv. Math. **86**, 357–370 (1986)

[OV 90]   Okonek, C., Van de Ven, A.: Stable bundles, instantons and C^∞-structures on algebraic surfaces. In Several Complex Variables VI, Encyclopaedia of Mathematical Sciences, 197–249, Springer 1990

[Pe 86]   Peternell, T.: Algebraic structures on certain 3-folds. Math. Ann. **274**, 133–156 (1986)

[Pe 93]   Peternell, T.: Moishezon manifolds and rigidity theorems. Preprint 1993

[PSW 92] Peternell, T., Szurek, M., Wisniewski, J.: Fano manifolds and vector bundles. Math. Ann. **294**, 151–165 (1992)

[PT 92]   Pidstrigach, V., Tjurin, A.: Invariants of the smooth structure of an algebraic surface arising from the Dirac operator. Math. USSR Izv. **52**, 179–371 (1992)

[Re 87]   Reid, M.: Young person's guide to canonical singularities. Algebraic Geometry, Bowdoin 1985. Symp. Pure Math. **46**, 345–414 (1987)

[Sh 80a] Shokurov, V.: Smoothness of the general anticanonical divisor on a Fano 3-fold. Math. USSR Isv. **14**, 395–405 (1980)

[Sh 80b] Shokurov, V.: The existence of a straight line on Fano 3-folds. Math. USSR Isv. **15**, 173–209 (1980)

[Sh 86]   Shokurov, V.: Theorem on non-vanishing. Math. USSR Isv. **26**, 591–604 (1986)

[Si 92]   Simpson, C.: Higgs bundles and local systems. Publ. Math. I.H.E.S. **75**, 5–95 (1992)

[Siu 89] Siu, Y. T.: Nondeformability of the complex projective space. Crelles J. **399**, 208–219 (1989)

[SY 80]   Siu, Y. T., Yau, S. T.: Compact Kähler manifolds with positive bisectional curvature. Inv. Math. **59**, 189–204 (1980)

[Sm 61]   Smale, S.: Generalized Poincarés conjecture in dimensions > 4. Ann. Math. **74**, 391–466 (1961)

[Ta 73]   Takemoto, F.: Stable vector bundles on algebraic surfaces. Nagoya Math. J. **47**, 29–48 (1973)

[Uh 82]   Uhlenbeck, K.: Connections with L^p-bounds on curvature. Comm. Math. Phys. **83**, 31–42 (1982)

[UY 86]   Uhlenbeck, K., Yau, S.-T.: On the existence of Hermitian-Yang-Mills connections in stable vector bundles. Comm. Pure Appl. Math. **39**, 257–293 (1986)

[Ve 86]   Van de Ven, A.: On the differentiable structure of certain algebraic surfaces. Séminaire Bourbaki 667 (1986)

[Vi 80]   Viehweg, E.: Klassifikationstheorie algebraischer Varietäten der Dimension 3.
           Comp. Math. **41**, 361–400 (1980)
[Wi 87a]   Wilson, P.M.H.: Towards classification of algebraic varieties. Bull. London Math.
           Soc. **19**, 1–48 (1987)
[Wi 87b]   Wilson, P. M. H.: Fano fourfolds of index greater than one. Crelles J. **379**,
           172–181 (1987)
[Yau 77]   Yau, S.-T.: Calabi's conjecture and some new results in algebraic geometry. Proc.
           Nat. Acad. Sci. USA **74**, 1798–1799 (1977)
[Yau 78]   Yau, S.T.: On the Ricci curvature of a compact Kähler manifold and the complex
           Monge-Ampère equation, I. Comm. pure and appl. math. **31**, 339–411 (1978)
[Zh 90]   Zheng, F.: On semi-positive threefolds. Thesis Harvard 1990

Printing: Druckerei Zechner, Speyer
Binding: Buchbinderei Schäffer, Grünstadt